Contemporary Measurement Concepts

ISBN: 978-0-578-11732-4

Physical Variables, Measurement Hardware and Control Principles for Process Automation and Experimental Testing

About the author

Jonathan Lambert is a professor emeritus of engineering technology at Black Hawk College in Moline, II, a senior member of the International Society of Automation (ISA) and a thirty-year member of the American Society for Engineering Education (ASEE). Jon holds an AAS degree in Electrical/Instrumentation Engineering Technology from Black Hawk College, a BS degree in Electrical Engineering Technology (EET) from Bradley University and an MS degree in Industrial Engineering from the University of Iowa. He began a career in electronics, sensors and controls repairing traffic signals for the City of Quincy, Illinois in 1970 and has held positions as a test engineer for Motorola Corporation, a chief broadcast engineer for Black Hawk College Educational Television/WQPT Quad-Cities Public Television and served for 35 years as a professor, lead instructor and department chair of engineering technology at Black Hawk College. Jon has provided technical assistance, engineering consulting and product development testing services for numerous Quad-Cities regional employers including the Harvester and Seeding divisions of John Deere Worldwide Product Development, Roth Pump, Small Newspapers, the Moline Dispatch, Martin Engineering, Chemplex Corporation, the US Army/Edgewood Chem/Bio Lab, River Stone Group, Moline Consumers Company, Ipsco Steel, Kewanee Boiler and numerous other regional firms.

Jon is currently providing contract engineering services and adjunct engineering technology instruction at Black Hawk College in Moline, Illinois.

Chapter 1 – Mechatronics, Automation and Control: Systems and Components

This chapter introduces the components and functions of Mechatronics. Automatic control, direct and supervisory control and computer integration is also introduced. Included topics follow.

Objectives:
Upon completion of this chapter, you should be able to:
- Explain the operation of common control loops
- Explain the contribution of each component in a control loop
- Identify and diagram each component within an industrial control loop
- Suggest techniques for control loop fault diagnosis
- Suggest calibration techniques for common control loops
- Reference variations on common closed-loop control systems

Introduction

Wikipedia defines Mechatronics as being centred on mechanics, electronics and computing which when combined make possible the generation of simpler, more economical, reliable and versatile systems. The portmanteau "mechatronics" was first coined by Mr. Tetsuro Mori, a senior engineer of the Japanese company Yaskawa, in 1969. Mechatronics may alternatively be referred to as "electromechanical systems" or as "instrumentation" or "control and automation engineering".

A typical mechatronic engineering degree would involve classes in engineering mathematics, mechanics, machine component design, mechanical design, thermodynamics, circuits and systems, electronics and communications, control theory, digital signal processing, power engineering, robotics and usually a final year thesis involving a project. Mechatronics is fairly new to the US but exists in university curricula throughout the world

Automation is the contribution of any activity that makes work less tedious. Automation systems range from computers running application specific software to self-regulating process controllers. Automation does not always require sophisticated electronics. As an example, a simple beam and fulcrum in a torque balance arrangement provides a means for error comparison and subsequent correction. Such components are commonly employed within pneumatic, hydraulic and other mechanical control systems.

This chapter will review the methods and systems to perform industrial automation. Systems to measure, assess, analyze and control will be investigated and applied to common industrial processes. Techniques to determine system operation and component specification will also be examined. For application purposes, the text will assume a manually operated process is in existence and the process will be automated for increased productivity, improved product quality or to allow the operators the opportunity to concentrate their efforts elsewhere in less tedious functions. History has proven automation for the sole purpose of employee elimination is not feasible, for even the most reliable automation requires supervision and maintenance.

Figure 1 – Pneumatic process controllers utilize torque, force and pressure mechanisms to achieve automatic control.

Reasons to Automate

As mentioned previously, increased productivity, enhanced control, improved product quality and to allow the operators the opportunity to concentrate their efforts upon less tedious functions are the expected results of process automation. Even the smallest increase in productivity can translate into profit, often with rapid payback.

For a multitude of reasons many processes are manually operated. Usually the operator of the process, in order to maintain control and consistency, operates the process at a rate somewhat less than the original design specifications prescribe. To an extent this is to be expected, since process operators are usually required to perform numerous functions concurrently. By alleviating the operator of routine multiple tasks, a self-regulating, automatically controlled process results in increased production rates since the automatic control system responds to variation more rapidly and accurately than the operator does.

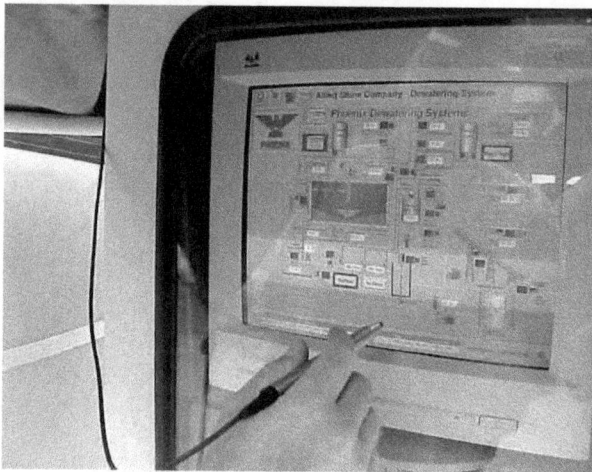

Figure 2 - A plant operator accesses process and control loop information via an interactive panel display.

To remain competitive today, most companies place a large emphasis upon quality. Quality procedures almost certainly require a degree of automation even if only a desktop PC is used to crunch numbers. As tolerances tighten and variation is reduced, automatic sensing, data acquisition and process analysis becomes more attractive. Thus, another reason to automate becomes obvious.

Automation Prerequisites

In reviewing the processes that the author has automated, two functions require emphasis. Procedurally, the first function is to become familiar with the existing process. The better one prepares for automation the smoother the implementation. It would not be exaggerating to contend that one must intuitively feel the process in its operation in order to correctly and successfully automate. In developing an intuitive feel, one should know the operating characteristics of all existing process components to the greatest extent possible. Even if a review of the internal construction and operation of each existing mechanism is necessary.

Remember also that the prescribed automation will be performing the routine functions presently provided by the operator, and that every detail no matter how minute must be observed, documented and understood. Consequently, your relationship with all involved personnel should be positive.

Secondly, an awareness of all available technology to perform the automation is also necessary. Information on computers, software, sensors, transmitters, controllers and any other instruments should be at your disposal. In addition, a means to remain technologically current is also required. Periodicals, professional society involvement, plant tours and courses are only a few methods by which one remains technically current.

In a de facto manner, all students of technology are students of automation. For technology is an integral component of automation. Mechanical aptitude, problem diagnosis and resolution, creative thought processes, continuous and organized documentation (like a lab journal), a genuine interest in how things work and a positive, patient and forgiving attitude will contribute to your success in automating processes. Again, of greatest importance is your working relationship with all involved people. All should be openly willing to participate, communicate and openly share in the successes and the inevitable disasters that can accompany changes of this type. Certainly, a coordinated team approach is in order.

Industrial Instrumentation
Instrumentation, in the broadest definition, is the study of measurement and control, apparatus and procedures. Industrial Instrumentation focuses upon measurement and control of physical quantities within the industrial arena called *process*

variables. Commonly used in manufacturing, typical industrial process variables are temperature, pressure, flow, level, position, velocity, humidity, density, composition and so forth.

Figure 3 - Open-pit mining automation involves sensing and controlling pressure, level, flow rate, position and numerous other process variables.

Industrial measurements of process variables are performed for primarily two reasons; to acquire information about the quality of a manufactured product and secondly, as a first step in automatically controlling a process in a manufacturing operation. This text will introduce industrial measurement and control devices, their specifications, operation and application but will not limit coverage to electronic devices. In order to effectively design, analyze or maintain industrial measurement and control systems, an understanding of many physical-operating principles is necessary. Hydraulics (fluids under pressure), pneumatics (compressed air), electronics and mechanics are common to the manufacturing, experimental testing and chemical processing industries and will be covered as required in this text.

Automatic Control Systems
As indicated earlier, a measurement is necessary to automatically (without human intervention) control, or regulate a process variable. A process variable is a physical quantity that is representative of, or has an influence upon the quality of a manufactured product. Among the most common process variables are pressure,

14

level, flow and temperature. The measurement device is referred to as a <u>sensor</u> and is often a *transducer*. A transducer accepts one energy form on its input and outputs another energy form. Transducer is a very general term that describes all types of components such as heaters, motors, lamps, microphones, speakers, buzzers, solar cells, etc. When an electronic computer is expected to control a temperature process, a transducer is required to convert the thermal (temperature) energy into an electronic signal for use by the computer. Another transducer is required to convert the computer output into thermal energy for control of the temperature process.

Figure 4 - The Fisher Controls type 546 I/P (pronounced "I to P") is a common transducer in the automation industry. Through a combination of electronic and mechanical components, the 546 converts a 4-20 mA input signal into 3-15 psi of pressure and is often mounted on a valve actuator. In this capacity transducers are often called servo valves.

The *sensor*, or sensor/transmitter, is the first of four essential components required to automatically control a process variable such as pressure, level, flow, or temperature (the four most popular process variables). The other three required control system components are the *process*, the *controller* and the *final control element*. All four components are coupled into a closed-loop configuration as shown in figure 5. The signal flow is also indicated. Although the set point is not a system component, it is usually displayed with the controller in system diagrams.

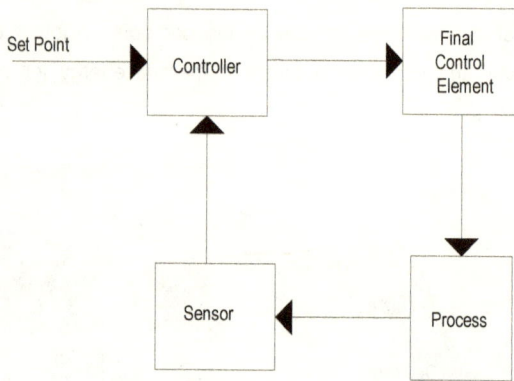

Figure 5 – Essential components for self-regulating automatic control.

The Process

The process contains the process variable and is a portion of, or an entire manufacturing operation. The process variable should be some quantity that is representative of a manufactured products' quality. The process variable is assigned the letter "C" and is also referred to as the "controlled" (for obvious reasons) or "dynamic" variable since it is capable of changing from external or *disturbance variables*. The process provides "C" to the sensor and receives input from the final (control) element. Manufacturing processes can assume many forms and degrees of sophistication, from ovens baking bread to reactors processing chemicals.

Figure 6 - Examples of manufacturing processes in the open pit quarry mining industry. This plant is operated using network communication from a handheld PC to the control tower in the background.

Two things should be remembered when considering the process. First, the process will usually contain a product input and output which may not be indicated in the previous diagram (figure 5). Also, remember that the rate at which the product is applied to the process may be capable of varying. This changing product feed rate is referred to as the load on the process. A larger or increasing load is understood to require more energy to the process from the final control element.

Figure 7 - A reservoir holding chemicals as part of a batch mixing process.

The Sensor

The sensor, or *primary sensor*, is typically located at the process and is often a transducer. The sensor output is commonly connected to a transmitter. Since the sensor is normally in direct contact with the process it is given the name *primary element*. The transmitter, or *secondary element*, output is assigned the symbol "Cm" representing the <u>measured</u> value of the controlled variable, "C." As was indicated previously, the sensor/transmitter's output will commonly assume a different form than its input, hence the description "measured value," and the designation, Cm, or measured value of the controlled variable "C."

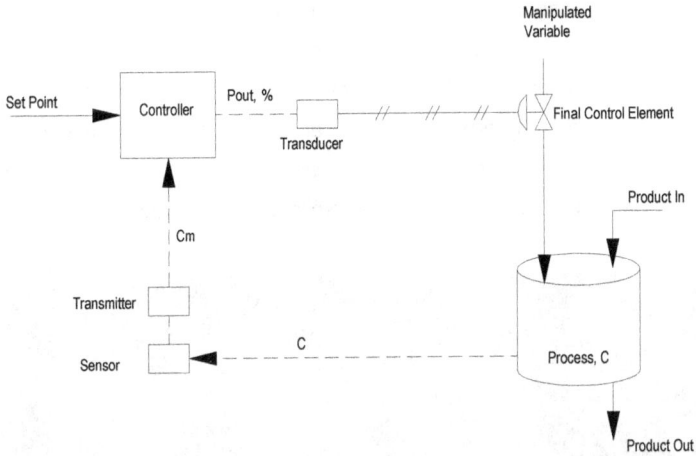

Figure 8 - A block diagram of common control system components.

Transmitters (abbreviated "Xmtr" throughout this text) convert the sensor's output into a compatible, standardized output signal span. Commonly available "standard" signal spans are 4 ma to 20 ma, 1 to 5 volts, 1 to 9 volts and in pneumatic systems 3 psi to 15 psi or 6 to 30 psi. Of all the electrical standard signal spans, the *current loop* (4-20ma) is the most common since there is no attenuation (loss of signal strength) over long distance signal lines, like there would be with a voltage span. The following diagram illustrates.

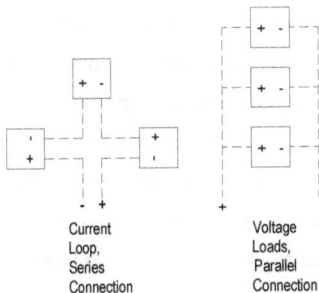

Figure 9 - Current versus voltage connections. Note the connection polarities.

Note that the sensor/transmitter's output is a span of values (4 to 20 mA) representing a span of process values such as 100 to 300 degrees, 5 to 25 gallons per minute or 30 to 60 inches of liquid. Although a control system is intended to maintain the process variable at a constant value, it is designed to operate over a span of process values. Designing a control loop to operate over a span of values allows fine-tuning of the process for improved product quality.

Figure 10 - A flow transmitter is considered a "secondary" element since it converts the signal generated at the primary element into an industry standard 4-20 milliamps. In this case the primary element is an orifice type flowmeter (not obvious in photo).

Sensor/Transmitter I/O Characteristics

When working with control loop components, one often needs to apply graphical concepts to associate the input and output quantities. Most of the current generation of control components input/output characteristics can be represented by straight-line, *linear* functions. Linear characteristics use a straight line X-Y graph to illustrate the relationship between X-axis (input) and Y-axis (output) quantities. A few non-linear sensors exist but are commonly made linear by mathematical functions performed in the transmitter. Non-linear sensors are addressed in later chapters.

As an example of a linear characteristic, assume a temperature process uses a sensor/transmitter combination to output 4 mA to 20 mA over a 100° to 500° Fahrenheit-temperature span. Since the relationship between temperature and current is linear, a proportional relation exists. Therefore, a process temperature of 50% of the input span (300°) would result in a transmitter output of 50% of the 4-20 mA output span, or 12 mA.

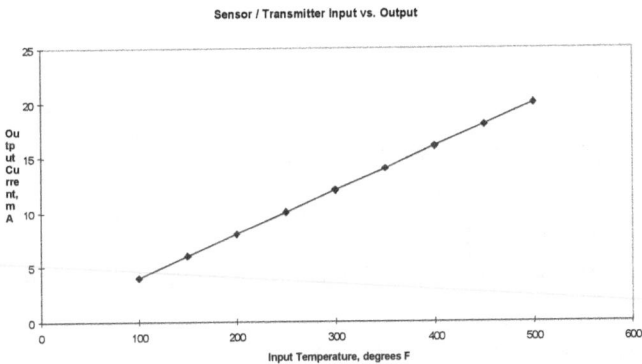

Figure 11 -Sensor/Transmitter Input/Output characteristic. In this example an input of 100 to 500 degrees generates an output span of 4 to 20 milliamps.

The previous graph and following equations describe the relationship between the process variable, C, at the sensor input and the measured value of the controlled variable, Cm, at the transmitter's output. Using the straight-line equation, Y=mX+b, the relationship between the process variable being sensed and the output span can be mathematically defined. The associated input/output values can then be accurately determined.

To determine the required mathematical quantities, again assume a sensor/transmitter is calibrated to output 4 – 20 milliamps over an input temperature span of 100° - 500° Fahrenheit. The mathematical characteristics of the sensor and transmitter combination (hereafter abbreviated as sensor/xmtr) can be determined using the slope and intercept computations associated with the straight-line, Y=mX+b equation. The characteristics can be determined as follows.

Given, Y=mX + b (general linear equation)
Where,
Y = Sensor/xmtr Output, units of milliamps (mA)
m = Sensor/xmtr characteristic slope or "Gain", units of mA/F
X = Sensor/xmtr Input, units of degrees F.
b = Sensor/xmtr characteristic y-axis intercept value or "Offset", units of mA

After subbing the operating terms into the Y=mX+b equation, the sensor/xmtr operating equation becomes,

$$Y = (m * X) + b$$

Sensor/xmtr Output = (Sensor/xmtr Gain)(Temp.Input) + (Sensor/xmtr Bias)

The specific values of slope (often called Gain or "Span") and intercept (often called Offset or "Zero") can now be determined. Solving for Gain (m) and Offset (b) using slope (m) and intercept (b) equations given the previous input/output characteristics yields,

Finding "Gain" -
Using the "Rise over run" slope equation to find sensor/xmtr "Gain" -
Sensor/xmtr Gain = Δout / Δin = 4-20 ma / 100-500 F
Sensor/xmtr Gain = .04 ma/F

Finding "Offset" -

Rearranging the Y=mX+b equation to solve for the Offset or Y-axis intercept ("b") yields,

$$b = Y - mX$$

Substituting corresponding input temperature (500° for X) and output current (20 mA for Y) values to solve for Sensor/xmtr "Offset" –

Sensor/xmtr Offset = Output - (Gain * Input)

Sensor/xmtr Offset = 20mA - (.04 ma/F * 500°F)

Sensor/xmtr Offset = 0 mA

Using the computed values for gain and offset, any corresponding input/output quantity can be determined. For example, the output current at an input temperature of 150° F. is,

Output = (.04 ma/ F) (150 F) + 0 ma

Output Current = 6 ma

Note the units, called *dimensions*, in the previous computations cancel to yield the required milliamp current. Canceling of the units is referred to as *dimensional analysis*. Dimensional analysis provides a convenient method for checking to see if the mathematical mechanics of problem solution have been performed correctly. Dimensional analysis should be used wherever possible to assist in proper mathematical manipulations and as an aid to recalling equations.

In actual industrial sensor/transmitter combinations, the gain and offset quantities emphasized in the previous equations are made variable to provide a means of *Calibration*. The slope and intercept adjustments are typically labeled "Span" (for Gain, or slope) and "Zero" (for offset, bias or intercept) respectively. Calibration of the Span and Zero adjustments provides a means of accurately determining sensor/transmitter operation and also allows the sensor/transmitter combination to be used over a completely different span of temperature if desired.

The current state of transmitter technology provides isolation (for noise immunity) and linearization (when required) circuitry for the majority of sensor/transmitter combinations, making the previous Y=mX+b equation worth remembering. With a little practice the computations can be performed without paper and pencil, and will provide tremendous insight into the calibration process.

Figure 12 - A pressure transmitter (left) and two temperature transmitters (right). Recent developments in integrated semiconductor technology have allowed the development of disposable transmitters. The span and zero potentiometric adjustments are obvious on both transmitters.

The Controller

Working our way around the closed-loop system from the process, we have taken a process variable such as pressure, level, flow or temperature, sensed it, and transmitted its *Analog* (corresponding measured value) to the controller's input as a standard signal span. The controller is the next component in the loop and is responsible for maintaining the process operating value at the desired set point.

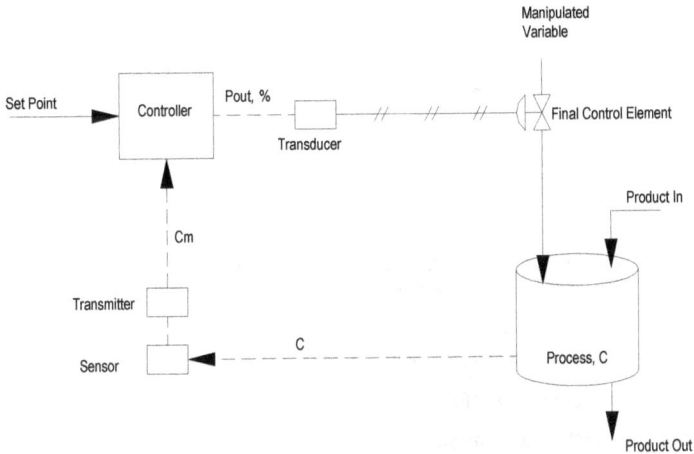

Figure 13 – The controller receives an analog (Cm) of the process variable (C) generated by the sensor/transmitter combination.

The controller detects changes in the process and manipulates its output (Pout, or %Pout) to keep the process at, or near Set Point. To accomplish the automatic control function the controller performs two basic functions. The first function is to compare the "actual" process value (Cm, from the sensor/xmtr) to the desired operating value or "set point", (SP), which is commonly entered manually by an operator. This comparison function is referred to as *error signal generation* and is another example of the ubiquitous comparison technique used throughout the field of instrumentation to observe any difference between a standard (in this case the Set Point), and an unknown (in this case the process variable, C) value.

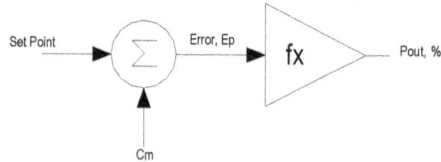

Figure 14 – The internal functions performed by a controller involve comparison (subtraction) and mathematical manipulation (Fx) of the difference (error).

The difference between the set point and the actual process value is the *Error* and is assigned the designation "Ep" for process error. An equation can be written that describes the function performed by the error signal generator.

$$Ep = Cm - SP$$

The above equation is understood to describe the operation of a "direct acting" (DA) controller since the error follows the input (Cm) from the sensor/xmtr, directly. In other words, as the input increases the error increases. "Reverse action" is usually a switch or software selectable feature on most single loop, commercial controllers and is described by the following equation;

$$Ep = SP - Cm$$

In this equation, it can be seen that as the controller input (Cm) increases, the controller generated error moves negative. - Hence the term, reverse action.

It should be understood that the *controller input is provided by the transmitter as Cm*, and not the set point, SP. The error generator or "summing junction" function can be performed a number of ways by using a computer software algorithm, an operational amplifier, a resistive bridge network or mechanical apparatus. Error generation, or comparing actual to desired values, can also be performed using pneumatics.

The error, or process error, can be expressed in terms of the process variable or as a percentage of the calibrated transmitter span. As an example, assume a set point of 200° F., and an actual process value of 160° F. When the transmitter is calibrated to output a standard span over an input of 100° F to 500° F., a direct acting controller would indicate an error of -40° F., or -10%. The following equation converts error into percentage of error.

$$\% \ Ep = [(C - SP) \div (Cmax - Cmin)] \ * \ 100 \ \%$$

Where,

Ep = Process error

C = Actual process value

SP = Set Point, or desired process operating value

Cmax = Maximum calibrated process value

Cmin = Minimum calibrated process value

Note that percentage of Error is simply the error quantity in relation to, or divided by the span of possible process values. The previous equation is provided as an aid since process controllers commonly display operating values such as Set Point, process value, Error and controller output as percentages.

After SP is compared to Cm, the controller's second stage performs the second function of determining the quantity and type of corrective action to be taken. The type of corrective action is contingent upon the relative sophistication of the controller but is ultimately determined by the operating tolerance with which the process must be maintained. Ideally, the controller should be "tuned" to return the process to the set point value as soon as possible after a process disturbance, load change, etc. The response of the process to a controller's corrective attempts is a function of every component in the loop but is mostly determined by the sophistication of the controller itself.

Controller operational modes, or the mathematical manner in which the various types of controllers respond to an error signal, are discussed later in this text and are the subject of Black Hawk College's ET 253 course. For now, assume that the controller's response to an error results in a returning of the process to set point, *although it should be stated that this is not always the case.* Some processes are allowed to operate within a tolerance around the set point. Such is the case with a home heating process, where the temperature is continually increasing and decreasing as the heater cycles on and off.

Figure 15 - Pneumatic controllers are commonly used where an electronic spark cannot be tolerated and in locations where electric power is not available.

In performing the comparison and corrective functions, the controller is referred to as the "brains" of the closed-loop control system. The controller output is normally expressed as a percentage, and is designated "Pout" to represent percentage of controller output. The controller output signal is commonly 4-20 mA or other standard signal span. Photos of common digital, analog and pneumatic industrial process controllers are included throughout the chapter.

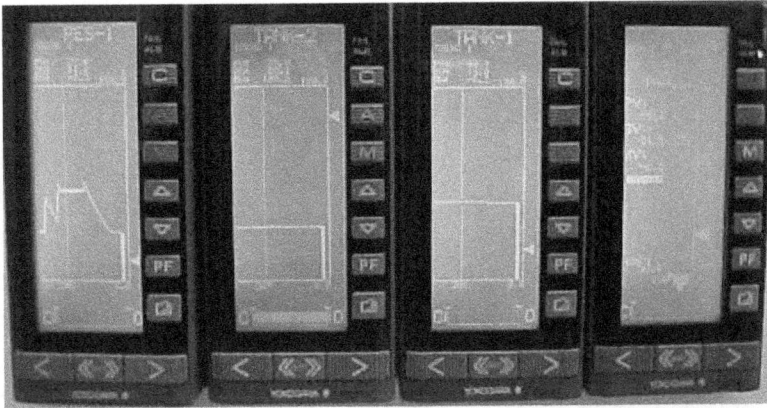

Figure 16 – Four Yokogawa YS-170 single-loop programmable process controllers. Set point, process quantity, process error and controller outputs are visible from the front panel display.

Final Control Element

The *final control element (FE)*, or final element, is responsible for regulating the *manipulated variable (MV)*, as dictated by the controller output value. The manipulated variable is commonly a raw energy source such as electrical power, natural gas, compressed air, mechanical force, rotational velocity and so forth. The manipulated variable will have a positive influence on the process variable. In other words, the controller "tells" the final element how much energy should be introduced to the process, based upon the process error, in attempting to maintain the set point.

Figure 17 - A bank of I/P's (pronounced "I to P") are utilized to convert the electronic signal from a controller output into a pneumatic signal span to position a proportional control valve. In this case a 4-20mA signal is converted into 3-15 psi. I/P's are also called servo-valves and are included as part of a control loop's final element.

The final element is frequently a valve but may also take the form of a motor, heater, pneumatic actuator or pump. Any required "drive" circuitry such as relays, SCR pulse circuits, or pneumatic gain/biasing components is also associated with the final element. The final element is often used in conjunction with a transducer. The transducer converts the generally electronic signal from the controller into some mechanical format such as air pressure, position, rpm or whatever energy form is required to position the final element. Common transducers found in this portion of the control-loop are current to pressure, (I/P), voltage to pressure, (E/P), digital to pressure (D/P) and variable frequency drive (VFD) AC motor components. Air pressure is a very common transducer output at the final element since large forces, created by relatively low pressures, are often required to position large industrial control valves. Final control elements appear in the photos following.

Figure 18 – A final control element consisting of a valve (bottom) and actuator (top).

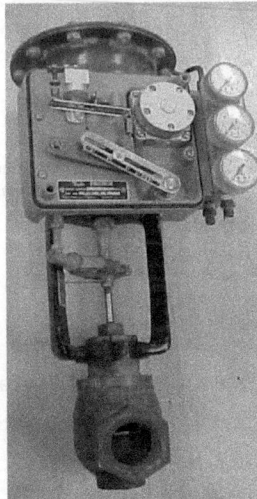

Figure 19 – A valve, actuator and "Precisor" (with pressure gauges attached) for precision control of valve position. The actuator is the top three-fourths of the figure with the valve at the bottom.

When the previously described control-loop components are connected in a closed-loop configuration, a self-regulating automatic control system is formed. The specific interconnection and setup of the various components are presented throughout the text. Each will require some form of setup (calibration) except the process, and additional components for signal conversion and compatibility purposes are usually included as shown in the following figure.

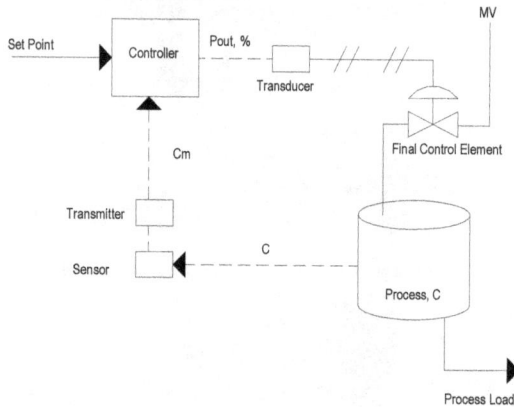

Figure 20 – The automatic control system with common transmitter and transducer. Note the signal path designation for electronic (dashed) and pneumatic (crossing) lines.

Open-Loop Control

Automatic control systems don't always assume the self-corrective, closed-loop configuration of the previous figure. In fact, if all the variables around the process were capable of remaining constant (especially the disturbance variables) then the controller and sensor would not be required to detect changes in the process. *Therefore, a closed-loop control system is necessary only if the process is expected to undergo load changes, or if any of the process related variables are susceptible to external influences.* In short, automatic control is required only if the process is susceptible to change. In cases where load changes and disturbances are relatively small and few or non-existent, the sensor/transmitter and corresponding feedback connection to the controller is not required. Eliminating the sensor and transmitter hardware would result in an "open-loop" control system as shown in the following figure.

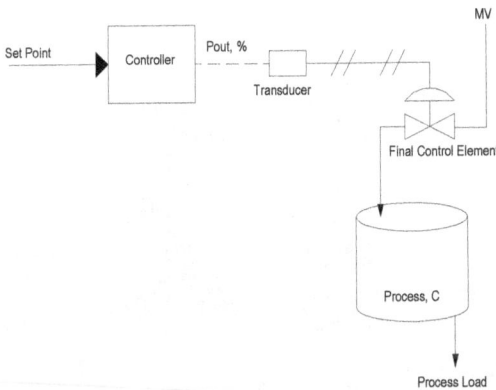

Figure 21 – The sensor and feedback is eliminated in an open-loop control system

An open-loop control system is understood to contain no summing junction (error generator in the controller), and no feedback from the process. Feedback is a term used to describe the function of the sensor and transmitter in providing information about the process "back" to the controller. The forward progress of the signal is understood to be from the controller to the final element, and on to the process. In most texts and courses, control systems are mathematically modeled in this manner. Consequently, the sensor/transmitter signal is the "feed-back" to the controller.

Examples of open-loop control systems include non-actuated traffic signal controllers, older generation washing machines, space heaters, some water heaters and so forth. Although many of the busier street intersections are currently "actuated" using inductive proximity detectors to detect the presence of vehicles, many are operated without a sensor. Traffic signals initially used a segmented cam rotating at a constant speed to change the traffic lights. Contact fingers rode upon the cam and were intermittently allowed to connect with a 120VAC bus bar to supply power to the 300-Watt lamps in the signal heads. The contemporary traffic controller uses a "drum" sequencing form of programmable controller algorithm to cycle through identical commands in a predetermined timing sequence.

Figure 22 - An older generation of traffic controller. Light cycle time was determined by one of the three synchronous motor timers. The contact "Fingers" on the timers would index the cam-switches in the following photo.

Figure 23 – Traffic light switching mechanism for early generation traffic controllers.

In the case of the washing machine, the drum-sequencing controller is expected to dictate a sequence of operations such as fill with water, agitate, drain the water, spin and repeat the cycle for a second or third series. The washing machine assumes that the water temperature and flow rate is correct and has no way of detecting if they aren't. Nor does the open-loop traffic signal controller have any way of detecting traffic patterns or frequencies. Controllers in open-loop systems are commonly timers, drum sequencers, or programmable logic controllers (PLC's). Although sophisticated PLC's are capable of accepting inputs from sensors, many of the control functions are performed open loop.

As an example of the susceptibility of open-loop control, take young Poindexter acting as a disturbance variable. Should someone decide to take a long, hot shower when you want to do the laundry, the washer will operate at a reduced water temperature, and possibly reduced water flow rate and level. The quality of wash will undoubtedly suffer. Another example of the limitations of open-loop control occurs when filling a bathtub. Experience has taught us that when the hot and cold-water faucets are set to a predetermined position(s), we can return in a few minutes to a warm bath or shower. However, should someone else have dishwashing or toilet flushing intentions elsewhere in the house, then we may find ourselves stepping into a tub of cold or excessively hot water. Certainly anyone who has lived in a large dormitory realizes the shortcomings of open-loop control. However, once you detect the error and attempt to take corrective action on the water temperature, the "Loop" becomes closed.

Closed-Loop Control

Corrective action is inherent in negative feedback, closed-loop control systems. Closed-loop describes the system configuration or interconnection of the components. Negative feedback refers to the manner with which the sensor's output (the feedback signal) and the set point are "summed" at the controller's summing junction (error signal generator). Since the Set Point and Cm are subtracted, they are "summed" out of phase, or of opposite polarities. Polarities are assigned to the summing junction input quantities in a reverse acting controller as indicated in the following figure. A direct acting controller (+ Cm, - Set pt) can be created with software or by a hardware switch that switches the set point and Cm inputs, a common feature in industrial controllers.

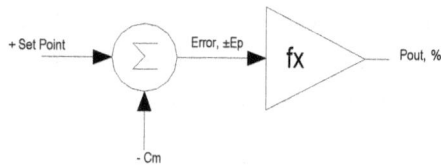

Figure 24 – A reverse-acting controller characteristic, Ep = SP - Cm

Again, the out of phase polarities associated with the summing junction is representative of negative feedback where corrective action is achieved in the following manner. Assume a temperature process whose temperature has risen as a result of a changing disturbance variable. Further assume a reverse acting controller (input goes up, output goes down) whose summing junction performs the following equation.

$$Error = SP - Cm$$

The temperature increase would be felt at the controller as an increase in Cm resulting in a negative-going (not necessarily decreasing, but building in a negative direction of polarity) error signal and decreasing controller output, % Pout. If the final control element is direct acting, as Pout decreases, the manipulated variable feeding the process through the final element also decreases, starving the process of energy. Under these conditions, the process would be forced to cool, canceling or correcting the initial temperature increase. In this way, corrective action and stability are inherent to negative feedback systems regardless of whether the controller is direct or reverse acting or whether the system is to control, condition, amplify, compare or compute. As long as the components are assembled to provide out of phase signals at the summing junction, the system will respond to an upset by driving the resultant error towards zero, or in a zero direction.

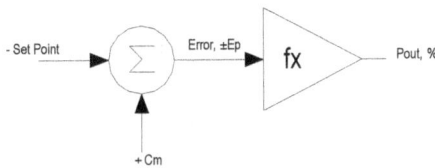

Figure 25 – A direct-acting controller characteristic, Ep = C – SP.

Negative feedback is used in electronics, pneumatics, mechanics and controls for the same reasons and with the same results. The error may not always be reset to zero but will always be driven in a minimizing direction around the Set Point. It should also be noted that all automatic closed-loop control systems contain the four essential components of process, sensor, controller and final control element. Moreover, a well-tuned automatic control system will attempt to correct an error in precisely the same manner, regardless of whether the control system is regulating temperature in petroleum refining processes or controlling the location of a guided missile.

Ironically, closed-loop control is not *perfect* control since the system can only respond to a generated error. When an error exists, product quality suffers. However, open-loop control with no sensor or feedback is capable of achieving perfect control, as long as none of the involved variables change. Once the appropriate recipe and processing sequence is determined, a "perfect" product will be manufactured as long as every required material and involved process remains constant. Should any of the variables be susceptible to change, then that variable will have to be regulated, or automatically sensed and controlled using a closed-loop system.

Computer Control

Computers were initially employed in automatic control systems in a direct-digital configuration but unfortunately; computers weren't reliable in the sixties and early seventies. Likewise, whenever the computer crashed, the control system stopped, along with production. Single loop, stand-alone analog controllers became ubiquitous during this period. Computers would return in various capacities in the eighties but not in the DDC configuration. Computers would be integrated with the existing single-loop controllers to form a *Supervisory Digital Control* configuration, as described following. Eventually computers would become quite reliable and would return to the industrial arena in the form of Direct Digital Control, Distributed Control Systems (DCS) and Supervisory Control and Data Acquisition (SCADA) systems.

Computers are integrated into closed-loop control systems in one of three basic ways, Direct Digital Control, Supervisory Control and Distributed Control. In *Direct Digital Control* (DDC) the computer is operating in the capacity of the closed-loop controller as shown in the following figure. A DDC computer usually replaces multiple single-loop process controllers.

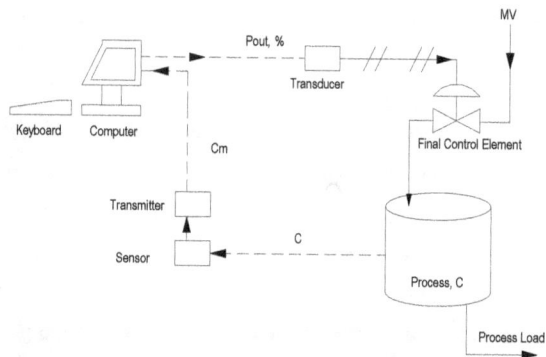

Figure 26 – A computer, keyboard and associated analog I/O replaces the closed-loop controller in a Direct Digital Control (DDC) configuration. This configuration is common in heating, ventilating and air conditioning (HVAC) energy management systems and experimental testing systems.

Distributed Control Systems (DCS's) typically utilize a single, centralized mid-sized or large-scale computer to control partial or entire processing and manufacturing operations. Present day DCS's are usually used in conjunction with operator consoles and multiple monitors displaying all acquired process variables. Although a DCS can be used without process controllers, and with all analog inputs and outputs connected directly to the DCS, many DCS's are networked to communicate with programmable controllers, single–loop controllers and analytical instruments. Thus distributing the control functions to the required location(s) throughout a plant. Most large controls manufacturers providing control components also support distributed control systems. Fisher-Rosemount, Bailey and Honeywell are common names in the controls industries providing DCS's.

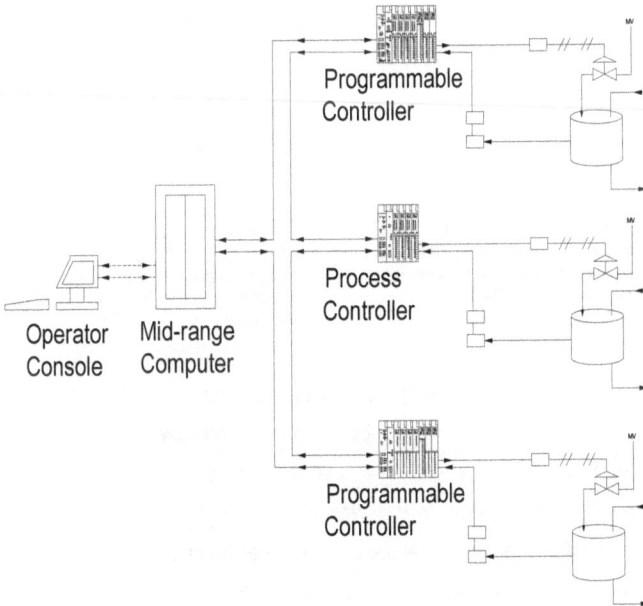

Figure 27 - Diagram of a distributed control system where a mid-range computer distributes control to PLC's and other single loop controllers.

Supervisory Digital Control (SDC), or Supervisory Control and Data Acquisition (SCADA) require all four of the essential components of a closed-loop system in addition to a computer. The computer becomes a fifth component that dictates set point information to the controller based upon the process status. To accomplish this, the remotely located single-loop or programmable controller must have a "remote set point" option, or other form of analog input which allows set point information to be transferred to the remotely located controller. A supervisory digital control (SCADA) system is shown following.

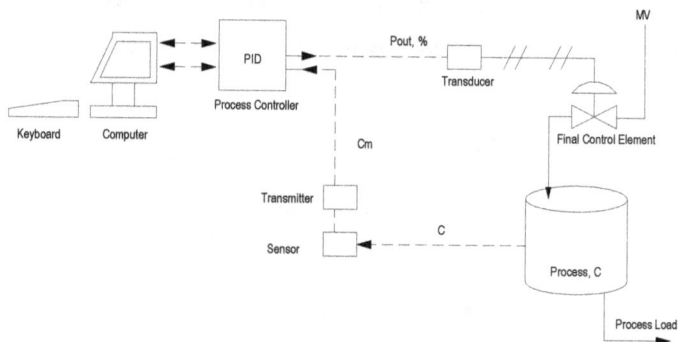

Figure 28 – A supervisory control and data acquisition system requires all of the components of a common closed-loop, control system and a computer to supervise the operation. Although not indicated in the above diagram the computer normally acquires process information independent of the process controller via the transmitters' output current loop. Radio frequency communication between the computer and the control loop is also common.

In order for the computer to perform an assessment of the process, the transmitted output from the sensor must be connected to the computer and the controller. This is usually performed using standard 4 to 20 milliamp current-loop signals to negate the effects of line resistance. In this way the computer monitors the process and makes subtle adjustments by manipulating the set point of the controller, just as a human operator would in place of the computer. Since the controller continues to monitor the process through it's sensor/transmitter connection, the controller maintains direct control of the process while the computer supervises operations through measurement and manipulation. The benefit of supervisory control is the isolated yet centralized monitoring and control of multiple loops, possibly including billing, inventory, shipping and other functions as well. Integrated plant automation or, *Enterprise* level control has found a number of companies providing factory automation hardware and software.

Each of the previous digital control systems have advantages and disadvantages. As with all measurement and control systems, the best system for any given application is dependent upon many factors such as expense, degree of control/tolerance, integration with existing system components, required maintenance and so forth.

Homework

1 – Name the four essential components of a closed-loop control system.

2 – Briefly identify the difference(s) between open loop and closed-loop control.

3 – Explain the 2 functions performed by a common industrial closed-loop controller.

4 – Identify the function(s) of a common open-loop controller.

5 – A sensor/transmitter combination is calibrated to output 4-20 milliamps over an input span of 100 to 300 degrees Fahrenheit. Determine the output current at,

    a. 200 degrees

    b. 150 degrees

    c. 250 degrees

    d. Less than 100 degrees

    e. Over 300 degrees

6 – A closed-loop process controller is connected to the transmitter of the previous problem. If the controller input is rated at 4-20 milliamps and the Set Point is set to 200 degrees (50% of the input span) Fahrenheit, determine the <u>percentage</u> of process error (Ep) when the transmitter output is,

    a. 12 milliamps

    b. 8 milliamps

    c. 16 milliamps

d. 18 milliamps

e. 5 milliamps

7 – Given the sensor/transmitter and controller of the previous problems (5 and 6), determine the <u>percentage</u> of process error (Ep) when the transmitter input is,

a. 200 degrees

b. 150 degrees

c. 250 degrees

d. 100 degrees

e. 300 degrees

8 – Explain the purpose of the Span and Zero adjustments on a typical transmitter.

9 – Explain the difference between a reverse-acting and direct-acting controller.

10 – Explain the purpose of a final control element.

11 – Identify 2 final control elements not presented in the text material.

12 – Identify 4 process variables not presented in the text material.

Chapter 2 – Standards and Symbols

This chapter introduces the concepts of comparison, calibration and standards. Calibration and standards are used to assure measurement and control accuracy in process automation. Common symbols, process and instrument diagrams are also presented. Included topics follow.

Objectives:
Upon completion of this chapter, you should be able to:
- Explain the purpose and use of reference standards in measurements
- Demonstrate the application of common instrument symbols and diagrams
- Interpret common instrument, balloon and process diagrams
- Assemble a routine instrument diagram
- Suggest calibration techniques for common sensors and measurement systems

Introduction

Measurement and control of Industrial process variables such as pressure, level, flow and temperature relies heavily upon the process of comparison. Comparison of an accurately known *Standard* quantity to an unknown or questionable one allows investigation and if desired, elimination of the difference. Any difference between a standard and a specimen under test is often called an error and can signify the difference between the actual process value and the desired set point, as well as the difference between a meter indication and the actual quantity being measuring. The importance of the comparison concept cannot be overemphasized.

Measurement Standards

The National Institute for Standards and Technology (NIST, formerly the National Bureau of Standards, NBS) and similar institutes around the world are responsible for the ongoing documentation, distribution and inventory of measurement standards and calibration procedures. Standards institutes commonly provide calibration services, procedures and reference data to labs and states within their jurisdiction. The measurement standards and services generated in technical institutes such as NIST, Canada's Institute for National Measurement Standards (INMS), Europe's Euromet and New Zealand's Measurement Standards Laboratory (MSL) are provided to meet the needs of state and local governments, federal agencies, industry, and the scientific community for *traceability*, at necessary levels of accuracy, to national standards. Traceability allows a measurement made in the lab of a university to be very close in quantity to the same measurement of the same specimen made with different measurement apparatus in a different part of the world. The measurement services afforded by standards institutes also provide expert guidance and services regarding legal metrology to state and local governments, business, and industry to ensure measurement procedural uniformity. The end result of national standards institutes is to assure equity in domestic and international commerce.

Standards take many forms, from procedures, to components, to observing the response of materials when subjected to some physical quantity such as pressure or temperature. A measurement standard is a resource for comparison that allows differences, or errors in measuring instruments to be quantified and labeled or corrected. For measurement standards to be of value, they must periodically be compared with standards of tighter tolerances and higher stature. *Traceability* is the term used to describe comparison among standards of higher order. *Working* standards are commonly found in labs and industry and are used for periodic calibration of common measurement instruments. Higher-level *tertiary, secondary* and *primary* standards facilitate comparison between regions, geographical locations and among countries internationally and assure that all measurements fall within controlled tolerances independent of location and measurement conditions. The following diagram depicts the measurement standard hierarchy of traceability.

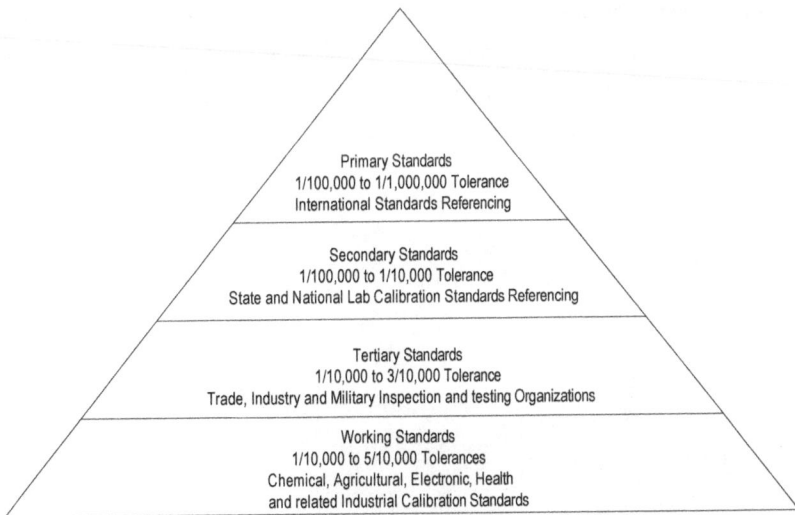

Primary Standards
1/100,000 to 1/1,000,000 Tolerance
International Standards Referencing

Secondary Standards
1/100,000 to 1/10,000 Tolerance
State and National Lab Calibration Standards Referencing

Tertiary Standards
1/10,000 to 3/10,000 Tolerance
Trade, Industry and Military Inspection and testing Organizations

Working Standards
1/10,000 to 5/10,000 Tolerances
Chemical, Agricultural, Electronic, Health
and related Industrial Calibration Standards

Figure 1 - The traceability pyramid of standards. Commonly used working standards are at the bottom. Each layer above represents tighter tolerances and higher stature in the comparison process.

At the top of the measurement hierarchy reside primary standards. National institutes such as the National Institute for Standards and Technology (NIST) in the United States, Canada's Institute for National Measurement Standards (INMS), Euromet and New Zealand's Measurement Standards Laboratory (MSL) administer primary standards and calibration related services. Standards residing at national institutes are used for comparison against standards of other countries and against secondary standards within the country.

Everything that is measured has a standard and all countries conform to the International System (or, SI for System Internationale – named after an international agreement on measurement units) standards of units. The International System (SI) of Units comprises seven clearly defined primary standard quantities. The quantities are length, mass, time, electrical current, thermodynamic temperature, amount of substance, and luminous intensity. From these quantities the base units of meter (m), kilogram (kg), second (s), ampere (A), kelvin (K), mol (mol), and candela (cd) are derived. Other units, known as derived units, can be determined from the seven base units. The following diagram illustrates.

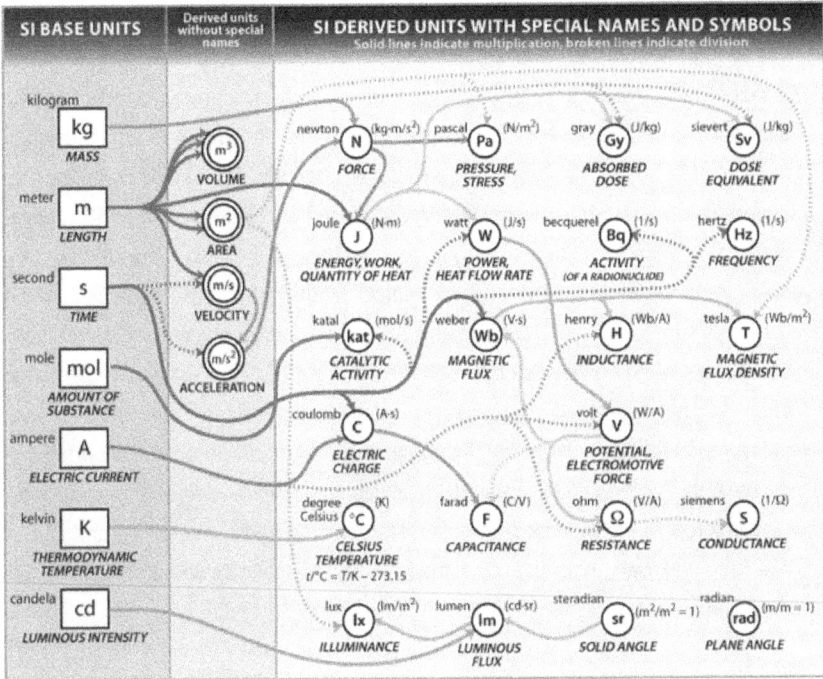

Figure 2 - The SI system of units. Combinations of each other unit define each of the 7 basic units of measure. Diagram courtesy of NIST.

Working standards are compared to tertiary standards, tertiary standards to secondary standards, and secondary standards to primary standards. Working standards such as manometers, standard (weight) masses, portable bridges and alcohol thermometers are most commonly accessible and are the only standards that are used "in the field." Tertiary and higher-level standards are generally expected to remain within a controlled environment.

Calibration

The ISA (the Instrumentation, Systems and Automation society), ANSI (American National Standards Institute, IEEE (Institute for Electrical and Electronic Engineers) and related scientific and industrial institutes develop procedural standards for use

47

in all aspects of industry. The ISA defines *calibration* as the procedure of evaluating, quantifying and/or correcting the error that is inherent in every measurement instrument. As might be expected, the underlying concept behind calibration is comparison of an accurately known standard value to an unknown or questionable value. Calibration consists of periodically using measurement apparatus such as voltmeters, ohmmeters, ammeters, thermometers, pH meters, etc., to measure a range of known working standards. The results are documented and if differences are noted adjustments are made within the instrument to cancel, or *Null* the error. If no calibration adjustment is available, observation of any difference between the standard and the "unknown" allows the error to be quantified, documented and compensated against in future measurements. Once calibrated, instruments are labeled with a calibration tag that denotes the date and if required, specifies the known errors and conditions. Calibration is at the foundation of the measurements and controls field and carries over into the daily activities of virtually every scientific "student" regardless of discipline.

Calibration labs, also called Metrology labs (*Metrology* is the field of study of measurements, standards and calibration), testing and standards labs, are located regionally throughout the world and use tertiary standards for calibrating and quantifying error in precision measurement instruments and working standards of all types. Calibration labs should ultimately be able to trace their standards to a national standards institute such as NIST. State divisions of "Weights and measures" routinely calibrate, inspect and tag measurement apparatus associated with commerce. Grocery store scales, grain elevators, automobile odometers, taxicab meters, gas pumps, and medical offices are mandated to perform routine calibration activities. The process is as simple as taking a standard gallon vessel and checking to assure that gas pumps are accurate. Next time you visit the gas pump look for the calibration sticker.

A variation on standards and yet another application of calibration is used at assembly and manufacturing plants. Test labs located within manufacturing facilities store individual standard parts, products or materials for each of the components assembled or produced at the plant. A plant assembling televisions will have each individual piece of every part of every assembled model within the plant stored in the test lab. Each of the individually stored components will be a working standard, used for measurement and comparison purposes whenever a manufactured or assembled part is deemed questionable.

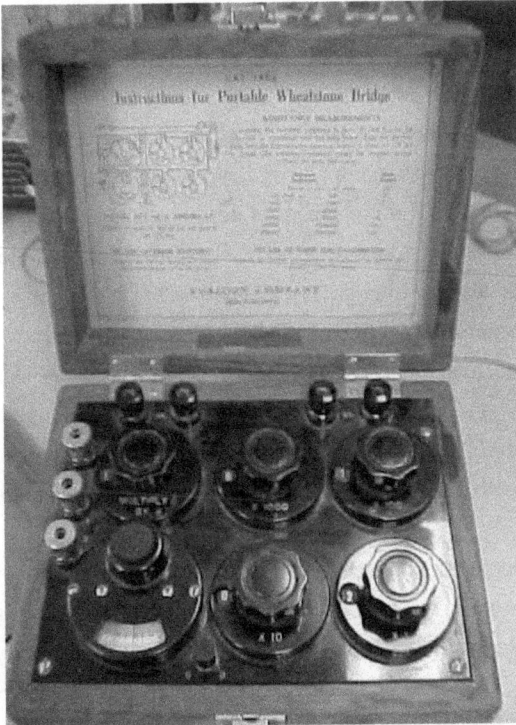

Figure 3 - An early portable Wheatstone resistance bridge. Wheatstone bridges are used to measure a wide span of resistance values but can also used be used as a precision resistors when required. The portable Wheatstone bridge is considered a working standard.

The following is an example of how standards are applied in manufacturing. Assume an assembler working in an assembly plant finds the individual components do not fit together as well as in the past. The assembler will call a technician to resolve the problem. The technician will observe the assembly process and decide if the problem lies within the process or the products. Assuming the assembly process is acceptable, the technician will make measurements of the assembled pieces and initially compare the measurement results with diagrams of the involved parts. Comparison of the measured value with a schematic diagram may resolve the problem. If not, the technician will take the part(s) in question to the test lab for comparison against the archived standard components. After measuring the standard and the component a difference should be obvious and a correction can be made. I've personally experienced this process on many occasions while employed as a quality engineer for Motorola. The 30-acre plant assembled automotive audio systems (AM-FM-8-Track, yeah, a long time ago), black and white monitors, and three models of color television sets. The test lab within the plant had archived a standard component for every part of every model for the occasions described previously. Inductors, resistors, integrated circuits, picture tubes (CRT's), deflection yokes and chassis components were inventoried for just the occasion described. Indeed, manufacturing standards of numerous forms must exist in order for a personal computer to communicate with any manufacturers printer.

The following photos illustrate common working standards. A Hewlett-Packard DC transfer standard, 100 Ohm and 1K Ohm wire-wound resistors, and a 1.019v Weston standard unsaturated cadmium cell are shown. The components are located in the Instrumentation Lab in building 2, room 110 at Black Hawk College and are used to provide periodic checks of accuracy of the multi-meters in the lab. Each of the components are designed to minimize the effects of temperature, humidity and age. The 100 Ohm and 1K Ohm wire-wound resistances sit in an oil-filled, insulated-wall container and are to be used at a temperature of 77 F (25 C) and a power dissipation of .01 Watt.

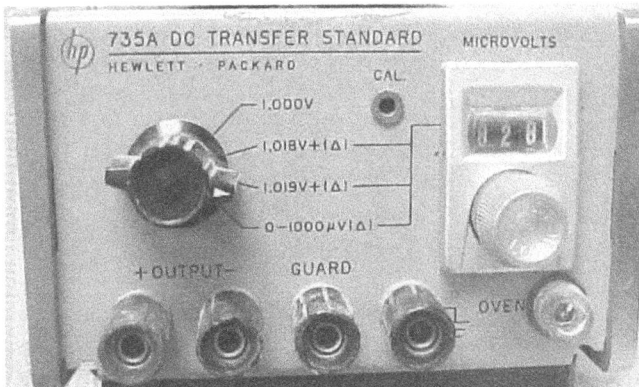

Figure 4 - A Hewlett-Packard Transfer Standard is used to generate very accurate micro - and millivoltages for the purpose of measurement instrument calibration.

Figure 5 – Two standard resistors of 100 and 1K Ohms used to calibrate resistance measurement apparatus.

Figure 6 – Internal construction of a working standard resistance. Great care is taken to maintain the value of standard components. The resistance in this photo is composed of fine gauge insulated wire. The coiled wire sits in an oil bath within the housing to reduce chances of having the resistance value change, or "Drift" as a result of the surface tarnish.

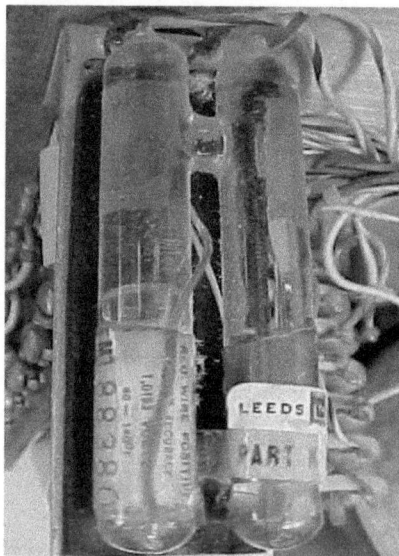

Figure 7 – Looking more like a couple of test tubes, the Westin standard cell holds a constant 1.0193 Volts (look closely at the left label in the figure, the voltage is indicated) for extended periods and loses the voltage quite rapidly upon aging. The Westin standard

cell is used in within portable potentiometric measurement bridges to calibrate the bridge as part of the setup procedure prior to use.

Terminology

The following NIST definitions apply to the hierarchy of standards.

Primary standard: A standard that is designated or widely acknowledged as having the highest metrological qualities and whose value is accepted without reference to other standards of the same quality.

International standard: A standard recognized by an international agreement to serve internationally as the basis for assigning values to other standards of the quality concerned.

National standard: A standard recognized by a national decision to serve, in a country, as the basis for assigning values to other standards of the quantity concerned.

Reference standard: A standard, generally having the highest metrological quality available at a given location or in a given organization, from which the measurements made there are derived.

Transfer standard: A standard used as an intermediary to compare standards.

Traveling standard: A standard, sometimes of special construction, intended for transport between different locations.

Working standard: A standard that is usually calibrated against a reference standard and is used routinely to calibrate or check material measures or measuring instruments. Working standards may also at the same time be reference standards. This is particularly the case for working standards directly calibrated against the standards of a national standards laboratory.

P&ID's and Balloon Symbology

Diagrams and symbols represent another form of industrial standard. In industrial instrumentation many forms of diagrams are available. The *process and instrument diagram* (P&ID) is among the more popular and is commonly used to represent the components, their functions and interconnections in chemical plants, petroleum refineries and power plants to perform a process operation. Process and power plants routinely use high pressure, high temperature and high energy to perform processing functions. As such, safety is an utmost concern. Process and instrument diagrams aid the safety effort by allowing one to determine the effect a change to one control loop or component can have on another loop in another portion of a plant.

P&ID's, their associated balloons and tag symbols are standard, meaning all companies performing similar processing functions with similar equipment will use the same symbols and conform to a similar set of standards associated with process and instrument diagrams.

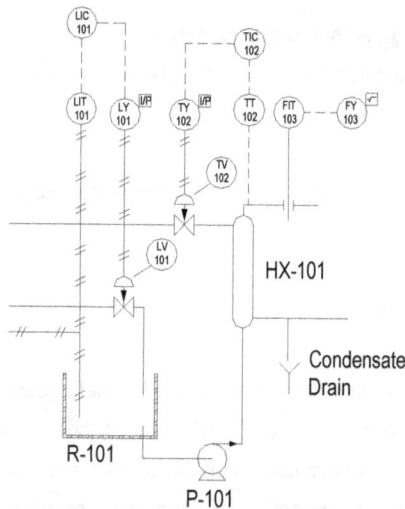

Figure 8 - An excerpt form a process and instrument diagram, or P&ID. Note the connection lines between symbols and the appearance of circular "Balloons."

Each of the symbols in the previous diagram represents a component's function, its location and the required interconnections to instrument a specific manufacturing process. The circular symbols are called "Balloons" and are used to represent transmitters, transducers, controllers, functional components (usually mathematical functions) and alarms. Within the balloons are 2, 3 or 4 letter designations to allow identification of the component's function, and a loop/component "Tag" number to facilitate determining the component's location within the overall plant. A components' loop number is unique. No two components within a given plant will have the same designation.

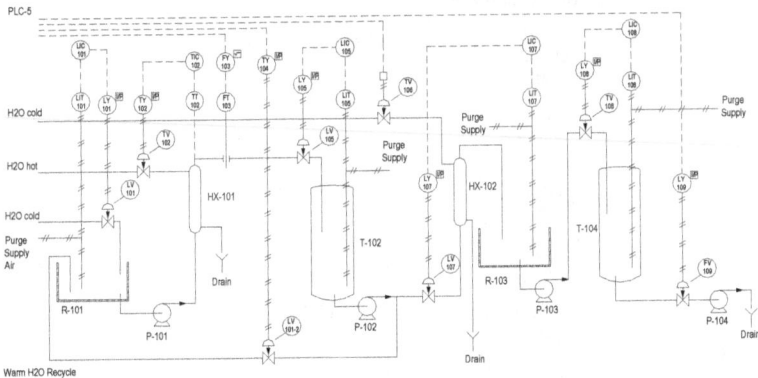

Figure 9 - Process diagram of the simulator in the instrumentation lab at Black Hawk College. More detailed diagrams of this system are located at the end of the chapter.

Within the individual balloon designations, the non-numerical tag references are usually in a group of letters representing the component function and the physical process variable being measured or controlled. A letter representing a process condition or description may also be present if required, as with alarms.

Determining the vast majority of balloon functions is fairly simple if one remembers the four most common process variables of pressure, level, flow and temperature. The first letter will designate the process variable type being measured or controlled. The letters most commonly used are P for pressure, L for level, F for flow rate and T for temperature. The second letter designates the function of the instrument represented by the balloon. Recalling the common control loop components of sensor (transmitter - T), controller (C), and final element (valve – V), the majority of component identifying designations can be determined. The letter "I" is used to designate an indicating device such as a pressure gauge, alcohol thermometer or digital panel meter, and "Y" represents a computational function such as a totalizing or square root mathematical function.

The following diagram illustrates balloon symbols, their characteristic function designations and loop locations.

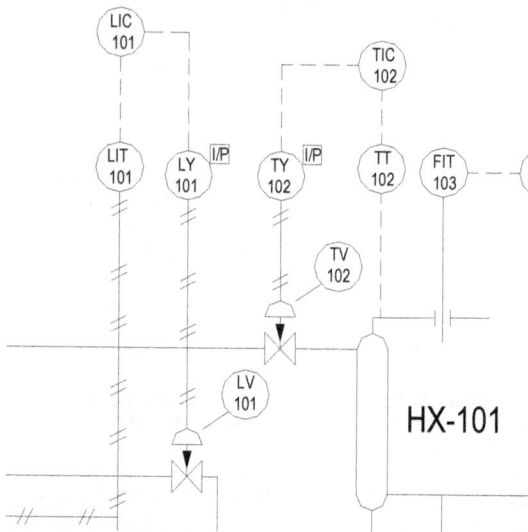

Figure 10 - Balloons represent common control loop components and contain tag information about the function and location of the component.

As an example of balloon tag interpretation, focus on the loop containing a component labeled LIT-101 in the previous diagram. Recalling the 4 common process variables of pressure, level, flow, temperature, and common control loop elements of transmitter, controller, recorder, indicator, alarm, etc., the first letter represents a level process; the second letter represents an indicating instrument and the third letter a transmitter. This component is a level transmitter in control loop #101 that is providing some form of local (meaning at the process) material level indication. LV-101 represents a level valve in control loop #101. TV-102 designates a temperature valve in control loop #102. Some designations are a little abstract. LHA represents a level high alarm; Y designates a general-purpose function and usually has an associated descriptor such as with the square root extractors in the previous diagram. Unusually exotic or out of the ordinary components are also designated on the process diagram through the use of a legend.

	Pressure	Level	Flow	Temperature
Transmitter	P T	L T	F T	T T
Indication	P I	L I	F I	T I
Controller	P C	L C	F C	T C
Valve	P V	L V	F V	T V
Indication - Transmitter	P I T	L I T	F I T	T I T
Indication - Controller	P I C	L I C	F I C	T I C
Indication Recorder	P I R	L I R	F I R	T I R
Alarm - Low	P A L	L A L	F A L	T A L
Alarm - High	P A H	L A H	F A H	T A H
Computation	P Y	L Y	F Y	T Y
Analysis - Analyzer	A P	A L	A F	A T

Figure 11 - Examples of balloon tag designations.

Balloon Symbol Variations

Instrument symbols assume sixteen basic configurations depending upon the location and type of instrument represented in the process diagram (P&ID). As a result of computer integration into automatic control systems, standards have been developed to provide symbols that represent analog and digital control system components. ANSI/ISA-S5.1 and ISA-S5.3 contain definitions of process terms associated with digital technology and should be considered when working with computer-integrated process measurement and control systems. Some of the symbols and terms addressed within the standards have been included following.

The four types of symbols are discrete instruments, shared controls or displays, programmable logic controllers and computer functions. Each can be either field mounted, board mounted, auxiliary mounted or mounted behind the board. The symbol types are detailed following the diagrams.

 Field mounted instruments are located at the process location, which is commonly outdoors or "In the field." Board mounted instruments are on the front of the control panel board and are considered accessible by the plant operators. Auxiliary mounted instruments are usually accessible to the operator but not located on the front or rear of the control panel. And behind the board devices are located behind the control panel but are not considered to be accessible by the operator.

Field Mounted Devices - Usually located near process vessels,
the point of measurement or final control element

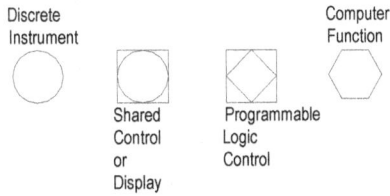

Discrete   Computer
Instrument   Function

        Shared   Programmable
        Control   Logic
        or   Control
        Display

Board Mounted Devices - Usually grouped with other
instruments near operator access

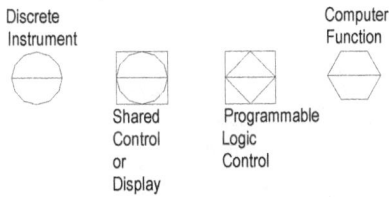

Discrete   Computer
Instrument   Function

        Shared   Programmable
        Control   Logic
        or   Control
        Display

**Figure 12 - Field and board mounted devices will assume the symbol forms indicated in
this diagram. The identification letters and tag number designations are normally included
within the symbol.**

Auxiliary Mounted Devices - Usually accessible to the operator
Discrete   Computer
Instrument   Function

        Shared   Programmable
        Control   Logic
        or   Control
        Display

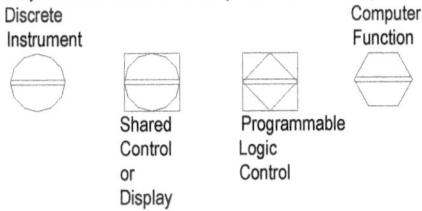

Behind the Board Devices - Normally not accessible by the
operator

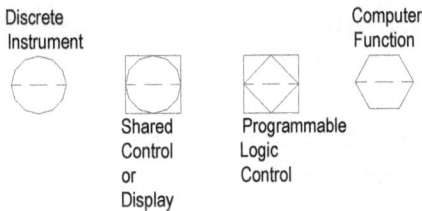

Discrete   Computer
Instrument   Function

        Shared   Programmable
        Control   Logic
        or   Control
        Display

Discrete instruments are common control loop components. Single-loop controllers, transmitters, analytical instruments, indicators, computational components and transducers are examples of discrete instruments represented by balloon symbols and the balloon symbol variations.

Discrete Devices

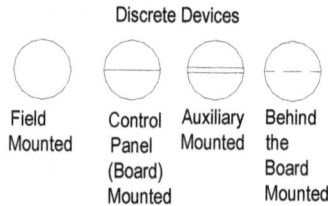

| Field Mounted | Control Panel (Board) Mounted | Auxiliary Mounted | Behind the Board Mounted |

Shared controls or displays include monitors or similar informational and control devices capable of interfacing with multiple process variables. Commonly a computer or digital system running a pre-programmed algorithm, a shared display can monitor and control an entire process from a remote location using a single device.

Shared Control or Display

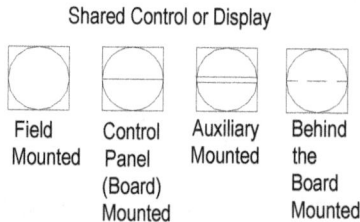

| Field Mounted | Control Panel (Board) Mounted | Auxiliary Mounted | Behind the Board Mounted |

Programmable logic controllers (PLC's) are commonly used to distribute control functions throughout a manufacturing operation and are interconnected with multiple inputs and outputs of an analog and digital nature. PLC's perform process measurement and control functions of an analog and discrete (digital) nature but usually do not contain monitors or input devices such as keyboards at the PLC location. Often PLC process data is provided to shared displays.

Programmable Logic Controls (PLC's)

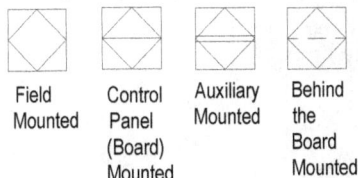

| Field Mounted | Control Panel (Board) Mounted | Auxiliary Mounted | Behind the Board Mounted |

Computer functions involve computers and related data transmission, network components, and programming devices commonly found in computer-integrated process operations. Servers, switches, routers, handheld programmers, dedicated data acquisition devices, dumb and smart terminals, and related hardware devices are included as hexagon symbols on process diagrams when appropriate.

Computer Functions

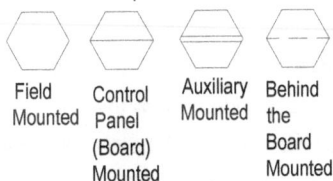

| Field Mounted | Control Panel (Board) Mounted | Auxiliary Mounted | Behind the Board Mounted |

Again, each of the four types of component symbols can be mounted in four different locations. Field mounted instruments are located at the process location which, commonly outdoors or "In the field." Board mounted instruments are on the front of the control panel board and are considered accessible by the plant operators. Auxiliary mounted instruments are usually accessible to the operator but not located on the front or rear of the control panel. And behind the board devices are located behind the control panel but are not considered to be accessible by the operator.

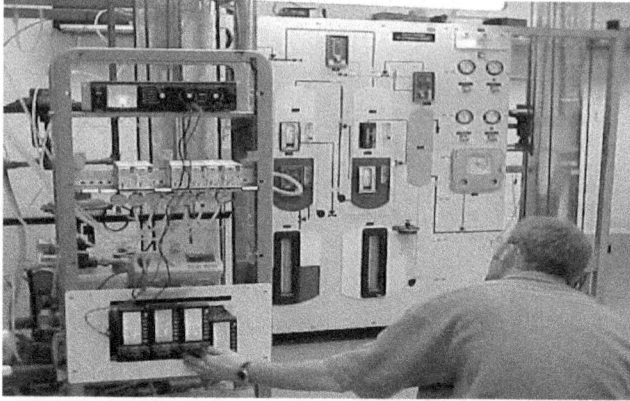

Figure 14 - The control panel board, is obvious in the center of this photo. Instruments are located on the front and back of the panel as well as on the process vessels "in the field." Balloon variations allow one to view a P&I diagram and determine where a component is located. The operator in the photo is making adjustments to an auxiliary located device.

Signal Line Symbology

Balloon symbols on P&ID's are interconnected by a series of line symbols. Line symbols provide a means of identifying the various forms of signals and communications between components. A heavy solid line indicates a mechanical process piping connection. A narrow solid line represents an instrument piping connection as between the process piping and FIT 101 in the following diagram.

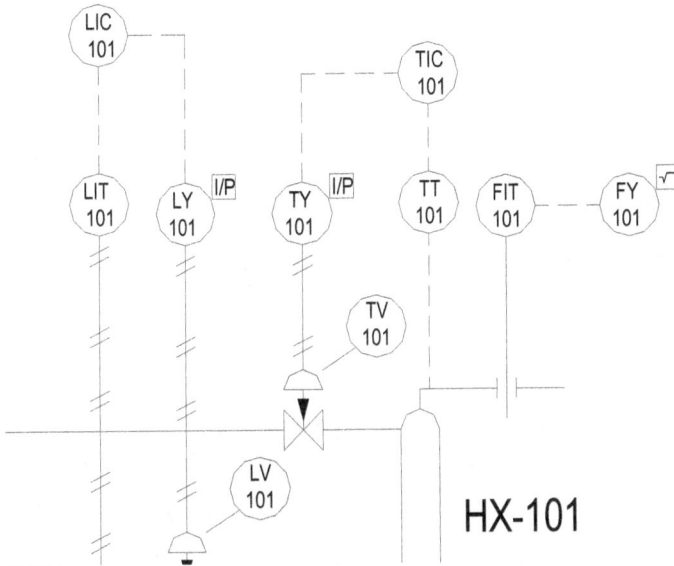

Figure 15 - Connection lines on process and instrument diagrams represent the various means of communicating a signal between components.

Solid lines with short diagonal pairs such as found in the lower left of the previous diagram represent compressed air pneumatic signal lines. Pneumatic lines are commonly connected to valves, as proportioning valves require a span of compressed air values for positioning. The dashed lines in the diagram represent electronic signal connections. Electronic symbol lines usually carry 4 to 20 milliamps but may also be 1 to 9 volts, 1 to 5 volts or 10 to 50 milliamps.

Other forms of line symbols exist for hydraulic, sonic, computer data, capillary and similar line symbols. Most are included on the following diagram.

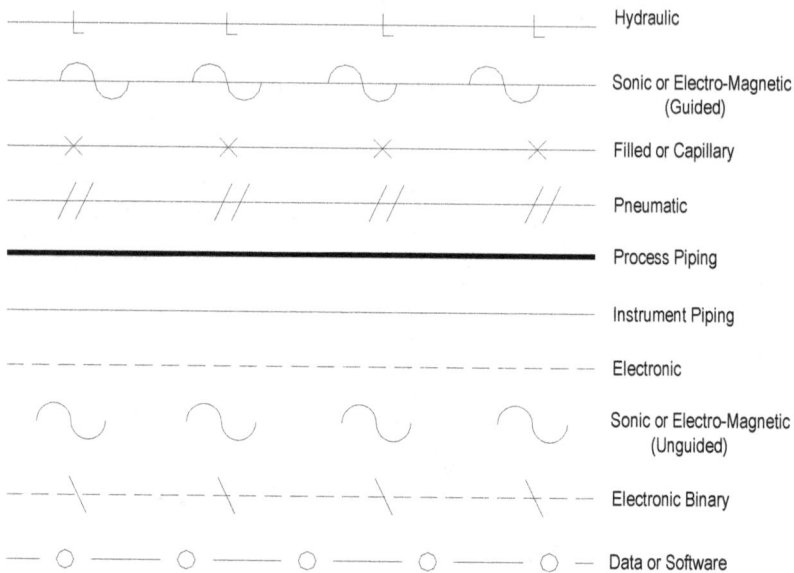

	Hydraulic
	Sonic or Electro-Magnetic (Guided)
	Filled or Capillary
	Pneumatic
	Process Piping
	Instrument Piping
	Electronic
	Sonic or Electro-Magnetic (Unguided)
	Electronic Binary
	Data or Software

Figure 16 - An overview of common signal communication line symbols. Note: These symbols are standard to United States process diagrams and may differ internationally.

Valves – Anatomy, Actuation and Symbology

Since every closed-loop control system contains a final control element, usually in the form of a valve and since valves do not conform to the balloon symbol convention as other discrete control loop components, a word about valve symbology is in order.

Final control elements are commonly composed of two basic parts, the actuator and the valve itself. Valves are available in many types but among the more popular are on/off and proportioning valves. On/off valves are also called solenoid valves and are capable of assuming two operational states, open or closed.

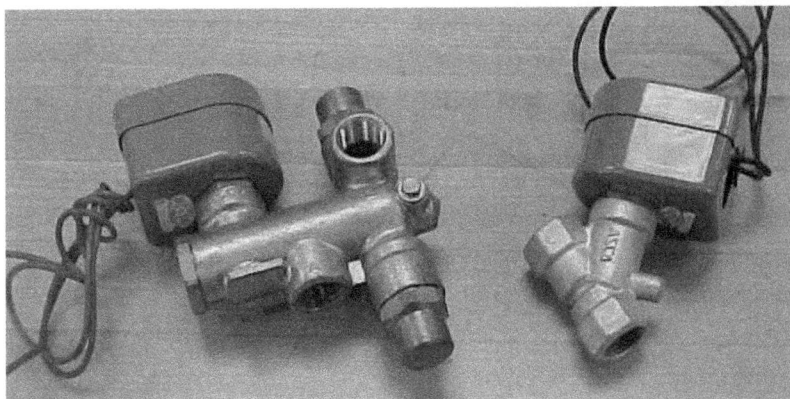

Figure 17 - Two solenoid valves, the actuators are the rounded rectangular boxes with leadwires connected at the top of the devices. The valve itself is the component in contact with the process material. The solenoid valve on the left is a "3-way valve" however; the three process connections are not all obvious in the photo. The device on the right is a two-way valve.

Proportioning valves are capable of assuming any position between fully open and completely closed and require an input signal span (such as 3-15 psi) to be applied to the actuator for positioning. Proportioning valves are either <u>air-to-open</u>, meaning an increase of input air pressure to the actuator will cause the valve to open, or <u>air-to-close</u>, meaning an increase in the applied actuator air pressure will cause the valve to close. More recently however, the air-to-open and air-to-close references have been replaced fail-closed (FC) and fail-open (FO) respectively. "Fail" in this context means a loss of air signal or failure of the actuator hardware that would not be able to position the valve. The FC or FO designation is usually associated with final control elements on P&ID's and is significant if one wishes to anticipate problems. Again, safety first.

The following air-to-close (FO) valve and actuator diagram assists with understanding valve symbology and operation by labeling the common components of a proportioning control valve. Proportioning valves open, or close in proportion to the amount of signal (usually 3-15 psi of compressed air) applied to the actuator. The actuator applies the compressed air to a diaphragm area and spring arranged in a force balance assembly. In operation, an air pressure increase at the actuator input increases the force generated by the diaphragm area; the spring compresses proportionally and strokes the valve in a direction consistent with the applied air pressure. The valve will stroke downward if the air is applied to the upper portion of the actuator as in the following diagram. Valves will stroke upward if the air pressure is applied below the diaphragm seal, the balancing spring will be located above the diaphragm as noted in two of the following diagrams.

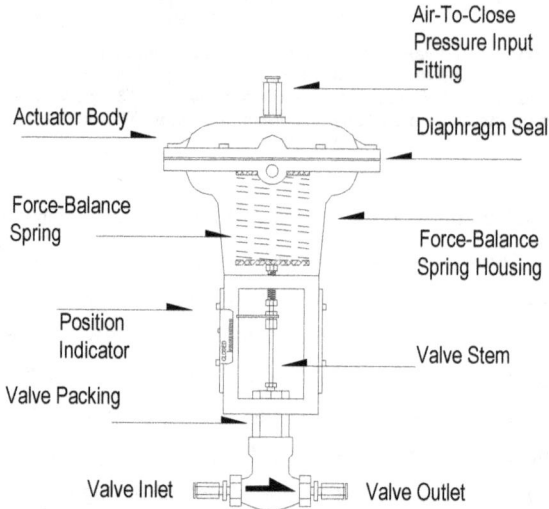

Air-To-Close Pressure Input Fitting

Actuator Body

Diaphragm Seal

Force-Balance Spring

Force-Balance Spring Housing

Position Indicator

Valve Stem

Valve Packing

Valve Inlet

Valve Outlet

Figure 18 - An air-to-close (Fail Open) proportioning control valve.

Figure 19 - An air-to-open (Fail Closed) proportioning control valve. Note the input air pressure signal is applied to the lower portion of the diaphragm. The balancing spring is located in the circular area above the diaphragm seal.

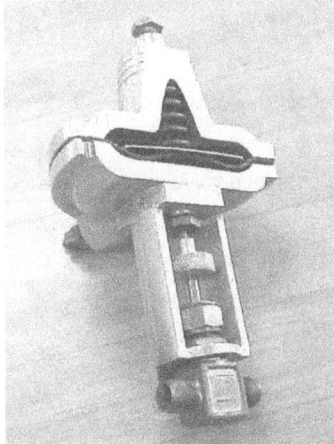

Figure 20 - A cutaway view of an air-to open proportioning valve. Note the diaphragm seal, location of the balancing spring in the upper portion of the actuator, and the air input fitting (back of actuator) below the diaphragm.

Additional actuator symbols are included in the following diagram.

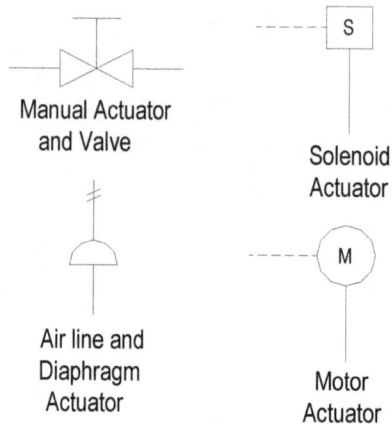

Manual Actuator
and Valve

Solenoid
Actuator

Air line and
Diaphragm
Actuator

Motor
Actuator

Figure 21 - Actuator symbols are always used in conjunction with a valve symbol.

In review, actuators are responsible for positioning control valves and the valve itself is in direct contact with the manipulated energy or process material.

Valve Symbology

Valve symbols greatly reflect the construction and operation of a control valve. To address the numerous types of control valves available to the field of instrumentation, a series of unique valve symbols exist. Control valves are better defined on P&ID's than balloon symbols so that specific valve types can be located in the field with relative ease. Following is an overview of some of the more popular forms of valves and symbols.

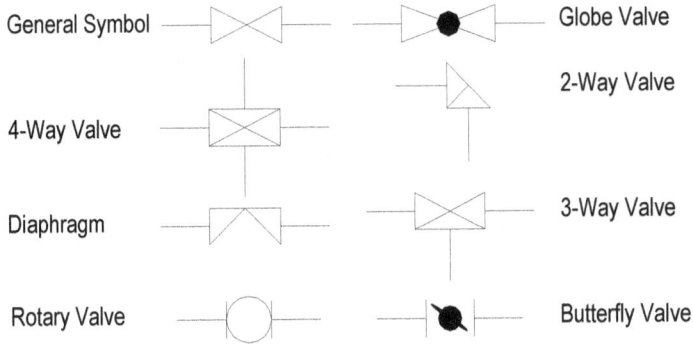

General Symbol ——————▷◁——⊢—— ———▷●◁——⊢—— Globe Valve

——————△—— 2-Way Valve

4-Way Valve ——————▷◁◁—————

3-Way Valve
Diaphragm ——————▷△————— ——————▷◁——⊢—— 3-Way Valve

Rotary Valve ——————◯—————— ——————●◁—⊢—— Butterfly Valve

Figure 22 - Common valves symbols without the corresponding actuator symbols. Each valve is uniquely constructed and is assigned a unique symbol.

Each of the preceding valves finds unique applications in instrumentation control systems depending upon the task to be performed. Globe valves are used in applications where a tight valve closure is required. A two-way valve is a common two position, on/off solenoid valve. Three-way and four-way valves have multiple connections and are used for diverting fluids. Angle control valves are designed to be located in turns and bends in piping. Butterfly valves use a rotating vane to provide closure. A rotary valve uses a turning motion to provide valve closure, similar to a home water faucet.

Proportioning valves and their associated actuators are symbolized as in the following diagram.

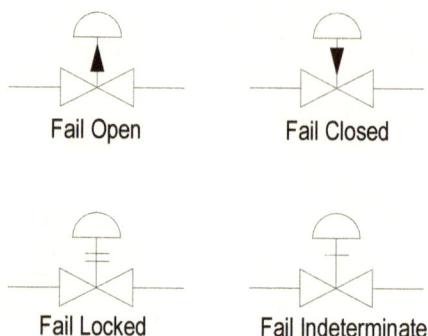

Figure 23 - Common proportioning valves with failure modes indicated.

As can be seen in the previous diagram, the symbols provide failure information that can also be translated into an operational description. Fail Open valves require an increasing air pressure signal to close. Fail closed valves require an increasing air pressure to open. Two other types of valve failure modes are Fail Indeterminate (FI), and Fail Locked (FL). A Fail Locked valve will hold the last position value when the power (input signal) failure occurs. A Fail Indeterminate valve may assume any position after a failure.

Electronic Loop Diagrams

Process and instrument diagrams (P&ID) depict the instrument components, process vessels, their functions and interconnections within a manufacturing process control system. *Loop diagrams* illustrate the electrical connections between the instrument and control components included within the P&ID. Pneumatic loop diagrams are available when pneumatic control systems are utilized and show the interconnections of all pneumatic signal lines and components. Loop diagrams provide valuable troubleshooting assistance by facilitating connection terminal location and as a result, signal tracing.

The loop diagram includes a small portion of a process diagram, just enough to allow one to focus upon the components of a single control loop. Hence the name, loop diagram. As an example of how loop diagrams are used, consider the temperature loop in the following process diagram.

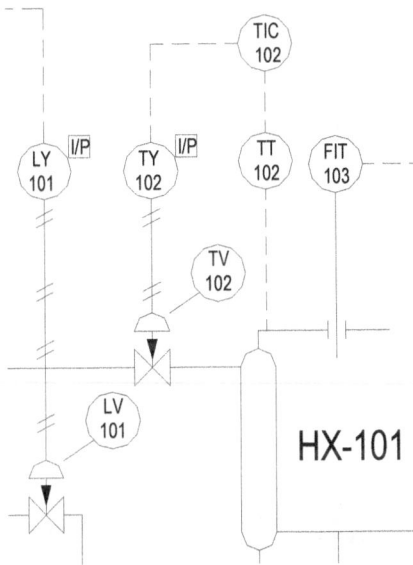

Figure 24 - As a first step in interpreting a loop diagram, locate the temperature control system 102. The loop is composed of TT-102, TIC-102, TY-102 (an I/P transducer) and TV-102.

The temperature control loop is composed of TT-102, TIC-102, TY-102 (an I/P transducer) and TV-102. From the line symbology it can be determined the sensor is an electrical/electronic component on the input to TT-102, and the output of TT-102 is also an electrical/electronic signal, probably 4 mA to 20 mA. The transmitter output is provided to an electronic controller, TIC-102, and the controller output is connected to TY-102, the current to pressure transducer used to convert the controller output signal (again, probably 4 mA to 20 mA) into 3 to 15 psi to position the actuator and valve. Although the P&ID shows all of the interconnected components, it does not show the junction boxes and connection terminals to electrically connect each of the components together into a control loop. The loop diagram is required to run-down or "Ring-out" the loop connections.

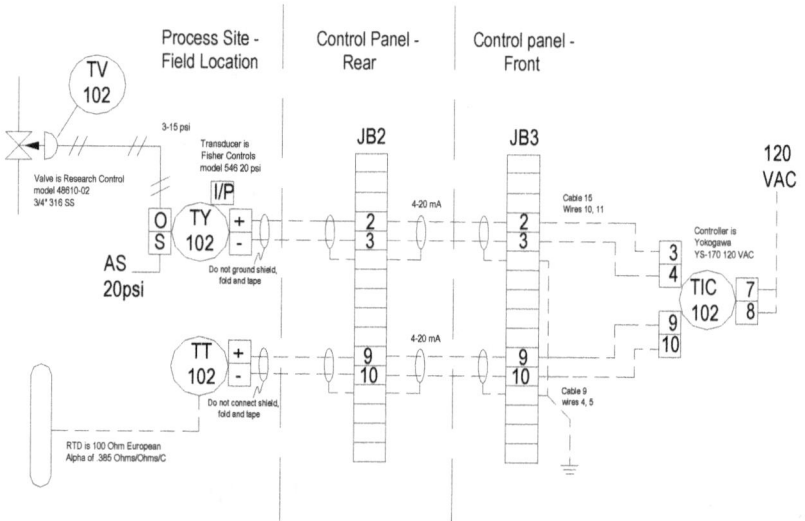

Figure 25 - In its basic form, the loop diagram provides connection information among the components for a single control loop.

Interpreting a loop diagram is performed by first locating the transmitter. Then, follow the transmitter connections through the junction boxes and terminal strips to the controller in a manner similar to the signal flow path. The controller output can then be located and the signal flow to the transducer and final element can be determined. Note the junction box and component locations in the field, and on the rear and front of the control panel. An actual loop diagram would also specify the individual component manufacturers and model numbers, cable bundle and leadwire numbering, and any other appropriate information to assist in locating and if necessary, replacing components within the loop.

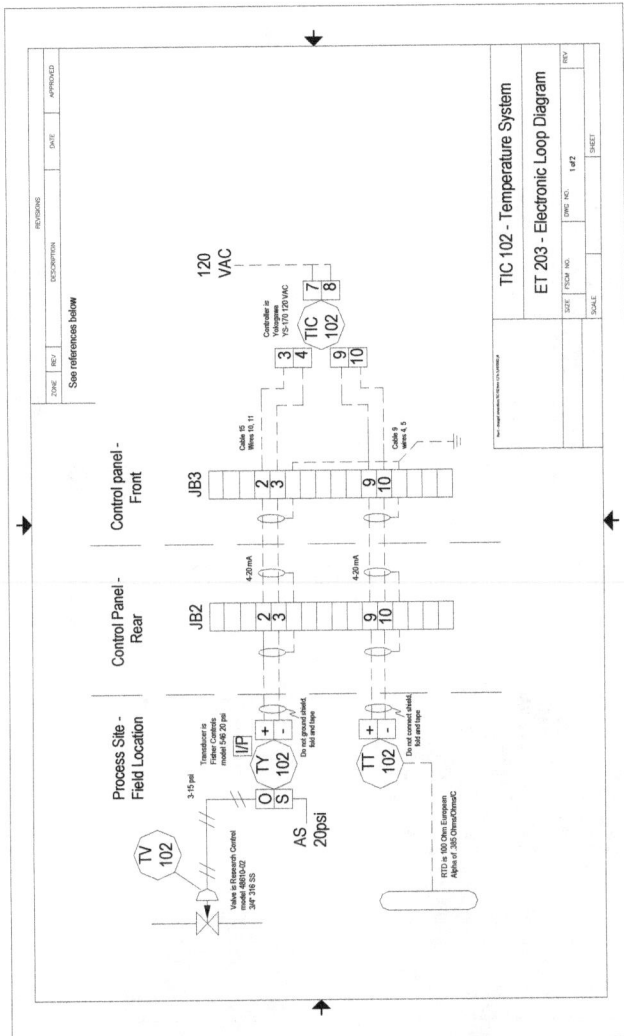

Figure 26 - Loop Diagrams contain information regarding the components and corresponding interconnections of a given control loop. (ANSI B title block.)

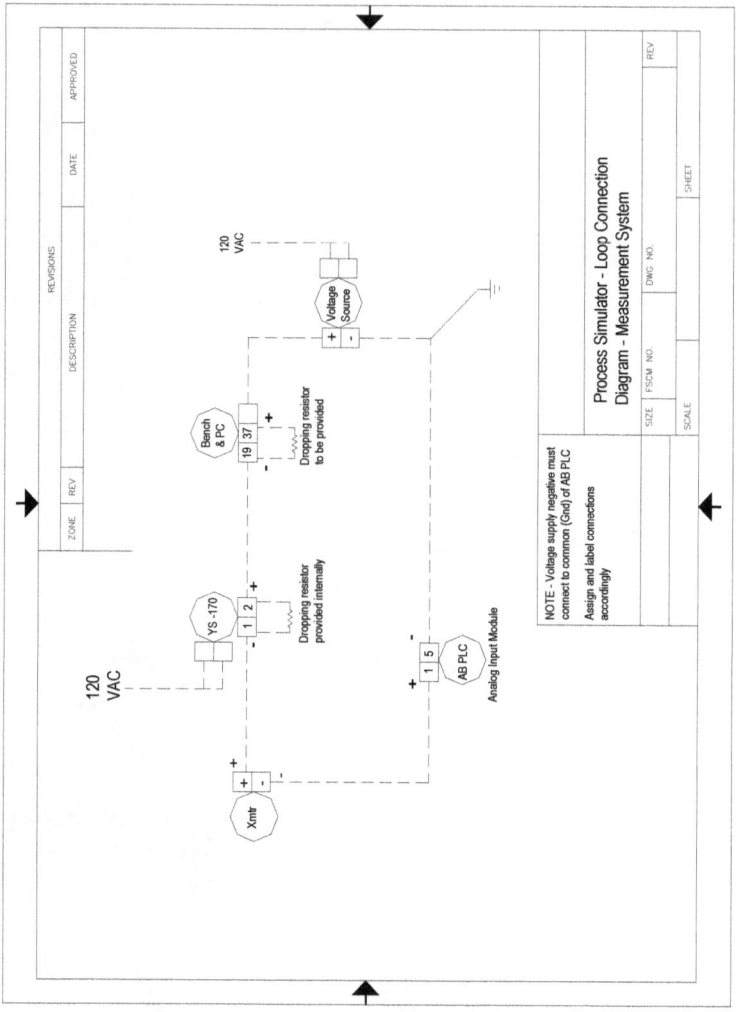

Figure 27 - A loop diagram of the process measurement system in the lab. Each of the six loops on the process simulator is connected as indicated.

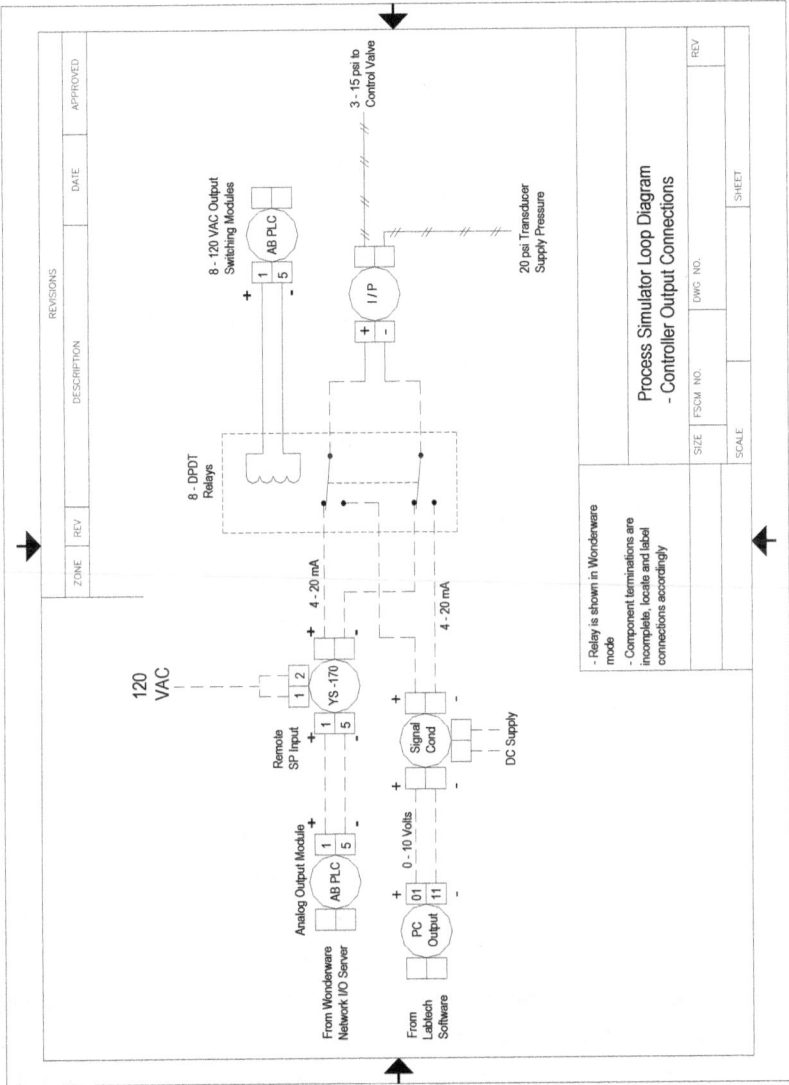

Figure 28 - A loop diagram of the output connections in the lab. Each of the six loops on the process simulator is connected as indicated.

Pneumatic Loop Diagrams

As was seen in the previous example, loop diagrams provide information about the components that compose a specific loop. Component function, tag name and number, type and model number, power requirements, signal span and connection information are clearly labeled on loop diagrams. Pneumatic loop diagrams are similar to electronic loop diagrams except that all of the components within the control loop are operated using compressed air, or pneumatically. Little or no electrical/electronic component information is contained within pneumatic loop diagrams.

The following represents a pneumatic loop diagram.

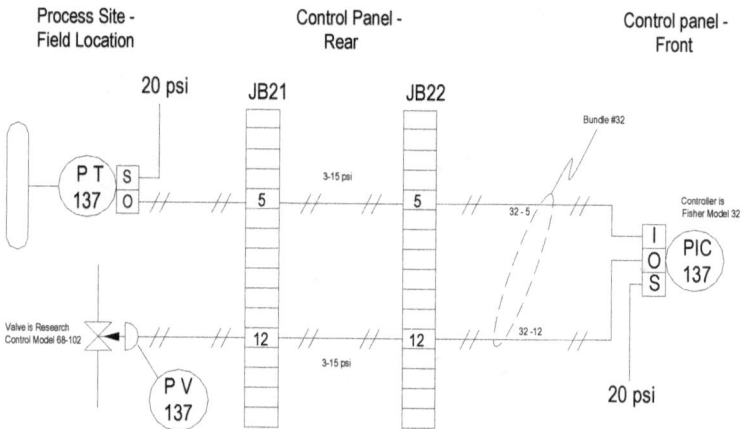

Figure 29 - A pneumatic loop diagram.

Interpreting the pneumatic loop diagram is identical to the electronic loop diagram. Locate the process or transmitter and follow the diagram through each of the control loop components and junction boxes to the final element. The diagram is placed within an American National Standards Institute (ANSI) "B" title block following.

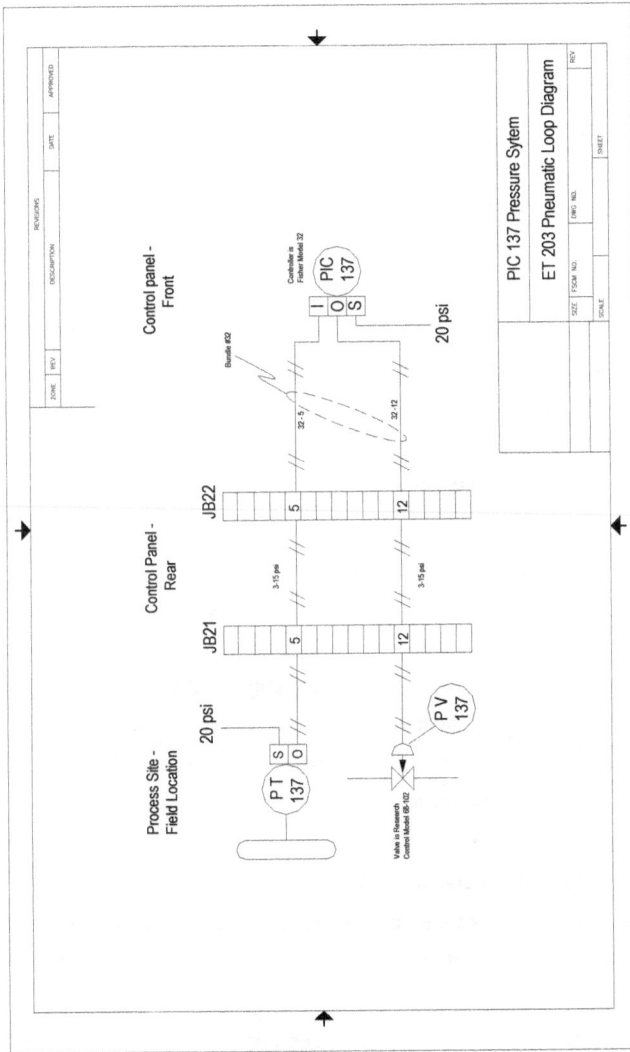

Figure 30 - A pneumatic loop diagram and ANSI B title block.

Project Methodology

In order to assure that all individuals become introduced to the sequence of events required to automate an existing process, a project methodology is being provided in the form of 10 Lab Performance Exercises. The project methodology is currently composed of ten distinct steps. Over the course of the next semester each team will be required to address and present each of the following items. The methodology steps will be assessed as quizzes. The steps include:

1 -System familiarization

> What components are involved and how are they interconnected? What is the function of each?

2 -System quantification

> What are the operational quantities including signal spans, power and pressure supplies, volumes, temperatures, rates, etc?

3 -Connection "Ring-out"

> Determine the system integrity including continuity checks, signal injection and operational verification of individual components.

4 -Assessment

> Determine the operational status of each component within the loop based upon the information acquired in the previous step(s). Determine the operational status of the existing system and become familiar with the manual-mode system operation.

5 -Proposal

> What needs to be done to make the existing system operational in a manual mode? What needs to be done to make the system operational in an automated mode? What additional components are required?

6 -Development

> The hardware and software is assembled and checked in a sequential and organized manner. Concurrently, the system continues to be operated and assessed in a manual mode.

7 -Documentation

Documentation to support the development and implementation stages is assembled. The documentation should be adequately detailed to support the implementation and also any future maintenance, fault diagnosis, repair, modification and system operational understanding. Remember that you are the contractor, not the client

8 -Implementation

An organized and sequential phase-in of the proposed and approved automation is initiated. Each step of the phase-in is assessed before proceeding. The system continues to operate in a manual mode.

9 -Operational Assessment

Each loop is individually brought "On-line." The loop is cycled throughout its range of operation, assessed and approved before continuing to subsequent loops.

10 -Review and follow-up

The complete system is assessed as a single, collective process. Long-term concerns are anticipated and addressed. Documentation is updated and approved by all involved parties.

ET 203

Methodology Performance Exercise #1 – 30 Points

Methodology Steps 1 and 2 - System Familiarization and Quantification

This will be an assessed performance evaluation worth 30 points. Within the next ten days, your team will be asked to provide evidence of the indicated methodology items. A brief presentation is expected of all teams.

Objective:

To locate each component within the assigned process loop, and to determine the function, capabilities and capacities of each component within the loop.

Procedure:

1 - Investigate the assigned process system by locating all components including the process vessel, sensor, transmitter, controller (or computer if appropriate), transducer, final control element and other ancillary hardware.

2 - Determine the corresponding signal, supply and other required connections to each component within the process loop. Discretely label all required connections.

3 – Determine the amount and type of required supply voltage(s), pressure(s), and other required material(s) to be used in conjunction with the process system. Measure and record all significant information including vessel dimensions, volume, material weights and densities, etc.

4 – Assemble a detailed diagram of your system. Use ISA symbology and include component labels and quantities from the previous item.

5 – Organize and prepare an informal briefing for the other teams.

ET 203

Methodology Performance Exercise #2 – 30 Points

Methodology Steps 3 and 4 – Connection Ring-Out and Assessment

This will be an assessed performance evaluation worth 30 points. Within the next ten days, your team will be asked to provide evidence of the indicated methodology items. A brief presentation is expected of all teams.

Objective:

Determine the connection and operational integrity of each component within the system.

Procedure:

1 – From the diagram assembled in the previous steps, determine the individual connections between each component in the system.

2 - Determine the appropriate pressures, currents, voltages and polarities of the connections.

3 – Using a visual inspection, assess the operational integrity of the components and corresponding connections. Assure that all connections are known, labeled and correctly documented.

4 – Using continuity or other appropriate measurement techniques, determine the connection integrity and document the measured results.

5 – Introduce an appropriate supply to the required components.

6 – Introduce the appropriate signal to each component individually (if possible). Closely observe the response of each component and most importantly, *realize the implications of your actions upon all related systems,* or you may spend the

remainder of the period mopping. Operate the components throughout their intended range while closely observing the response. Check for inconsistencies, hick-ups, non-linearity's and note ambiguities and discrepancies. Determine if the inconsistencies are system threatening. Determine and document the implications of all operational discrepancies. It's best to highlight or keep the discrepancies separate for future reference when troubleshooting. In short, anticipate.

7 – Assess the system operational integrity and become familiar with manual operation. Attempt to maintain a constant set point through process observation and manual manipulation of the final control element. Develop an intuitive feel for the process.

8 – Organize and prepare an informal briefing for the other teams.

ET 203
Methodology Performance Exercise #3 – 30 Points

Methodology Steps 5 and 6 – Proposal and Development
This will be an assessed performance evaluation worth 30 points. Within the next ten days, your team will be asked to provide evidence of the indicated methodology items. A brief presentation is expected of all teams.

Objective:
To assemble a proposal for automating the current manually operated process. To assemble an implementation strategy for a seamless transition to automation.

Procedure:
1 – Continue to assess the operation of the manually operated process. Document all significant observations. Perform and document any required repairs. Observe and document operational values of all components when the

process is manually operating at set point. Document gain and offset values of all components.

2 – Assemble a brief proposal addressing the purpose, justification (get creative, this will usually be performed prior to your getting the contract) and components of the automation project. Support your proposal with hardware and software diagrams. Include a proposed implementation strategy sequence. Have your "seat" together and present the proposal. Solicit approval from the client and other teams.

3 – While continuing to observe the system in manual operation, begin assembling the software, and acquiring the hardware components. Perform an operational checkout (similar to MP exercise #2) of all newly acquired automation components. Perform any required initial maintenance and calibration of newly acquired hardware items.

4 – Perform an incremental operational checkout of the assembled software using input and output simulations. Do NOT assemble all of the software and then perform the operational check. Proceed in operational modules or divisions. Integrate the hardware if appropriate.

5 – Review the implementation strategy. Anticipate and document perceived problems. Review previous discrepancies and check for solutions in the proposed automation strategy.

6 – Organize and prepare an informal briefing for the other teams.

ET 203

Methodology Performance Exercise #4 – 30 Points

Methodology Steps 7 and 8 – Documentation and Implementation

This will be an assessed performance evaluation worth 30 points. Within the next ten days, your team will be asked to provide evidence of the indicated methodology items. A brief presentation is expected of all teams.

Objective:

To assemble the documentation required to support the project implementation, and to begin the transition to automatic control.

Procedure:

1 – Minimize and organize the documentation to support the sequential conversion to project automation. Organize any support documentation such as component specifications, process and connection diagrams, I/O tables, operational characteristics (typical outputs when operating at set point, gain and offset values of signal conditioners, transmitters, transducers, etc.) and other relevant items.

2 – Begin a logical and sequential implementation. Check the operation of each component after installation. Do NOT assemble multiple components unless operation can be verified or substantiated as a functional unit. Avoid troubleshooting more than two components concurrently. Use caution when connecting or disconnecting components. Again, _realize the implications of your actions upon all related systems,_ or you may spend the remainder of the class period mopping. Continue to install the project hardware and software components individually.

3 –Once installation and operational verification of all components is complete, contact all other teams involved with processes related to yours and inform of

your current status. Once approved, bring the process up to operating set point manually. This is the _start-up_ phase of the automation.

4 – Observe the operation of all components while operating manually at a value close to, or equal to the set point. Once your team is confident that all components are operating correctly and the process is stable, switch the controller to automatic. Observe the response. If the process deviates rapidly from set point, place the process back into manual and re-acquire control of the process. Assess the operational discrepancies of placing the system into automatic control. Review any previously noticed discrepancies and determine their effect upon the automation.

5 – Attempt to isolate and repair any problems associated with the start-up automation. Repeat step 4 until the process holds a set point. Provide a briefing while demonstrating system operation.

Methodology Performance Exercise #5 – 30 Points

Methodology Steps 9,10 – Operational Assessment and Review/Follow-up

This will be an assessed performance evaluation worth 30 points. Within the next ten days, your team will be asked to provide evidence of the indicated methodology items. A brief presentation is expected of all teams.

Objective:

To observe the operation of the automated system and determine any inconsistencies and operational limitations. To assemble any support documentation.

Procedure:

1 – Complete any required troubleshooting begun in the previous MPE. If the problems appear to be irreconcilable, consider returning to the engineering aspect of the automation in the proposal step. The problems may not be associated with the hardware but with the design. If so, make the necessary modifications to the proposal and follow-through with the remaining steps.

2 – If the system appears functional, exercise the system by varying the set point throughout its intended range of operation. Assess system operation as a stand-alone loop and as a collective, multi-loop system.

3 - Review the documentation and make any required updates and modifications. Attempt to be as specific as possible. Provide as much information as possible. Ideally, you don't want to return to this site again. Organize and prepare the documentation for final form.

4 - Educate and assist the system technicians and operators.

5 - Prepare a final briefing and a large bill. Plan your early retirement.

6 - Visit the facility at a later date to assure long-term operation. Offer to make arrangements with the plant manager to play golf and drink heavily. Remember that one successful project usually leads to additional project opportunities.

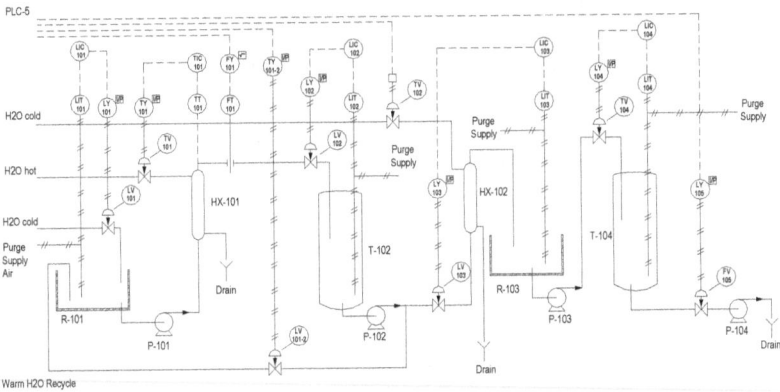

Figure 31 – The P-1 Process Simulator. Each student should become familiar with all systems and components noting any discrepancies or modifications. The components are not labeled to facilitate the familiarization process. The diagram is drawn as an AutoCAD drawing (.dwg) using ISA symbology. It is suggested that each student label the vessels, components and signal lines using ISA symbology. For assistance review chapter 1 of the manuscript. A larger copy of the diagram appears on the following page.

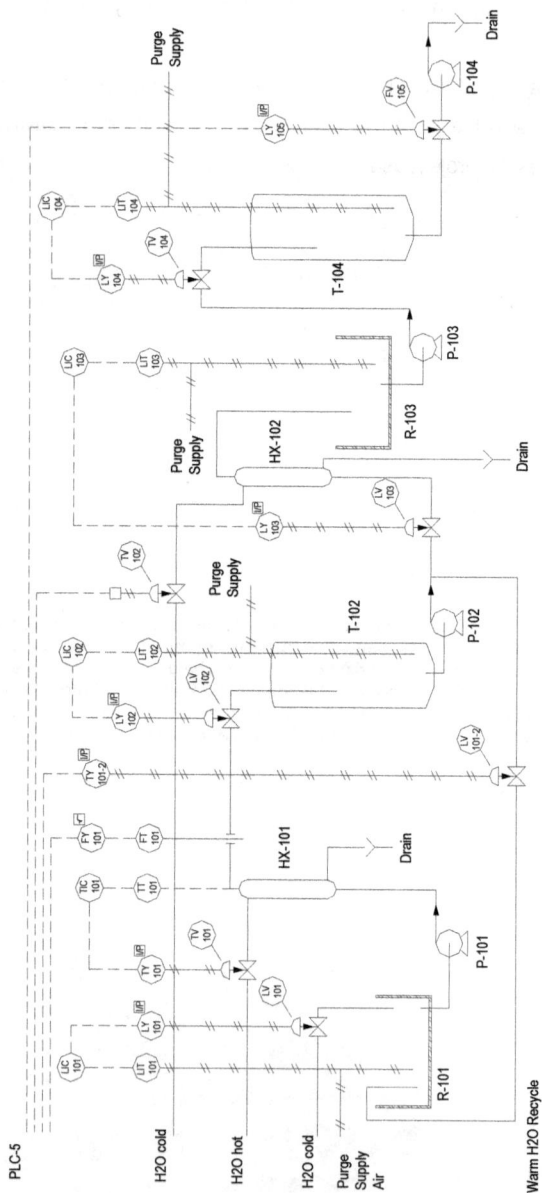

Homework

1 – Early in the chapter an example is provided of comparison and standards in an assembly plant. Suppose the comparison of the questionable part to the standard in the final portion of the example yielded no obvious differences. Where would you look for discrepancies next?

2 – Perform a web search on "Measurement Standards". Search for standards institutes from 10 foreign states and/or countries. Document the state or country, the purpose and function of the organization.

3 – Explain how standards maintain consistency of measurements made around the world.

4 – Refer to the P-1 process diagram. Identify the function of all balloon symbols.

5 - Refer to the P-1 process diagram. Identify the various forms of process and instrument connection symbology. Each of the types are represented in the following diagram.

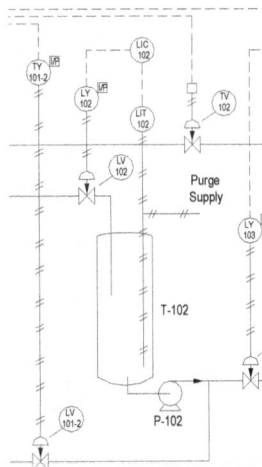

6 - Refer to the P-1 process diagram. Identify the final control elements and determine if each is Fail-Closed (FC) or Fail-Open (FO). All are similar to those in the following diagram.

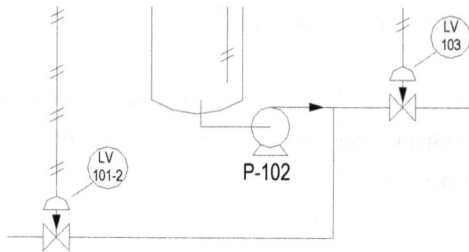

7 – Locate the cold-water source to R-101 in the P-1 process diagram. Attempt to determine the flow path of the water from R-101 to the drain located at the output of P-104. The water flow path should pass from R-101 through HX-101, T-102, HX-102, R-103, T-104 and into the drain.

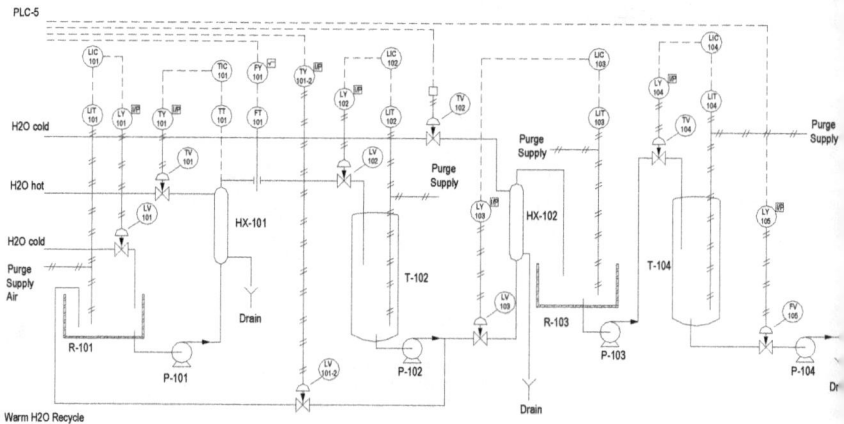

This page is intentionally blank.

This page is intentionally blank.

Chapter 3 - Physical Variables and Instrument Mechanics

This chapter introduces measurement and control variables from an applied physical perspective. Introductory physical concepts are investigated for application into measurement and control apparatus in subsequent chapters. Simple mechanisms are introduced and applications are demonstrated. The following topics are included.

Objectives:
Upon completion of this chapter, you should be able to:

- Explain common physical principles employed in measurement instruments
- Demonstrate introductory applied physical concepts
- Describe concepts of force-balance, pressure-balance and torque-balance
- Demonstrate linear concepts employed in measurement apparatus

Introduction

Force and Force Balance

Spring Rate

Energy and Work

Work

Weight

Weight Density

Specific Gravity

Torque

Torque Balance

Analysis of Moments

Pressure

Level

Hydrostatic Pressure

Absolute, Gauge and Differential Fluid Pressure

Linear Instrument Mechanisms

Homework

Introduction

The operation of measurement and control instruments and the processes they support can be best interpreted if a few basic concepts of mechanisms are applied and understood. Basic principles of force, density, distance, work, torque and pressure are applied throughout the controls industry in a myriad of instruments used to measure and control every conceivable physical quantity.

Applied mechanical concepts are examined to provide a framework to study industrial instrumentation and experimental data acquisition systems, and to investigate the operation of common industrial components, process control systems and the processes themselves. As will be demonstrated, even sophisticated industrial processes can be reduced to a collection of fundamental concepts.

Figure 1 – A clamp-on piezo-electric pressure transducer to measure muzzle pressure of prototype M-16's of varying barrel lengths. Shown is a 5.56mm suppressor on a 10.5-inch barreled M-16 in a lead-sled rest test fixture.

Force and Force Balance

Force is the application of energy, measured in units of pounds or kilograms that attempts to create motion or perform work. Note that the definition implies that motion, or movement need not result from an applied force. Note also the other key

94

words in the description of energy and work. When reviewing technical descriptions and especially technical equations, attempt to recognize the focus of the subject and all interrelations with other variables within the statement or equation. Reading into the description or equation to establish a mental image of the forces at work is very beneficial in developing an engineering method of thinking.

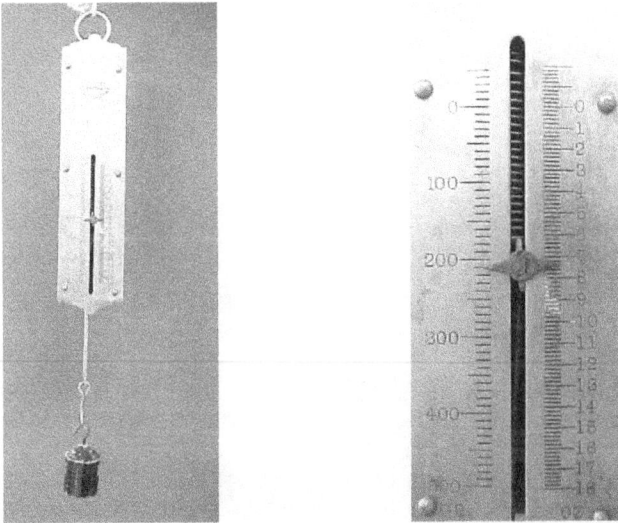

Figure 2 - Weighing scales are common examples of applied force-balance principles. The downward force created by the weight of the mass is suspended by a stretched spring.

Force Balance is the term used to describe multiple forces at work. In many cases observing the difference between two forces provides insight into the operation of scientific apparatus. Two applied and opposing forces are **balanced** if they result in a net force of zero, and no motion.

10# -->|<-- 10#

Measurement instruments are designed to operate in a balanced condition to achieve accuracy and predictable response. The instruments are only momentarily out of balance when their input variable (pressure, level, flow, temperature) is changing. The eventual result is a balanced condition. Balanced forces are demonstrated following.

20# --->|<--- 20#

Or,

5# --->|<--- 5#

Figure 3 - The actuator of a valve is a good example of a force-balance mechanism. Here, an input pressure is applied below a limp diaphragm to create a force. The force is then applied upward compresing the spring. The resulting displacement opens the air-to-open valve

Although two forces may be balanced, a unique value of mechanical displacement may be observed at each value of input force. When forces are unbalanced in measurement and control apparatus, the resulting displacement of a mechanical component is commonly used to activate some form of self-correction and an eventual balanced condition between the input and output forces.

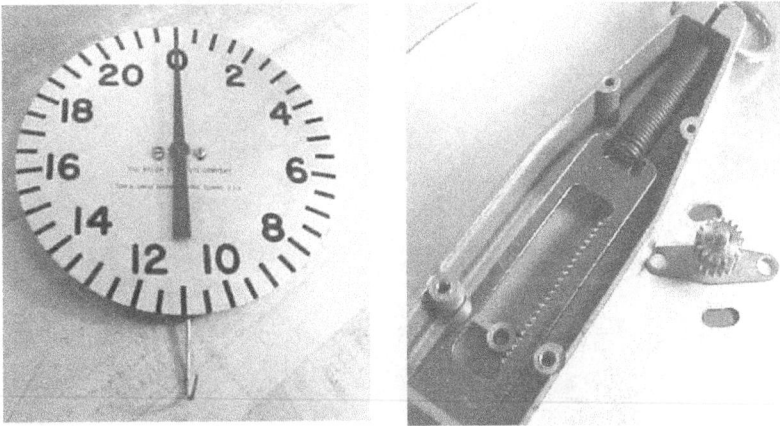

Figure 4 - A rotary spring scale (left) utilizes a rack and pinion in conjunction with a spring (right) for force-balance.

Springs commonly provide a balancing force against applied input forces. As can be imagined, a spring compresses or stretches to a different length with each unique value of applied force. The displacement distance of the spring-force interface can provide for a display of the input variable causing the force, or control the process causing the applied force. A simple mechanical indicating arrow pointing toward a graduated scale provides a process indication with an opportunity for self-regulation of the process. Imagine if this motion were used to position a valve, turn off a heater or start a motor and pump.

Figure 5 - A variation on the force-balance concept, air pressure is applied to a bellows to compress a spring. As the pressure increases the spring force increases to maintain a condition of balance.

Spring(s) opposing the process input, and the absence of levers recognize force balance in mechanical measurement and control apparatus. Force balance is most commonly applied in pneumatic components such as air relays, pressure regulators and process transmitters. Watch for this in pressure components and diagrams, the springs are easily recognized. Once noticed, the technician's thought processes in analyzing the operation of the instrument should be focused upon the force balance concept. A common spring assisted weighing scale is another application of the force balance concept.

Figure 6 - This pneumatic liquid level transmitter outputs 3-15 psi of air pressure as tank level varies and uses torque-balance and force-balance principles.

Spring Rate

Forces created by springs are capable of exhibiting linear or non-linear force-distance relationships. Linear springs are rated in pounds per inch (lbs./in.) or kilograms per centimeter (kg/cm) and as a linear device, springs follow a Y=mX+b operational equation. The offset (b) associated with springs is assumed equal to zero, resulting in only a gain (m) quantity to relate force and distance. Likewise,

$$\text{Spring Rate (SR)} = \text{Force} / \text{Distance}, \quad (\text{lbs./in.})$$
$$\therefore \text{Force} = \text{SR} * \text{Distance}$$

And,

Distance = Force / SR

Figure 7 - Experimental testing involves a considerable amount of data acquisition measurements but usually involves closed-loop control as well.

Energy and Work

Energy is available in two forms, **Kinetic Energy** or energy of motion (KE) and **Potential Energy** (PE) referred to as banked, stored or held energy. The exchange of potential energy into kinetic is required if work is to be performed. **Work** is defined as a force exerted <u>over</u> a distance, or W = F*D. To apply a force over a distance implies motion, which is required to perform work, and an application of kinetic energy. To develop kinetic energy requires a reservoir or bank of stored, potential energy. Voltage, compressed air pressure and temperature are common sources of potential. Electron flow, airflow and heat flux are the resulting kinetic currents. The motion of currents as required by the Work definition, is used to rotate turbines, motors and generators to perform industrial and scientific functions. The exchange of Kinetic into Potential energy is rarely transferred entirely. The amounts of kinetic energy resulting from the stored or reserved energy will always be less since some amount of energy is always lost as friction. Energy lost to friction appears as heat and/or sound.

An appreciation of force, energy, work and friction are requisite to an understanding of industrial controls, instrument mechanics and the operation of scientific apparatus.

Work

Work is defined as the application of force over a distance. From the definition, two primary ingredients are required to perform work, force and distance. The resulting equation demonstrates.

W = F * D, Measured in foot*pounds or kilogram*meters

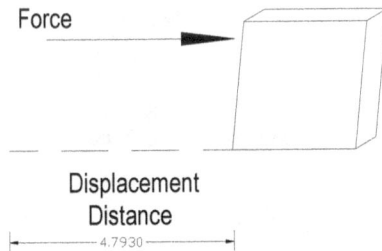

Figure 8 - Work involves force and a distance of displacement.

Weight

Weight (Wt.) is a downward force created by gravity, and measured in pounds or kilograms. For purposes of analysis, weight should be seen as a downward generated force. Gravitational acceleration coefficients will not be included in our computations. Therefore,

$$Wt. = F\downarrow$$

Weight Density

Weight Density is the ratio of a material's weight to its volume. Measured in pounds per cubic foot or kilograms per cubic centimeter, weight density allows a rapid determination of which objects are "Heavier" than others. Water (62.4 $\#/ft^3$) is heavier than wood (50 $\#/ft^3$). Mercury (848.6 $\#/ft^3$) is heavier than water. Weight density allows rapid determination of whether materials would mix or separate if combined. Materials of differing densities tend to separate when mixed with heavier materials sinking while lighter materials rise.

When a material's density is specified, temperature and atmospheric pressure are usually referenced as each can influence an object's volume. Temperature change causes expansion or contraction in solids, gases and liquids. The resulting change in volume changes the object's density. As temperature increases and volume increases due to thermal expansion, an object's density decreases.

Density, d = Weight / Volume, Measured in pounds per cubic feet, $\#/ft^3$

Density explains why ice, petroleum products and most woods float when placed in water.

Figure 9 - Due to its lighter density, 10-W-30 Motor oil floats on water.

Specific Gravity

The ratio of a <u>solid</u> or <u>liquid</u> density to the density of water is referred to as Specific Gravity (SG). Specific gravity is often used when working with materials of differing densities. Materials that are lighter than water such as alcohol, plastics, some woods and petroleum products have a SG less than 1.0. Materials that are heavier than water such as metals, a few woods and some liquids have a SG greater than 1.0. Water is always used as the reference material when determining Specific gravity. The ratio for specific gravity follows.

$$SG = d_x/d_{H2O}$$

Where,

SG = Specific Gravity

d_x = the density of the material

d_{H2O} = the density of water

Density and Specific Gravity should not be confused with viscosity, which is the measure of opposition to flow. Oil is a viscous substance but water is not.

Note that gases are not included within the definition of Specific Gravity. The ratio of gas densities at defined temperature and pressure conditions are referred to as Specific Gas and is addressed later in this text.

Torque

Torque is defined as a force applied <u>through</u> a distance that <u>attempts</u> to create a rotating or turning effect. At first, this definition appears similar to that of Work. However, upon closer inspection two key differences appear. First, the force is applied using a lever, or moment arm distance. Second, the applied force *attempts* to create motion. During a torque application, there exists no motion between the applied force, moment arm and the object being torque'd. Both objects may be in rotation as in the case of a motor driven shaft turning a load, but the relative motion between the motor and load is zero. The torque in this example is a twisting in the shaft.

$$T = F * D, \text{ measured in pound*feet or meter*kilograms}$$

Again, notice the key words of force, through a distance, attempts and rotating effect. Apply a force using a lever and torque is created when motion does not exist. Again, <u>Torque involves no motion</u>. This may seem foreign at first but if motion is involved then Work is performed, not torque. When a force is applied through a lever to generate a resulting force, Torque is developed. Many examples of torque, or torque balance, are common. Automobile jacks and tire-irons, torque wrenches and screwdrivers (especially when used for prying) are common torque generating instruments.

Since Torque also equals Force * Distance, T = FD, some confusion between Torque and Work is appropriate. Remembering their differences is in order, Work requires motion over a Distance whereas Torque does not involve motion, and again a mental image of each is in order.

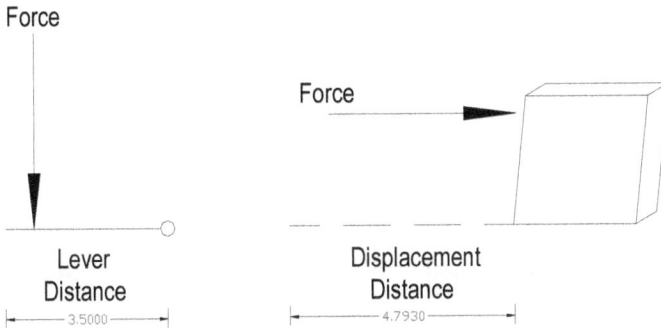

Force

Force

Lever
Distance
|——— 3.5000 ———|

Displacement
Distance
|——— 4.7930 ———|

Figure 10 – Distance and force are required for both Torque and Work. The difference is whether or not motion results. Torque is an applied force without motion.

Torque Balance

Torque-balance occurs when opposing forces are applied to the same lever. Usually the forces are applied to opposite sides of a lever pivoting around a fulcrum. Since torque is the product of force multiplied by distance, the forces and distances may not be equal. For torque-balance to occur no motion can be produced and two opposite and balanced torques, or *Moments* must be applied. A moment is the product of a force and a distance. With torque balance, the balancing torque may not always be obvious but two opposing torques, or Moments are always applied in opposing (clockwise and counter-clockwise) directions about or around a fulcrum, as in the following figure.

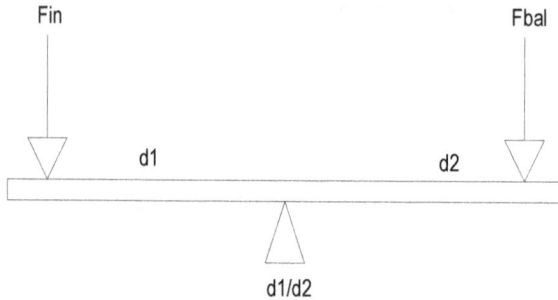

Figure 11 - An example of torque balance, the product of force and distance on one side of the fulcrum is equaled by the product and distance on the opposite side.

Balance occurs when the individual F*D products are equal. In a Torque balance condition, the CW and CCW moments (torques) are equal, although the applied forces and distances may be different. This means one force on the moment arm lever may be larger than other applied forces, resulting in an effective force amplification, or force attenuation (reduction). Motion distance and moment arm distances work in a similar manner, where moment arm arcs of rotation may vary substantially from each end of the lever. In industrial measurement and control applications, distances of arc motion are kept to a minimum to avoid errors associated with friction and bending. As such, the arc motion of the torque beam or moment arm will be inconsequential in our study of automation.

Analysis of Moments

The vast majority of mechanical measurement and control instruments employ a torque-balance operational concept and all torque-balance measurement and control instruments achieve a condition of balance in producing an output. As will be seen in subsequent chapters, measurement and control instruments always compare the output against the input. This is done to assure that the output is a precise representative of the input. Whenever a difference, or error exists between the input and the output, the mechanism detects the difference, adjusts the output and in the process of changing the output, self-corrects. This inherent self-correction process is used in transducers, transmitters and controllers of all types in instrumentation and control.

Figure 12 - The output pressure from this pneumatic flow transmitter is applied to a lever for comparison against the input force. The output will change until the two applied forces (actually torques, note the lever and thumbwheel fulcrum in the photo) are equal.

Applications of torque-balance can be analyzed relatively easily. A mathematical analysis of torque-balance always begins with a statement of equal clockwise and counter-clockwise torques. As an example, assume the balancing force, F_{bal}, is greater than the applied input force, Fin, In the following figure. This can be proven using torque balance techniques.

At balance,	$T_{CCW} = T_{CW}$
Since,	$T = F * D$
And,	$T_{CW} = F_{bal} * d2$
And,	$T_{CCW} = F_{in} * d1$
Therefore,	$T_{CCW} = T_{CW}$

Becomes, $F_{In} * d1 = F_{bal} * d2$

Or, $d1 / d2 = F_{bal} / F_{In}$

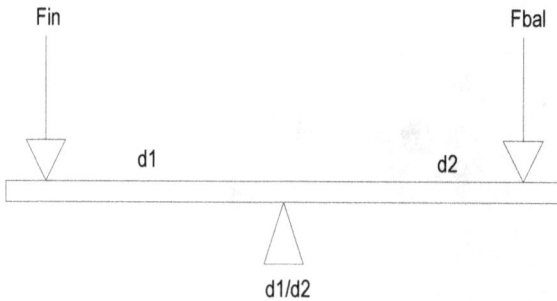

Figure 13 – Simple in appearance and operation, the torque balance mechanism finds applications in all aspects of measurement and control.

Using input and balancing force spans,

$$d1 / d2 = F_{bal} / F_{In}$$

When the ratio of d1 to d2 is greater than unity, the balancing force will be greater than the input force. From the equation, the ratio of the balancing force to the input force will be equal to the ratio of the input and output moment arm distances. The result is force amplification or, *mechanical advantage*. Force attenuation will occur at values of d1/d2 of less than unity. Similarly, positioning the fulcrum determines the ratio between the two applied forces. More specifically, positioning the fulcrum determines *Gain* by establishing the ratio of the *spans* of input and output forces. The following equation demonstrates.

$$d1/d2 = \Delta Fbal/\Delta Fin$$

Where the use of the Greek letter Delta, Δ, implies a change of quantity or span of force values. Assuming a fulcrum ratio (d1/d2) of 3:1 results in an output (balancing) force to input force ratio of 3:1. An input span of 1 to 4 pounds will result in an output force span of 3 to 12 pounds.

As in the following figure, Torque-balance principles and the associated force and distance (moment arm) components are usually easily recognized in scientific and industrial apparatus by the appearance of lever arms and fulcrums.

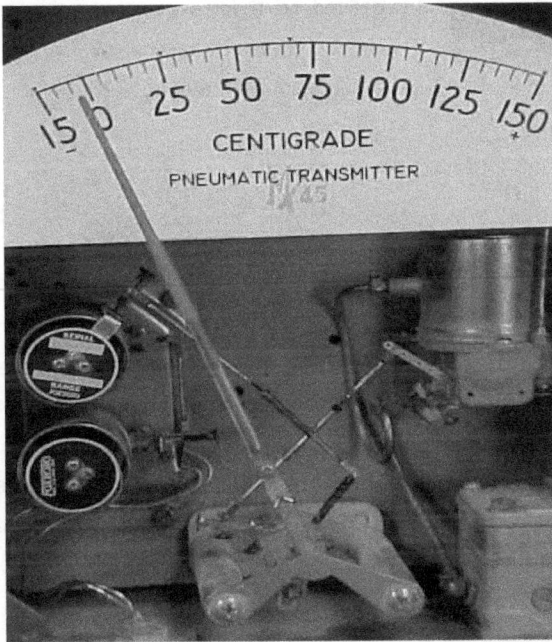

Figure 14 – The moment arms (levers) and pivot point (fulcrum) at the bottom center in this temperature transmitter indicate the application of torque balance principles. The transmitter outputs a 3 to 15 psi span of compressed air in proportion to the indicated temperature span.

Pressure

Pressure is always defined as a force applied over an area. The resulting equation is identical to the definition.

$$P = F / A,$$

Measured in pounds/square inch or kg/square centimeter, pressure is also measured in Pascal's, atmospheres, millimeters of Mercury, and numerous other abstract dimensions. All represent force over area.

Pressure defined as above can be difficult to visualize. It's difficult to imagine a contained, pressurized gas defined as a force over an area. Although this description is accurate, in many pressure applications, an algebraic re-arrangement of the previous pressure equation is more appropriate.

Since,

$$P = F / A,$$

$$F = P * A.$$

The resulting equation demonstrates the effect of applying a pressure against an area, a force is created. Imagine the effect of air pressure in automobile tires supporting the vehicle weight. Again, although fundamental and intuitive this is also an essential concept in measurement instrumentation and process control. If one were to study the evolution of automation it would be found the earliest self-regulating, negative feedback control systems employed this basic force-pressure concept universally.

Figure 15 - An example of force equals pressure multiplied by area. This pressure to dispacement (distance) mechanism uses air pressure to create a force, a spring to balance the created force and results in displacement of the shaft running through the center.

Pressure can assume three basic forms of liquid, gas and solid. Hydrostatic (standing liquid) pressure is created by the weight of a volume of liquid distributed over the areas (including the sides) of a surface such as a holding tank. Gas pressure is created when a gas, such as air, is compressed and contained within a vessel, like a balloon. In which case, the pressure is distributed and felt upon all three dimensions of the containing surfaces. Solid pressure is normally referred to as Stress, and is created whenever forces are acting upon a solid material. Forces on solid materials can be applied in a number of ways including tension, compression, shear and bending. Stress pressures although not always obvious will *always* result in a deformation of the solid object, or Strain. Stress and Strain are defined in a subsequent section in this manuscript.

111

Level

Liquid and solid level measurements are common process variables with a variety of sensing techniques and apparatus available. It should be understood that level isn't always measured for the sake of knowing the surface height or depth of material within a container. Level is measured to determine volume, to determine weight or to determine level itself. With level control systems the particular information desired may not determine the type of sensor utilized. A batch-mixing process performing a weight-based ratio of two products may be measuring level or pressure as an indication of each of the two mixed weights. A symmetrical container may allow the detection of liquid level to yield volume and/or weight information. The difference between the three measurement types may reside in the required information for the application and not in the sensor. This type of sensing principle where one variable is sensed to provide information about a related physical variable is referred to as an *Inferred or Indirect measurement* and is common in level control.

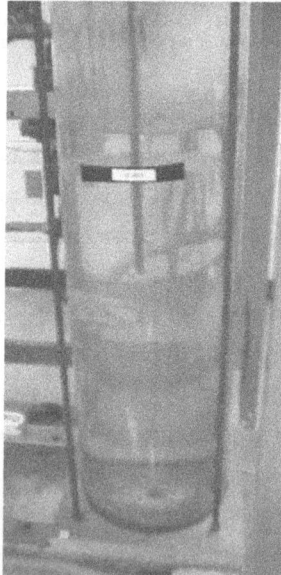

Figure 16 - An example of an indirect measurement, pressure is measured to determine water level in this tank.

When analyzing control system operations, the physics and unique characteristics of the process must be understood to the greatest extent possible. If not, the end-result of specifying a measurement or control component could be expensive and disastrous. In the case of level measurement, the process vessel and the contained material must be considered in conjunction with desired measurement result. A symmetrical process vessel may be able to utilize a simple float mechanism when sensing liquid level, weight or volume. But a float would yield less than desirable results if measuring crushed rock due to the rock's angle of repose (maximum slope before sliding occurs) and the possible inconsistency of material density.

Hydrostatic Pressure

Anyone that has been swimming to any depth has an appreciation of hydrostatic pressure concepts. Simply stated, hydrostatic pressure is the result of a liquid's weight (or force) distributed over an area, which results in pressure.

$$P = Wt / Area = Force / Area, \text{ measured in psi or kg/cm}^2$$

Since weight is a downward force, it stands to reason that pressure can be created by weight being supported by an area. This can be proven through algebraic re-arrangement of the weight density equation, where;

$$Density = Weight / Volume$$

$$Weight = Volume * Density$$

Since,

$$P = F / A = Weight / Area$$

And rearranging the density equation,

Weight = Volume * Density

Substituting for Weight in the pressure equation,

Pressure = (Volume * Density) / Area

Since Volume equals Area multiplied by Height,

Volume = Area * Height

Substituting for Volume in the pressure equation yields,

Pressure = (Area * Height) * Density / Area

Canceling areas yields,

P = Height * Density, measured in psi or kg/cm

Or,

$$P = H * D$$

Assuming the density of the stored liquid is constant, it can be seen from the derivation that the pressure above a certain measurement location is directly proportional to the height of the liquid, *regardless of the shape or area of the storage vessel*.

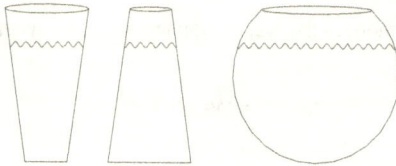

Figure 17 - The pressure at any depth below the liquid surface is the same in each of these vessels.

This is partially true for dry material as long if the material seeks a flat surface level and maintains constant density throughout the container. However, some dry material has a tendency to settle and compress under its own weight resulting in greater density at the bottom of a storage vessel than at the top. This would make a pressure measurement for the purpose of volume determination unreliable.

In liquid measurement and controlled processes, the P=HD relationship is fundamental. This type of liquid measurement is often referred to as a 'Head' of pressure, referring to the liquid above the measurement point as creating a "*head of pressure.*" Many liquid level pressures are expressed as inches of water (In H2O) or inches of Mercury (In Hg), the different material densities being responsible for different pressure ranges.

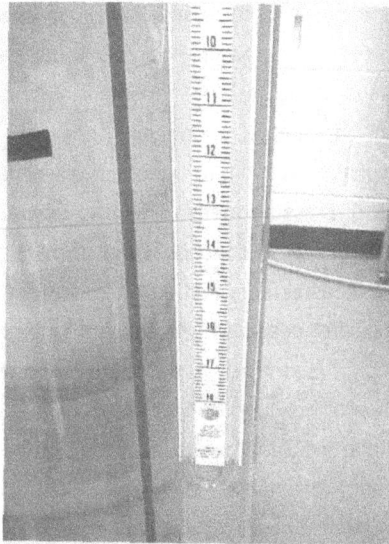

Figure 18 - A U-Tube manometer operates according the principle of pressure-balance and P=Ht*D. Pressure applied to one side forces the liquid to elevate up the other side. The difference in liquid elevations equals the applied pressure.

Absolute, Gauge and Differential Fluid Pressure

Being a potential energy form, pressure will always be a differential measurement. Pressure, like voltage or any other potential energy form will always be measured as the difference between two quantities. As an example, imagine measuring the pressure in your car tires. The pressure indicated on the gauge is the difference between the pressure within the tire itself, and the atmospheric pressure you're breathing. This type of pressure measurement, where atmosphere is used as the reference, is referred to as a *Gauge* pressure measurement. So called because the vast majority of pressure gauges work in this manner, indicating pressure above atmosphere.

Processes that require vacuum pressures (pressures less than atmosphere), or pressures close to atmosphere usually do not utilize the gauge-pressure measurement concept since the atmospheric pressure reference changes with the ambient weather conditions. Substantial error can be introduced by the always changing atmospheric pressure conditions as well as the possible mechanical shortcomings associated with repeatability, resolution and accuracy of the measurement apparatus. Normally pressures measured around atmosphere utilize a constant vacuum (or as practically constant as possible) as a reference. This type of pressure measurement is referred to as an *Absolute* pressure measurement. Similar in concept to the absolute temperature scales, absolute pressure is a measurement that includes atmospheric pressure in the indicated value, or pressure measured with respect to the absolute lack of any atmosphere, a vacuum reference. Atmosphere is normally assumed at approximately 30 inches of mercury, or 14.7psi *absolute* (abbreviated as psia). The following defines the individual pressure types and details the conversion between absolute and differential pressures. The following equations demonstrate the measurement concepts and conversions.

Differential Pressure,

$$PSID = P_{Hi} - P_{Lo}$$

Or,

$$PSID = P_{Test} - P_{Reference}$$

Gauge Pressure,

$$PSI = PSIG$$

Atmospheric Pressure,

$$Patmos = 14.7 \text{ psia}$$

Absolute Pressure,

$$PSIA = PSI + 14.7$$

Gauge Pressure,

$$PSI, \text{ or } PSIG = PSIA - 14.7$$

Most low pressure, meaning close to atmosphere, measurement applications specify pressure in terms of a liquid column. As an example, atmospheric pressure is normally expressed as 30 inches of mercury (30" Hg). Many pressure transmitters also utilize this approach, preferring to specify pressure ranges in inches of water (designated WC or In. W.C., as inches water column). To determine the numerical value associated with the specification, multiply the given range distance times the density of the indicated liquid as was discussed in the section on Hydrostatic Level measurement. As indicated,

$$Pressure = Height * Density$$

As an example, take the atmospheric pressure stated previous, thirty inches of mercury, or 30" Hg. Using the density of Mercury as 848.6 pounds per cubic feet, atmospheric pressure can be determined by multiplying the height (30") by the density of Mercury (848.6 lbs/ft^3).

$$Pressure = 30 \text{ inches} * D_{Hg}$$

Converting the density into pounds per cubic inch by dividing by 1728 in^3/ft^3 allows computing an answer in pounds per square inch,

$$Pressure = 30 \text{ inches} * (848.6 \text{ lbs/ft}^3 * 1/1728 \text{ in}^3/\text{ft}^3)$$

$$Pressure = 30 \text{ inches} * .491 \text{ lbs/ in}^3 = 14.73 \text{ PSIA}$$

Unless otherwise indicated pressure measurements are assumed to be gauge. Atmospheric pressure however by its nature would be absolute. Therefore,

$$30" \text{ Hg} = 14.73 \text{ PSIA}.$$

Some process measurements such as flow through an orifice plate or a Venturi tube, or liquid level under pressure utilize a true differential pressure measurement where the reference pressure is neither atmosphere nor a vacuum. Gauge pressure measurement applications (PSI or PSIG) are typically more common than either absolute or differential. Consequently, PSIA (absolute) and PSID (differential) are used to signify the appropriate type of measurement application, and PSI (or occasionally PSIG) is utilized for all others.

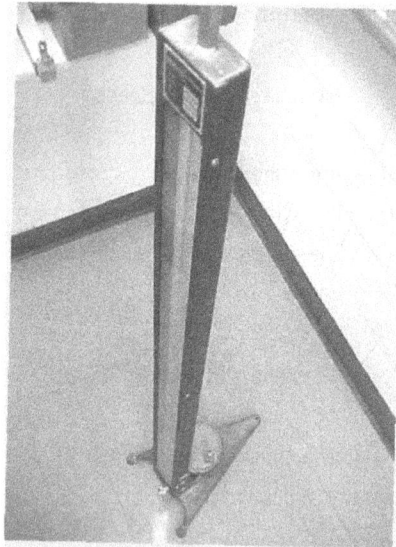

Figure 19 - A well-type manometer, the reference pressure is applied at the top and the test pressure is applied to the well (bottom of the photo) containing the indicating fluid in the rear of the unit.

Linear Instrument Mechanisms

The vast majority of sensors and mechanisms are linear. Linear describes proportional components that produce a straight-line segment when the input/output characteristics are graphed. The following graph illustrates.

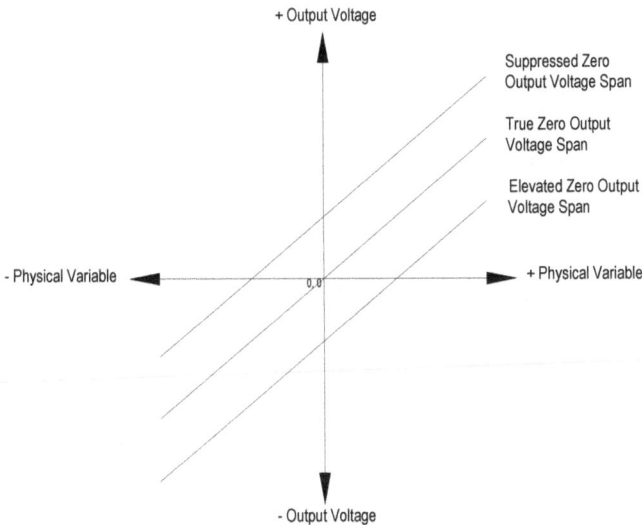

Figure 20 - True, elevated and suppressed zero output voltage spans.

In responding to a change in an applied physical variable a mechanism or sensor outputs a signal that varies in intensity. The output signal can follow, or track the input in a linear or non-linear manner. Due to measurement accuracy and control concerns linear sensors and mechanisms are desirable. Linear sensors and components follow the mathematical Y=mX+b model.

Where,

Y = the output variable

119

m = the slope of the input/output relation

X = the input variable

b = the Y-axis intercept value

In measurement and control terms Y=mX+b is often expressed as output (Y) equals gain (m) times input (X) plus offset (b).

The following is an example of a linear mechanism. Assume the fulcrum location, spring force and spring force direction is to be determined. The system uses a 5 to 15 pound input force span applied at Fin 1 to balance a 10 to 80 pound output force range applied from a mechanical load at Fin 2. Assume the beam is 8 inches in length.

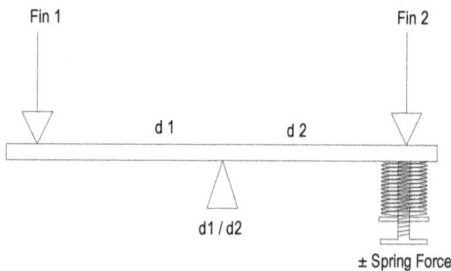

Figure 21 - Figure for the linear example problem.

Solution:

First, an equation must be written to represent the operation of the mechanism. Since the mechanism involves the application of forces through distances, a torque-balance (T=F*D) equation will be derived. To derive the equation, assume the spring force is pushing up on the right side of the beam. If this assumption is incorrect the resulting spring force will be negative indicating the direction should be reversed.

Second, observe the forces working on the beam. Three forces labeled Fin1, Fin2 and spring force are applied to the mechanism. Since the spring is initially assumed to be pushing up, Fin1 and the spring force (Fsp) aid each other and should sum in the soon to be written equation. Note also that each of the forces is applied to the fulcrum through the distances labeled d1 and d2.

Finally, recognizing the beam is a torque mechanism, that torque equals force multiplied by distance and that the beam will be balanced when the forces are applied, the equation can be written and the problem solved.

At balance, clockwise torque equals counterclockwise torque or,

$$T_{CW} = T_{CCW}$$

Since torque equals force multiplied by distance or,

$$T = F*D$$

The force and distance quantities can be substituted into the balance equation.

$$(Fin2*d2) = (Fin1*d1) + (Fsp*d2)$$

Or, substituting Fout for Fin2 yields,

$$(Fout*d2) = (Fin1*d1) + (Fsp*d2)$$

Dividing each term by d2 to solve for Fout yields,

$$Fout = (Fin1*d1/d2) + (Fsp*d2/d2)$$

In a form similar to Y=mX+b, solving the equation for Fout yields,

$$Fout = (d1/d2)(Fin1) + Fsp$$

With the equation expressed in this form it can be seen that d1/d2 performs the slope (gain) function and the spring force (Fsp) performs the biasing (offset) function. To find the specific numerical values associated with d1/d2 and Fsp, assume Fsp is set to zero in the previous equation and solve for d1/d2 using the provided input and output force ranges.

Since Fsp represents the Y-axis intercept value and does not influence the slope computation, it must be set to zero to find d1/d2. The resulting equation becomes,

$$Fout = (d1/d2)(Fin1)$$

And,

$$d1/d2 = Fout/Fin$$

Or, more specifically since Fout and Fin1 are both force spans,

$$d1/d2 = \Delta Fout/\Delta Fin$$

Where,

$$\Delta Fout = 80\# -10\# = 70\#$$

And,

$$\Delta Fin = 15\# -5\# = 10\#$$

Therefore,

$$d1/d2 = 70\#/10\# = 7{:}1$$

The distance of d1 is seven times longer than d2. Since d1 plus d2 equals eight inches,

$$d1 + d2 = 8"$$

From the previous computation

$$d1/d2 = 7 \text{ or, } d1 = 7d2$$

Subbing for d1,

$$(7d2) + d2 = 8"$$

$$8d2 = 8"$$

$$d2 = 1"$$

Therefore,

$$d1 = 7"$$

Knowing the d1/d2 ratio and the corresponding input and output forces, values can be substituted into the mechanism equation to find the spring force.

Since,

$$Fout = (d1/d2)(Fin1) + Fsp$$
$$Fsp = Fout - (d1/d2)(Fin1)$$

Substituting d1/d2 and corresponding input/output values into the equation yields,

$$Fsp = 80\# - (7/1)(15\#)$$
$$Fsp = 80\# - 105\#$$
$$Fsp = -25\#$$

Note the negative quantity implies the spring force direction should be reversed. The resulting spring force needs to be down at 25 pounds with d1 set to 7" and d2 of 1".

Proof:

Subbing the values into the mechanism and checking for a condition of balance at the input/output force extremes can verify the results.

Proving at minimum input force (5#) and minimum output force (10#).

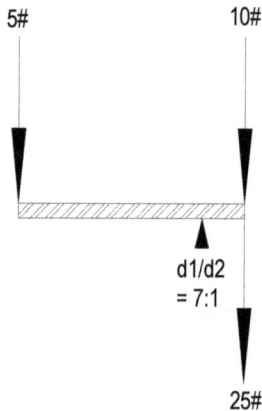

Figure 22 - Proving a condition of balance at minimum input and output values.

Summing the torque moments around the beam yields,
$$(5\#*7") = (10\#*1") + (25\#*1")$$
$$35lb\bullet in = 10lb\bullet in + 25lb\bullet in$$

$$35lb\bullet in = 35lb\bullet in$$

The balanced condition is verified at the minimum input/output values. The values are again proven at the maximum input and output values.

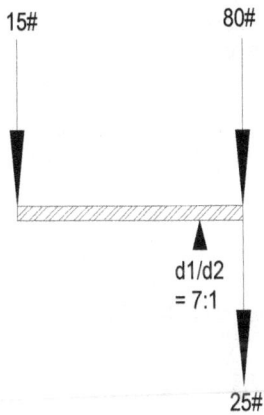

Summing the torque moments around the beam yields,

$$(15\#*7") = (80\#*1") + (25\#*1")$$
$$105lb\bullet in = 80lb\bullet in + 25lb\bullet in$$
$$105lb\bullet in = 105lb\bullet in$$

The balanced condition is verified at the maximum input/output values. The d1/d2 and spring force values are correct.

The previous computations are commonly employed when performing signal scaling and conditioning functions with electronic as well as mechanical systems.

Homework

1 – A 10-lb. weight is hung from a spring with a spring rate (SR) of 20 lbs./inch. Determine the distance of spring stretch.

2 – Determine the necessary spring rate to balance an applied force of 48 pounds with a displacement of .6 inches.

3 – A force of 135 lbs is applied over a half-mile distance to move a 3200-pound automobile into the nearest gas station. Determine the work performed by the poor slob doing the pushing.

4 – If a calorie of energy equals 3.088 foot-pounds of work, and 3600 calories equals 1 pound of human body flab (weight), how far is one required to push at a force of 20 pounds to lose 15 pounds of body mass?

5 – A cubic foot of Mercury (chemical symbol Hg) weighs 848.6 pounds. Determine the weight of one fluid ounce. Assume a cubic foot equals 7.48 gallons.

6 - A cubic foot of water (chemical symbol H_2O) weighs 62.4 pounds. Determine the weight of one gallon.

7 – Determine the specific gravity of Mercury (Hg).

8 - A cubic foot of salt water weighs 64.0 pounds. Determine the specific gravity of salt water.

9 – A two foot bar and fulcrum are used to support the weight of a 2400 pound automobile. If the auto's weight is applied to one end of the bar and the fulcrum is located 4 inches from the auto, determine the force required to balance the auto's weight.

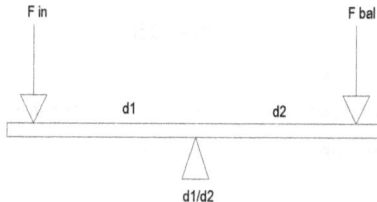

Figure 24 - Diagram to support problem #4.

10 – A spring (SR = 25lbs/in) pulls down on one end of a 10-inch beam. If the spring is stretched two inches, determine the necessary force applied to the opposite end to balance the beam. Assume the fulcrum is located 8 inches from the spring.

11 – Using the diagram of figure 20 and a 10-inch beam, assume the fulcrum is positioned 4 inches from the input force and the input force varies between 5 and 15 pounds. Determine the output force span at Fbal.

12 – Using the mechanism and forces of figure #20, assume a spring is added below the beam from Fbal and is pushing up at 10 pounds of force. Determine the output force span.

13 – Using torque-balance concepts, determine the equation for F_{out} in the following diagram.

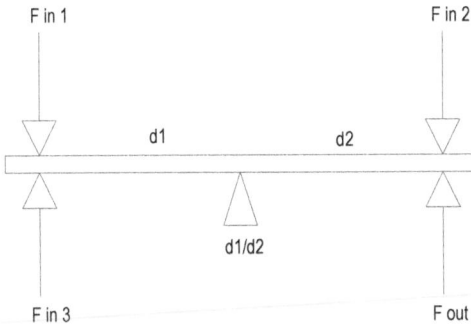

Figure 25 - Diagram for problems #11 and #12.

14 – Using the equation generated in problem #13, determine the two components in the previous diagram responsible for gain and offset adjustment functions.

15 – Determine the fulcrum location, and required spring force and direction using the following diagram to convert a 1 to 5 pound input force span applied at Fin 1 into 8 to 20 pounds of output force to be applied at Fin 2. Assume the beam is 8 inches in length.

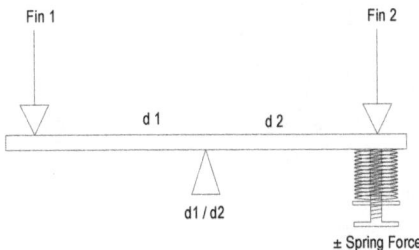

Figure 26 - Figure for problem #15.

This page retained intentionally blank.

Chapter 4 - Stress, Strain and Strain Gauges

This chapter introduces the subjects of stress, strain, strain gauges and related measurement considerations. Ductile materials are investigated and stress, strain and related strain gauge computations are provided. Strain gauges of various types are examined, and resistance bridges, signal conditioning and temperature compensation techniques are introduced.

Objectives:

Upon completion of this chapter, you should be able to:
- Explain the elasticity coefficient of ductile materials
- Differentiate between the physical variables of stress and strain
- Explain the purpose and operation of a strain gauge
- Perform tension and compression stress computations
- Perform tension and compression strain computations
- Determine the change in resistance of a strain gauge
- Determine the output voltage signal from a strain gauge bridge circuit
- Determine the required signal conditioner gain and offset to calibrate a strain gauge-based measurement system

Introduction

Pressure, stress and strain are common measurements in the construction, manufacturing, experimental testing, research and product development industries. The strain gauge, used to measure mechanical deformation, is the primary sensing element of numerous instruments and is possibly the most universally applied sensor in the industry. Used to measure stress, strain, pressure, vibration and numerous related physical variables, the strain gauge is also an intrinsically simple sensor if one is familiar with the basic concepts of resistance in conductors.

Physical Principles

The study of strain gauges begins with an examination of pressure, or *stress*. Elastic materials and their corresponding yield characteristics are then investigated in the subsequent pages. And finally, the strain gauge, it's associated computations and many of the required conditioning circuits are provided in the hardware portion towards the end of the chapter.

Stress

Pressure exhibited by, or within solids is often referred to as *Stress* and is symbolized with an upper case "S". Measured in the same dimensions of pressure as pounds per square inch (or Newtons/m^2), stress is also defined as a force exerted over an area. If one can imagine placing a cubed object of substantial weight over a support of fixed and known area, the resulting stress can easily be determined by dividing the material weight by the size of the supporting area. One can further imagine the supporting area yielding to the applied force of the object's weight. The change in physical dimension experienced by the supporting structure is the result of an applied stress, and is referred to a *Strain*.

To begin an explanation of stress, strain and strain gauges, an introduction to applications of both stress and strain is appropriate. As mentioned previously, stress is the application of a force that creates a resulting deformation or strain. As can be seen from the equation, stress is a pressure developed within solid materials,

$$Stress = Force / Area, psi$$

The applied force can assume the forms of tension (pulling), compression (squeezing), shear, bending or torque (twisting). For the sake of simplicity, and to assist in focusing upon the objective of strain gauges, only tension and compression forces will be considered in the chapter computations.

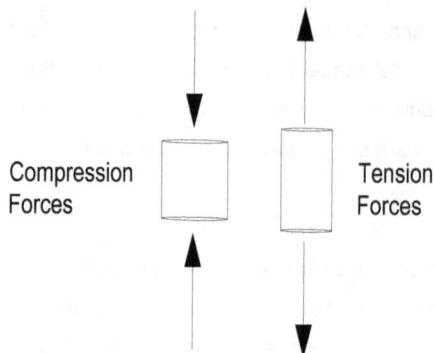

Compression Forces

Tension Forces

Figure 2 - Compression forces cause objects to become shorter and wider. Tension forces cause objects to become longer and narrower.

Ductile Materials

All materials will deform when placed under a load, or under an applied force. If the load is small and creates a stress that is within the material's "elastic limit," the material will return to its original dimensions when the load is removed. Ductile materials exhibit elastic characteristics if the applied force or "load" is relatively small. Metals, plastics, fabrics, wood, concrete and even glass are examples of ductile materials. All elastic materials conform to a predictable set of characteristics when destructively tested by being subjected to gradually increasing loads. The characteristics are obvious when graphed upon a stress versus strain characteristic plot.

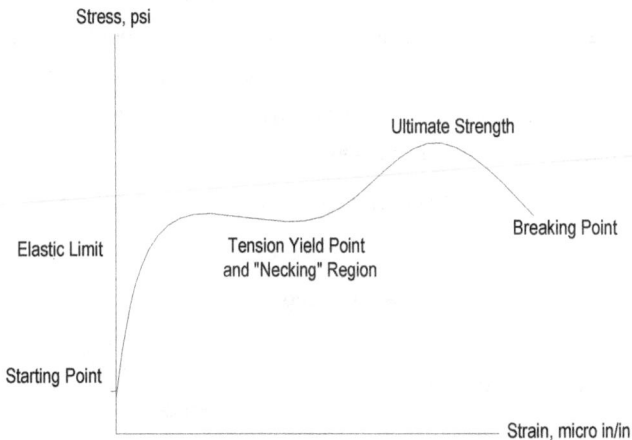

Figure 3 – Ductile (elastic) materials will exhibit common characteristics as indicated. Note the applied input force is plotted along the Y-axis with the resulting elongation is plotted along the X-axis. This is the only case in this course where input is plotted along the Y-axis and output is plotted on the X-axis.

Strain

Longitudinal Strain (symbolized with a lower case Greek letter sigma, ε_L), is defined as a single dimension (distance) fractional deformation and is measured in fractional units of change in length divided by the original length, unit dimensions are inches/inch.

Longitudinal Strain, ε_L = change in length / original length

Or,

$$\varepsilon_L = \Delta \text{ length / Lorig, } (\mu \text{ in/in})$$

The fractional longitudinal strain units of inches per inch are more commonly expressed as micro-inches/inch and are often termed micro-strain. Strain is not just a single dimensional change however. As an object's length changes under an applied stress, its area must also deform in the opposite direction. For example, as a sample length increases with an applied tension (pulling) force, the cross-sectional area decreases. And if a sample length decreases as resulting from a compression force, the cross-sectional area increases. The terms transverse (across) strain and longitudinal (lengthways) strain are used to represent physical deformation in the two dimensions.

Transverse Strain, ε_T = change in area / original area

Or,

$$\varepsilon_T = \Delta \text{ area / Aorig, } (\mu \text{ in}^2/\text{in}^2)$$

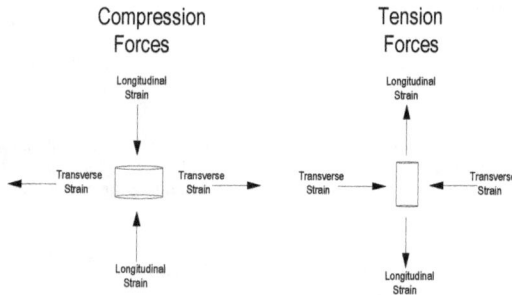

Figure 4 - Stress forces and the resulting strains.

Transverse strain, ε_T, is measured in units of micro-inches squared per inches squared (often called micro-strain transverse) since both the numerator and denominator contain area quantities. Transverse strain and longitudinal strain are related through a coefficient called the *Poisson Ratio*, symbolized as μ.

$$\mu = \text{Transverse Strain / Longitudinal Strain}$$

Or,

$$\mu = \varepsilon_T / \varepsilon_L$$

The previous description provides a good example of a basic property in instrumentation and experimental measurement. Stresses and strains are not always visually obvious in mechanical systems. When a large truck drives over a bridge the resulting stresses and strains are not obvious to the human eye. Yet, if one has ever stood upon a bridge when a large truck passes, one can surely feel the bridge yielding to the applied load. In instrumentation, it is often useful to remember that any applied force, independent of how large or small, will create a deformation, and the resulting deformation can therefore be used to determine the magnitude of the applied force. If one retains this fundamental principle, it will most certainly be observed repeatedly in practice.

Elasticity Modulus

As can be seen in the following plot, stress/strain graphs covering a span of zero to breaking stress exhibit five distinct points between the application of an initial force and the material breaking point. The starting point, elastic limit (also called the limit of proportionality), tension yield point, ultimate strength and breaking point (point of fracture) are encountered as one continues to apply force (stress) from zero through breaking of a ductile material sample. As long as the applied stress remains within a specific maximum limit, the sample being stretched or compressed will return to its original dimensions, acting much as a spring. Observing the stress versus strain characteristics of a ductile (elastic) material operating within its elastic limit exhibits a linear characteristic. The slope of a given materials stress/strain plot within the linear region is referred to as the Modulus of Elasticity (E) and is referred to as Young's Modulus.

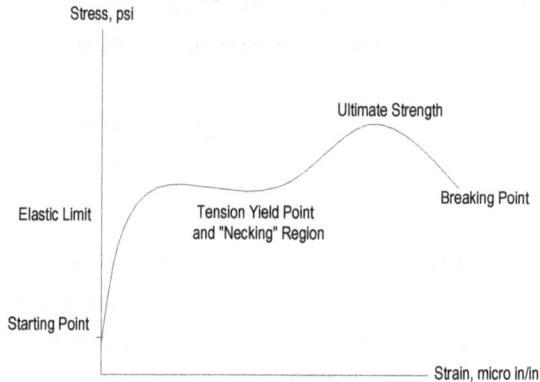

Figure 5 – A typical tensile (tension) characteristic for ductile (elastic) materials. Note the applied input force is plotted along the Y-axis with the resulting elongation is plotted along the X-axis.

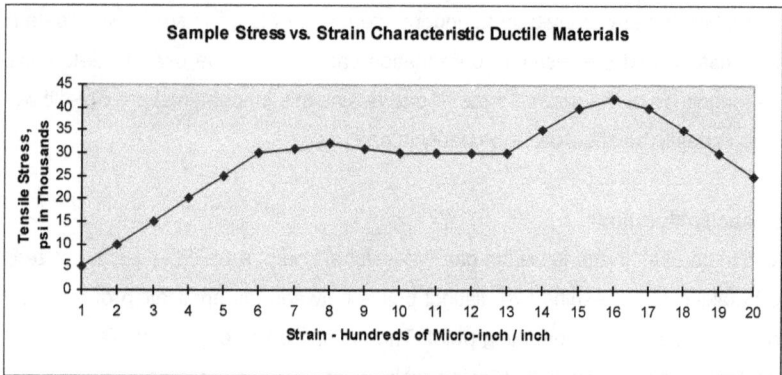

Figure 6 – Stress vs. Strain characteristic, data taken from a 1020 steel tensile test result.

E = Stress / Longitudinal Strain, within the elastic limit

Young's Modulus allows the determination of a materials strain if a given stress is known.

136

$$\text{Longitudinal Strain, } \varepsilon_L = \text{Stress} / E$$

Typical values of "E" are 31 million for 1020 steel and 10.7 million for T-3 aluminum. Should the elastic limit be exceeded however, the sample reaches the Tension Yield Point (TYP) and will appear permanently deformed with the material's elastic characteristics greatly reduced. Stressing a material beyond the elastic limit and into the Tension Yield Point results in "Necking." The material will continue to strain without any significant increase in stress, and the cross sectional area will become substantially and obviously narrower. Further increasing the applied stress beyond the necking region will exceed the material's Ultimate Strength (US) followed immediately by the sample's point of fracture, or breaking point. Commonly available materials handbooks contain values for Young's Modulus, the Tension Yield Point, the Ultimate Strength and other relevant information.

Figure 7 - A standard sample of 1020 steel with a strain gauge attached is ready for a tensile (tension) test.

The information provided in reference texts is usually supplied as fractional quantities to facilitate universal application to material samples of all shapes and sizes. However, standard samples are often used as well. A standard material sample for tensile testing is 2 inches long, ½ inch in diameter with threaded ends for gripping by a tensile testing machine. An initial force of 100 pounds, approximately 500 psi, is applied to assure the tensile testing machinery doesn't slip. Data is acquired in a manner similar to the following spreadsheet.

LOAD	STRESS	ELONG	STRAIN	MODULUS
(#)	(psi)	(in)	(in/in)	(psi)
0	0	0.0000	0.0000	
1000	4993	0.0002	0.0001	49926068
2000	9985	0.0004	0.0002	49926068
3000	14978	0.0007	0.0004	33284045
4000	19970	0.0010	0.0005	33284045
5000	24963	0.0014	0.0007	24963034
6000	29956	0.0017	0.0009	33284045
7000	34948	0.0020	0.0010	33284045
8000	39941	0.0023	0.0012	33284045
9000	44933	0.0026	0.0013	33284045
10000	49926	0.0029	0.0015	33284045
11000	54919	0.0032	0.0016	33284045
12000	59911	0.0036	0.0018	24963034
13000	64904	0.0040	0.0020	24963034
14000	69896	0.0044	0.0022	24963034
15000	74889	0.0053	0.0027	11094682
16000	79882	0.0068	0.0034	6656809
17000	84874	0.0103	0.0052	2852918
18000	89867	0.0214	0.0107	899569
18700	93362	0.0780	0.0390	123492
13700	68399	0.3350	0.1675	#VALUE!

E aver =   31619843

Figure 8 - Stress Vs. Strain characteristic for 1018 Cold Rolled Steel, lab data.

Example computations of the quantities included within the spreadsheet appear later in this chapter. Plotting the stress and strain data points yields an obvious linear region. The slope of the linear region is averaged to determine the

138

Modulus of Elasticity, or Young's Modulus. The Modulus was determined to be 31.6×10^6 psi/in/in.

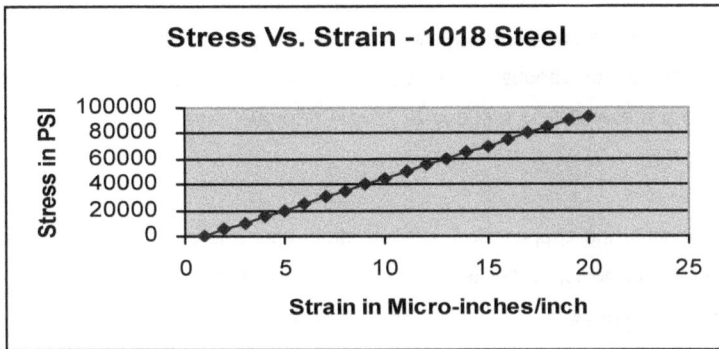

Stress Vs. Strain - 1018 Steel

Figure 9 - Stress vs. Strain 1018 Steel

Alloy, Treatment and/or Temper	Tensile Strength, Normalized	Yield Strength
1020 Steel	64 ksi	50.3 ksi
1050 Steel	108.5 ksi	62 ksi
1080 Steel	146.5 ksi	85 ksi
4150 Steel	167.5 ksi	106.5 ksi
2024-T3 Aluminum	70 ksi	50 ksi
2219-T42 Aluminum	52 ksi	27 ksi
5005-H12 Aluminum	20 ksi	19 ksi
7175-T66 Aluminum	86 ksi	76 ksi

A table of mechanical properties for common carbon and alloy steels and wrought aluminum alloys. KSI equals thousands of psi.

Stress/Strain Hardware

Being a resistor, the strain gauge can be applied and effectively used in a multitude of circuit configurations. By far, the series voltage divider circuit arrangement dominates strain gauge signal conditioning configurations. However, a host of non-ideal temperature and leadwire considerations must be addressed if the strain gauge is to perform as desired. The following addresses some of the application concerns when using strain gauges.

Strain Gauges

The Strain Gauge is a resistive sensor used to measure stress, strain and related physical variables. Composed of a single piece of copper alloy wire, (an alloy is used to control temperature effects) the strain gauge is shaped for sensitivity to applied forces of various types. Chevron, rosette and a multitude of other patterns are available to respond to forces of torque, tension, compression and bending.

In application, the copper-alloy, foil-grid strain gauge is glued to the surface of the sample or device under test such that the physical dimensions of the strain gauge will change in an identical manner to the material sample surface undergoing the applied stress. In this manner, the strain of the sample becomes the strain of the gauge.

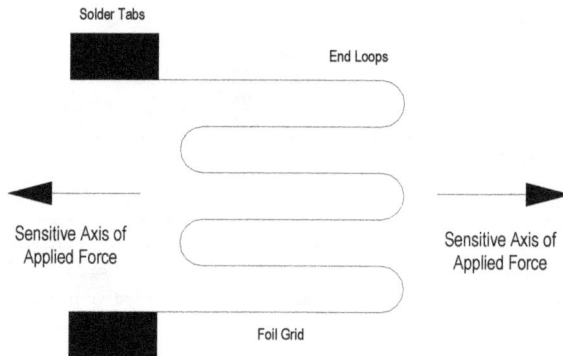

Figure 10 – A drawing of a strain gauge with a few locations indicated. Note the sensitive axis indicates the direction of the applied force. Forces can be applied as tensile (shown) or compressive with opposite results. Forces applied 90 degrees to the sensitive axis will result in zero or minimal change in resistance.

As the gauge's physical dimensions change, the wire's copper alloy grid dimensions also change. From fundamental electronics, the resistance of any wire equals the wire's resistivity coefficient (ℜho, determined by the wire material type) multiplied by the wire length divided by the wire's area.

Resistance = ℜho * Length / Area

If a strain gauge is glued to the surface of a metal column to be sensitive to longitudinal strain created by applied <u>tension</u> forces, the vertical stretch of the column will cause the strain gauge to stretch as well. The strain gauge, similar in appearance to the illustrations included in the following pages, increases its wire length and decreases its wire area resulting in an increase in the gauge's resistance according to the preceding equation.

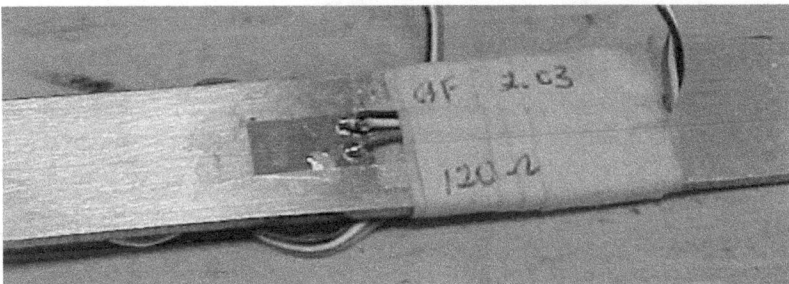

Figure 11 – A single strain gauge applied to measure tension and compression forces. Note the three wires coming from the strain gauge. Three wires provide a means of temperature compensating the leadwires against changes in resistance.

Figure 12 – The previous strain gauge observed closely. Note the really bad soldering results but before you laugh, keep in mind this is a very small area for soldering.

A compressive applied force will result in a decrease in strain gauge resistance since all of the previous conditions have reversed. Compressive loads cause the length of the sample and strain gauge to decrease which according to the equation for resistance, will result in the gauge resistance decreasing. In either case, the change in strain gauge resistance is a linear representative of the applied stress and resulting sample strain.

Gauge Factor

The Gauge Factor (GF) or Gain Factor of a strain gauge relates the output (resistance) change to the input (strain) change of the gauge. Since gain is computed as output divided by input, the gain of a strain gauge is computed as its fractional change in output resistance divided by the fractional change in input strain. In equation form;

GF = fractional output resistance change / fractional input strain change

Or,

$$GF = (\Delta R \ / \ Rgauge) \ / \ (\Delta Length \ / \ Lorig)$$

And,

$$GF = \Delta R \,/\, R_{gauge} \,/\, \varepsilon_L$$

Where,

      GF = strain gauge Gain Factor

      ΔR = strain gauge resistance change

      R_{gauge} = original gauge resistance before stress

      ε_L = Longitudinal Strain developed within the sample under test

Rearranging to determine the amount of resistance change given longitudinal strain, gain factor and original gauge resistance yields;

$$\Delta R = R_{gauge} \,{}^{*}\, \varepsilon_L \,{}^{*}\, GF$$

Typical values of resistance change are under .1 Ohms for a 120-ohm and 350-Ohm strain gauges however, as a material sample approaches its limit of elasticity the strain gauge resistance change will commonly approach .5 Ohms. For non-destructive testing, a strain gauge resistance change of approximately .5 Ohms represents the maximum change it will encounter.

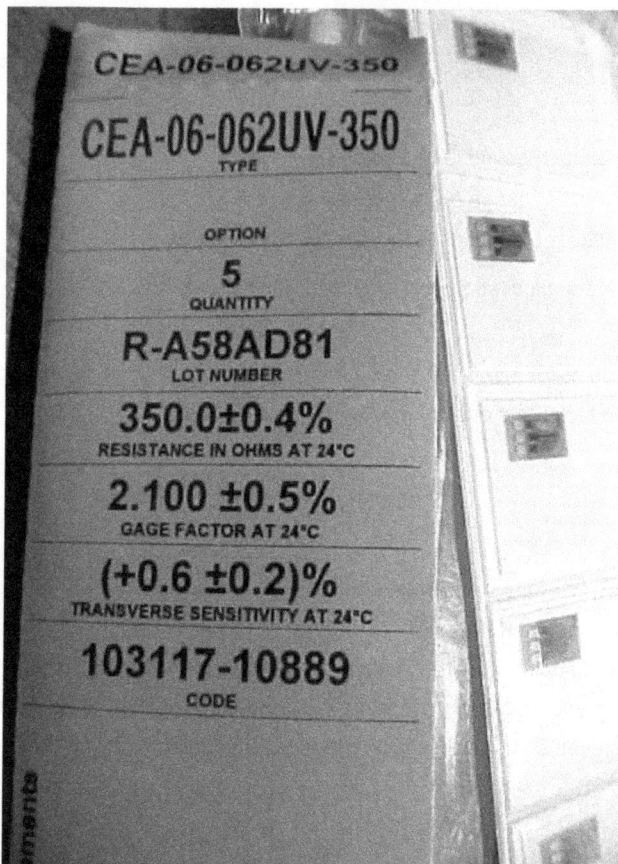

Figure 13 - A set of "Chevron" strain gauges used to measure torque and the enclosed specification sheet. Note the gauge resistance (350 ± .4%) and gauge factor (2.1 ± .5%) specifications.

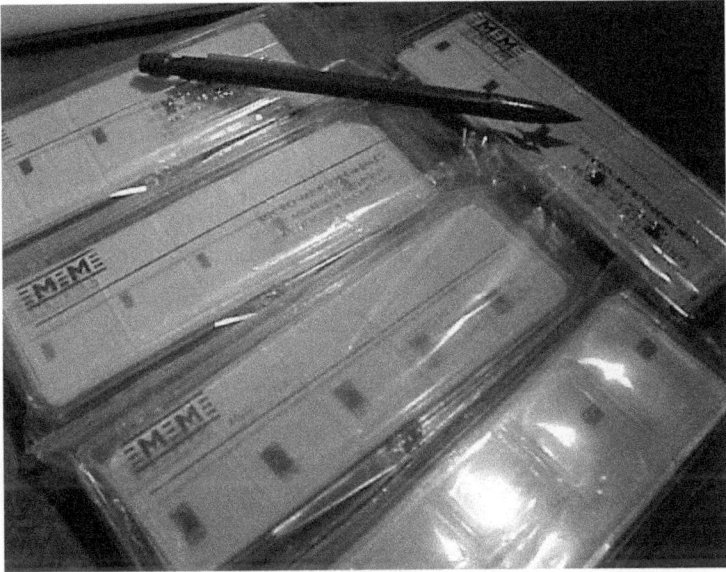

Figure 14 – Strain gauge sets of various types.

Bridge Conditioning

Strain gauge use is not limited to only stress and strain measurement. In fact the strain gauge is arguably the most popularly applied sensor, being used to also measure fluid (liquids and gases) pressure, weight, density, acceleration and an assortment of other physical variables. Simple in operation and application, the strain gauge appears universally within a variety of force and pressure measurement applications.

Acting as a variable resistor, the strain gauge must be incorporated into some form of resistance bridge circuit to convert the resistance change created by the applied stress and strain, into a voltage (millivoltage) signal for amplification. Two common conditioning circuits include the half-bridge and full, or Wheatstone bridge. Occasionally a one-milliamp constant current source is also used to generate an output voltage from the single strain gauge but this technique is somewhat rare.

In process automation and especially experimental testing applications, resistance bridge circuits can be found embedded as the active sensing element within a sensor, or in the "front-end" of a signal conditioner converting resistance changes into signal voltage for amplification. In bridge applications, the voltage difference across the bridge is considered the output signal.

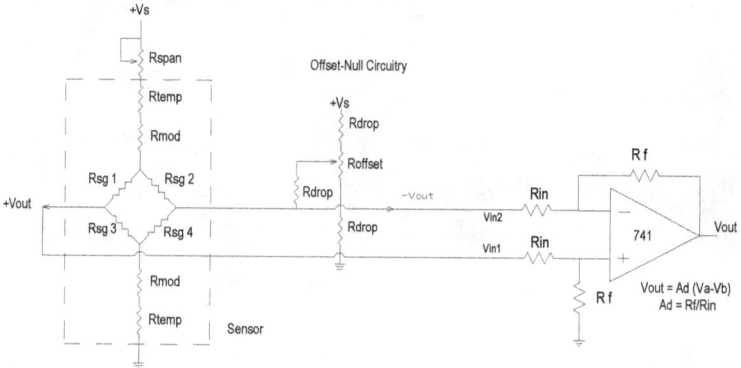

Figure 15 – A Resistive Bridge (diagonal components) and associated conditioning circuitry.

The bridge output is initially assumed to be "null-balanced" when zero physical force is applied. Null-balance occurs when each of the resistors in the bridge circuit are of equal value. Since the output voltage is taken by measuring the difference or subtracting the voltage available from each side of the bridge, if the resistor values are equal the voltages at each output are equal and the net output voltage will be zero, or *Nulled*.

The bridge gradually unbalances as the applied physical variable increases. As the bridge becomes unbalanced with the physical input changing, a *differential output voltage* is generated.

146

+ Vs

R1   R2

- Vout   + Vout

R3   R4

Ground or -Vs

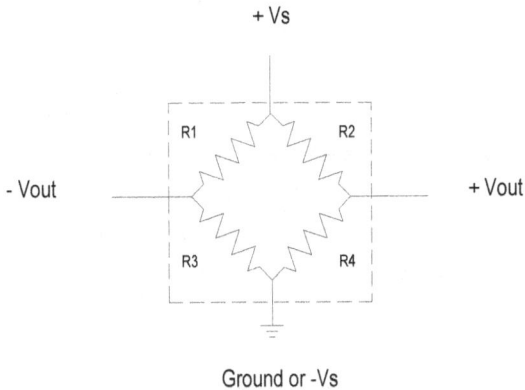

Figure 16 - A typical resistance bridge as used in sensor applications. Note the two outputs and corresponding polarities. In this example, the four resistor elements are included within a sensor housing indicated by the dashed line around the bridge. The diagram is representative of stress, strain and pressure sensors.

The differential output voltage is the result of having the voltage at each output (with respect to ground) moving in different directions when the physical variable is applied. This is caused by the manner with which the individual strain gauges are connected, and the physical variable applied internally to the individual strain gauges. Upon inspection of any four-arm bridge-type sensor, one will notice that each output has an associated polarity. The polarity indicates the direction that particular output will move when the input is applied and increased. Note this does not mean the polarity of the voltage at that output location, but the <u>direction</u> the output will change when the input variable is applied. Negative polarity indicates the output will be negative-going with an increase of the input variable, and positive implies the output will be positive-going with an increase in the applied physical variable.

The output polarities also imply something about the operational conditions of the individual resistors that compose the bridge. Typically, the resistors across each bridge output will change its resistance in a manner indicated by the output polarity.

If the polarity is negative, the resistor at the negative output (to ground or –Vs) will be decreasing in resistance value. At the positive output the resistor between the output and ground (or –Vs) will be increasing in resistance. The polarities associated with the resistor labels in the following diagram represent the direction the individual resistance changes when an input variable is applied and increased. As is the case with most housed, integral bridge sensors, all four internal bridge components are made sensitive to the input variable. In which case resistive sensors in opposite legs will vary their resistance in a corresponding manner, and resistors in adjacent legs will vary in an opposing manner.

The following diagram demonstrates the phase relationship among the four active, internal bridge components.

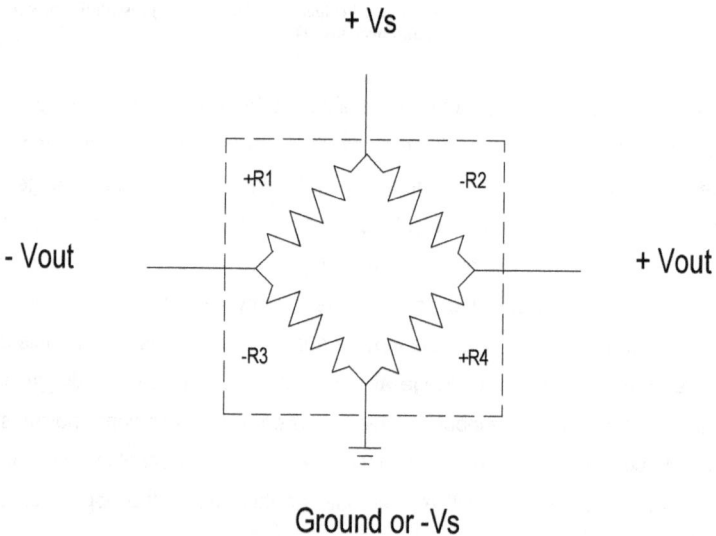

Figure 17 – A common bridge circuit representing the operation of many sensor circuits. The R1 through R4 resistor polarities represent the direction of the resistance change with an increase in the applied physical variable.

With all circuits employing differential outputs, neither output is connected directly to ground. Therefore, interpreting the output signal requires measuring (or amplifying) the *difference between the two voltages at the positive and negative output leads* independent of a reference to ground. Usually a ground connection is available but is not required to determine the output signal. Only a measurement of the voltage difference between the two outputs is required to determine the output voltage signal. Likewise, **when asked to measure the output from a bridge, the voltage measurement should be made <u>differentially</u> between the (positive and negative) outputs and not from either output to ground**. When used with strain gauges, the bridge output voltage is typically in the range of microvolts to millivolts and usually requires further amplification for the output voltage to be of any significant use.

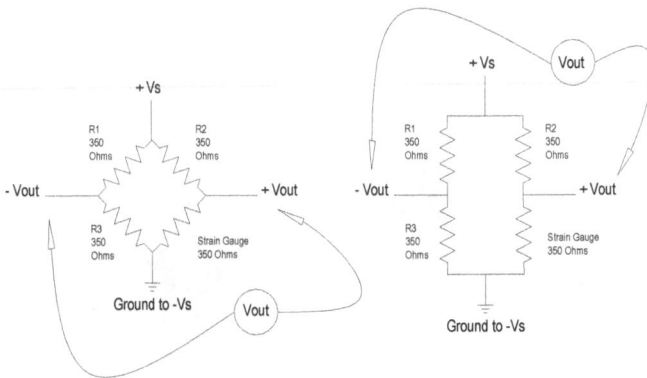

Figure 18 - Bridge circuits are composed of two side-by-side series circuits and may appear as either of these drawings indicates. The output signal voltage is measured between the differential (plus and minus) locations as indicated in both diagrams.

Resistive bridges may contain multiple (2, 4 or more) resistive sensing elements depending upon the desired sensitivity, application demands and the nature of the sensing elements. Bridges containing a single sensor are referred to as single active (sensor) element bridges or quarter active bridges, or simply quarter bridges.

149

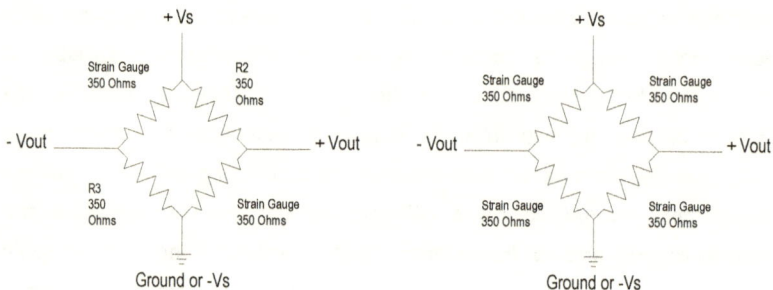

Figure 19 – A two-active element bridge (left) and a four active element or full-bridge (right).

Two and four active sensor bridges are commonly employed with strain gauges. The enclosed sensors discussed in the previous section were assumed to be full bridge circuits since all four of the resistive components were assumed to be sensing and changing with an applied input variable.

Figure 20 -A 3000 psig Viatran pressure sensor utilizes two strain gauges (barely visible in the upper right assembly) mounted on a metal diaphragm. The bridge completion resistors are visible on the adjoining circuit board.

Quarter bridges are rarely employed without additional bridge completion components since having only a single component in the circuit would result in an output voltage equal to the applied source voltage. When quarter bridges are used without additional bridge completion resistors, a *constant current generator* IC (integrated circuit) is placed in series with the single sensing element. The current generator develops a constant value (1, 5 or 10 milliamps is most common) of current, independent of imposed disturbance variables such as ambient temperature. The output voltage signal is taken directly across the sensor, as indicated in the following diagram.

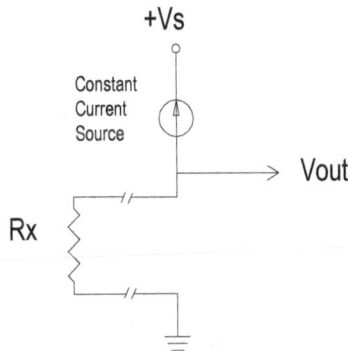

Figure 21 – A quarter bridge employs a single sensor Rx and a current generator. The output signal is taken across the sensor. Circuits of this type are common with RTD temperature sensors and single strain gauges. Temperature sensors for this purpose are available containing 4 leads, two for the current supply and two for the output to minimize the effects of leadwire runs.

Occasionally in bridge circuit diagrams the sensing element (shown as resistor Rx in the previous figure), is indicated as being removed from the bridge to place emphasis on the remote location of the sensor from the remainder of the bridge circuit. This is done to draw attention to the lead wire resistance. Bridges of this type often employ only a single resistive sensor such as a strain gauge at the point of measurement, and a single fixed resistance at a remote location (where the data acquisition system resides) to complete the bridge. It might be appropriate to again mention the original idea behind the bridge in sensor applications was to convert a resistance change into a voltage change. Therefore, bridge completion as referred to previous implies the insertion of an additional component(s) for the purpose of converting the resistance change into a voltage change.

Half-bridge circuits are common in experimental applications where a test fixture is assembled utilizing a single sensor to measure a specific characteristic associated with the device under test. The fixed resistor (R1 in the following figure) utilized in the half-bridge is selected to be a value equal to the sensors' resistance value at zero applied input.

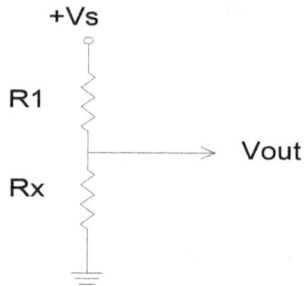

Figure 22–Schematic representation of a single active element half-bridge, or voltage divider circuit. The lower resistor labeled Rx is the single sensing element. The upper resistor is a precision, low temperature coefficient fixed value resistor. The circuit converts a change in resistance to a proportional change in voltage but exhibits an output-offset equal to half the supply voltage at the minimum input condition.

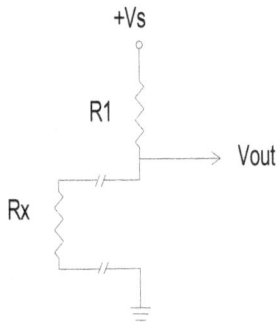

+Vs

R1

Vout

Rx

Figure 23 - A single active element employed in a half bridge. The sensor is indicated as being remote to emphasize the effects of the leadwire resistance.

As an example, a single 120-Ohm strain gauge will employ a 120-Ohm fixed resistor as the bridge completion component and a single 350-Ohm strain gauge would require a 350-Ohm fixed resistor. Occasionally a variable resistor is used in the location of R1 to set the initial bridge output condition.

Figure 24 – An inexpensive two active element load cell (top) with two forms of bridge completion resistors (center and bottom). The resistor in the center is 499 Ohms at .01% tolerance; the two resistors at the bottom are low temperature coefficient, 01% tolerance, 120-Ohm bridge completion resistors intended for use with strain gauges.

It should be noticed that using half-bridges would result in a minimum output value of one-half the supply when the applied physical variable is zero. Since the value of both bridge components is equal initially, the resulting output will equal one-half the bridge supply voltage. If the applied supply voltage is 5 volts, the initial output from the bridge is 2.5 volts. This relatively large amount of initial voltage, or offset, (2.5V) can impose a severe limitation on the amplifier gain following the bridge. For this reason the initial bridge output is usually eliminated, or nulled-out, before amplification when using half-bridges.

The following section provides an example of strain gauge bridge circuit analysis.

Sample Computation

The following provides an example of strain gauge related computations. It is provided as reference for student computation and to demonstrate the proportional relationship between the related items of stress, strain and gauge resistance variation.

-Given the following information, determine the included values:

-Given:
    Area = 2 in²
    Length = 10 inches
    F_{comp} = 5000#
    E = 31 E6 $_{PSI/in/in}$ (1020 steel)
    μ = .288 $_{in^2/in^2/in/in}$
    $R_{Strain\ Gauge}$ = 120 Ω
    $GF_{Strain\ Gauge}$ = 2.12
    Vs = 5 Volts

-Find:
A) Stress
B) Strain (ε_L)
C) ΔLength
D) Strain (ε_T)
E) ΔArea
F) ΔRsg
G) Bridge Vout

-Single active strain gauge element, full bridge configuration.

-Finding Stress:
    Stress = Force/Area = 5000# / 2 in² = _2500 psi_

-Finding Longitudinal Strain:

Since,

E = Stress / Strain

Strain = Stress / E = 2500 psi / $31*10^6$ PSI/in/in

Strain = 80.645×10^{-6} in/in or,

Strain = _80.645 μ Strain_

-Finding the change in length:

-Since,

$Strain_{Long}$ = ΔLength / Length

ΔLength = Strain * Length

ΔLength = 80.645×10^{-6} in/in * 10 in

ΔLength = _806.45×10^{-6} in_.

-Finding Transverse Strain:

Since,

μ = Transverse Strain / Longitudinal Strain

Or, $\mu = \varepsilon_T / \varepsilon_L$

$\therefore \varepsilon_T = \varepsilon_L * \mu$

Transverse Strain = $.288$ $in^2/in^2/$ in/in * 80.645×10^{-6} in/in

Transverse Strain = _23.22×10^{-6} in^2/in^2_

-Finding the change in area:

-Since,

$Strain_{Trans}$ = ΔArea / Area

ΔArea = $Strain_{Trans}$ * Area

ΔArea = 23.22×10^{-6} in^2/in^2 * 2 in^2

ΔArea = _46.44×10^{-6} in^2._

-Finding the change in resistance of the strain gauge:

-Since,

Gauge Factor, GF = ΔRsg / Rgauge / ε_{Long}

ΔRsg = GF * Rgauge * ε_{Long}

ΔRsg = 2.12 Ω/Ω/in/in * 120Ω * 80.645 x 10^{-6} in/in

ΔRsg = _0.020516088 Ohms_

-Finding the new resistance of the strain gauge:

New Rsg = Rsg $\pm$ ΔRsg

$\therefore$ Rsg = 120 - 0.020516088 (compression decreases Rsg)

Rsg = _119.979483912 Ohms_

-Finding Vout from the bridge using the voltage divider equation:

Vout = [5V (119.979483912 /(120+ 119.979483912))] – 2.5 Volts

Vout = [5V * (0.499957254496)] – 2.5 Volts

Vout = 2.49978627248 Volts – 2.5 Volts

Vout = -0.0002137275202188 Volts

Or,   Vout = _-213.728 μ Volts_ (negative output designates compression)

It is suggested that the previous computations be performed in the sequence provided.

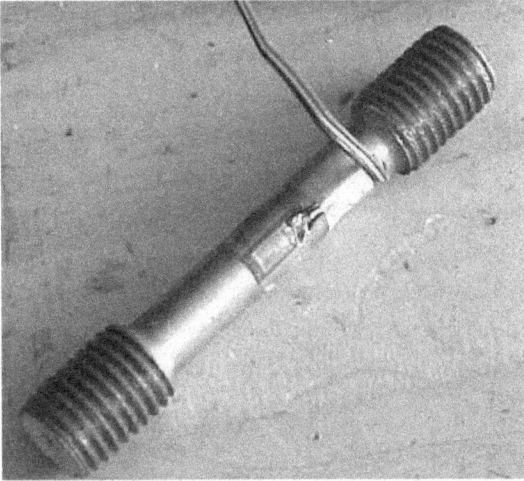

Figure 25 – A standard material sample with strain gauge applied. The previous example problem was derived from applications utilizing samples similar to this one.

The preceding computations were performed around a single strain gauged sample. Two or four gauges mounted on the same sample and connected within the same bridge increase sensitivity by proportionally increasing the output, and assist in decreasing sensitivity to thermal errors.

Gauge Compensations

After performing the computations of stress, strain, ΔR and bridge output voltage for strain gauge circuits, one develops an appreciation of the low-level signals being developed. Since low-level bridge outputs are usually amplified many thousands of times before being recorded, it is desired to keep the signal as "noise-free" as possible. When working with "micro-measurements," subtle influences due to component noise, warm-up drift, magnetically induced noise, temperature effects and thermocouple connections need to be considered and minimized.

Among the most prevalent effects causing an output to shift in strain gauge circuits is temperature. Even the smallest thermal effect can cause a substantial change in the output of a bridge circuit and especially a high-gain amplifier's output. To minimize the effects of temperature upon the strain gauges and leadwire connections, circuit configurations and strain gauge placement and mounting techniques have been developed.

Frequently, all four bridge resistance components are mounted at the sample under test site, and are all strain gauges. In this case two of the strain gauges are *active* or sensitive to the applied stress, and two are passive or "dummy" gauges, and are insensitive to the applied stress. Dummy gauges are actual strain gauges mounted on the sample under test with the (two) active gauges, but are usually rotated 90 degrees from the active gauges to be insensitive to the applied stress and resulting strain. Since the (two) dummy gauges are wired in to the full-bridge circuit in the location of the fixed resistors, the dummy gauges perform the required voltage divider functions. Since the dummy gauges are located in the same temperature environment as the active gauge, any temperature-induced change of operating characteristics will appear at all gauges in the bridge. The thermal effects will be negated due to the differential nature of the bridge output, and since all of the four bridge components are changing their characteristics together. A result of this application, full bridges with all strain gauges mounted at the application are understood to be inherently temperature compensated.

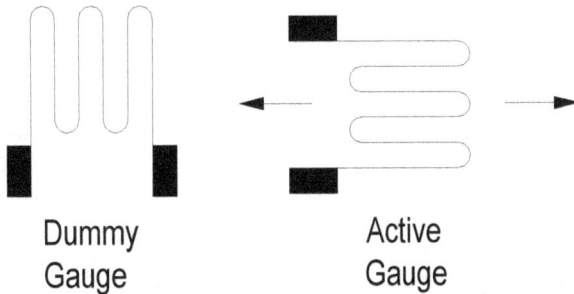

Dummy
Gauge

Active
Gauge

Figure 26 – Active and dummy strain gauge orientation. Both gauges would be applied in close
proximity on the sample. The applied force(s) would be in a horizontal direction.

However not all bridges contain two or four active strain gauges. Many contain
only a single active strain gauge, and usually the remainder of the bridge circuit
is located remote to the single strain gauge. For these circuits, the leadwire
resistance must be compensated. The length of leadwire between the bridge
circuit and the strain gauge may introduce considerable resistance into the bridge
circuit. If the resistance of the leadwire were constant, the effect of the wire could
be offset, or biased-out. Nevertheless, the effect of the wire resistance is rarely
constant, since as the temperature of the wire varies so does the resistance.
With the leadwire resistance capable of varying, it is conceivable that the bridge
output could vary in relation to the leadwire temperature even if the applied
stress or strain was to remain constant. This condition would yield a false stress
or strain indication, or what is occasionally referred to as an *apparent* microstrain.

Leadwire compensation is best understood if one is familiar with the differential
manner by which the bridge circuit generates an output voltage. As an aid in
understanding the following explanation, one should assume that for any single
leg of a bridge circuit there exists a single resistance in an opposite location and
two resistors in adjacent locations. Mentally visualize the bridge circuit and verify
the preceding statement before proceeding. The bridge output voltage is equal to

the difference between the two outputs, labeled ⁻V and ⁺V. Considering the output signal is the <u>difference</u> between each bridge output lead, any voltage common to each output will be subtracted. Any undesired effects introduced into adjacent legs of a bridge will cancel, having no contribution to the output signal. Leadwire compensation is accomplished by connecting one of the two leadwires to the strain gauge, in each of two adjacent legs of the bridge circuit. In this manner, the introduced leadwire resistances will establish equal voltages in adjacent legs of the bridge, canceling in the bridge output and effectively canceling the leadwire resistances. The following diagram demonstrates this condition.

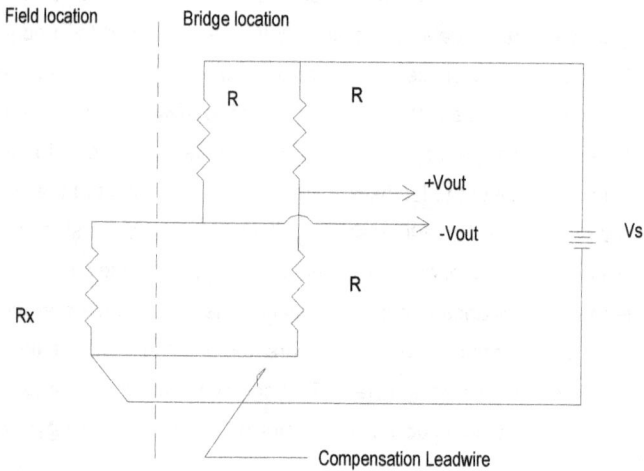

Figure 27 – Single resistance sensors commonly contained three wires connected to two terminals at the sensor site to provide thermal leadwire resistance compensation.

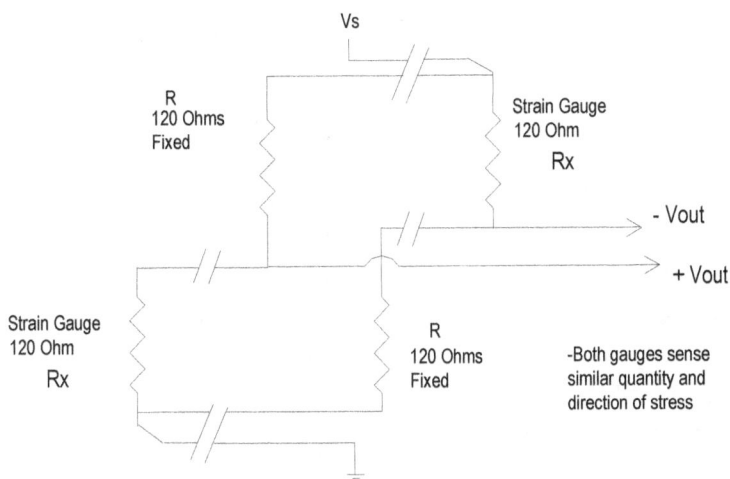

Vs

R
120 Ohms
Fixed

Strain Gauge
120 Ohm
Rx

- Vout

+ Vout

Strain Gauge
120 Ohm
Rx

R
120 Ohms
Fixed

-Both gauges sense
similar quantity and
direction of stress

Figure 28 – A bridge containing two remotely located and leadwire compensated sensors.

Bridge Gain and Offset

The Strain Gauge Bridge with 1, 2 or all 4 active gauges will often include provisions for gain and offset adjustment. Offset, zero, bias, or null adjustment is performed in one of two ways. The most direct method is to include a potentiometer within the bridge between any two of the bridge resistance components. With the potentiometer connected at one of the bridge outputs, the wiper will provide the output for the respective side of the bridge. As the wiper is adjusted, the voltage at that one side, and therefore the differential output voltage, will change slightly in a positive and negative direction. The potentiometer could also be connected between two components with the wiper to the supply or to ground. With the wiper connected to the supply or ground, the current in each leg of the bridge will vary in opposite directions as the wiper position is varied. In this method, the output at each side of the bridge will vary slightly positive and negative as the supply or ground potential is adjusted. Example diagrams of the techniques will provide additional insight. (Not included, 9/98)

161

Another common approach to bridge null adjustment is to include a voltage divider in parallel with either leg of the bridge circuit. This is accomplished with two fixed resistors and a variable resistance. The three are connected in series between the bridge supply and ground, with the potentiometer connected in between the two fixed resistors. This circuit provides a voltage that varies slightly above and below the voltage at one of the bridge outputs. The wiper of the potentiometer is then connected to the bridge output through a current limiting (usually around 100K resistance). Although sophisticated in appearance, the overall idea is to vary the voltage at either bridge output by only a few millivolts, enough to null any non-zero output voltage when the applied stress/strain is minimal. The following diagram illustrates the nulling circuit configuration.

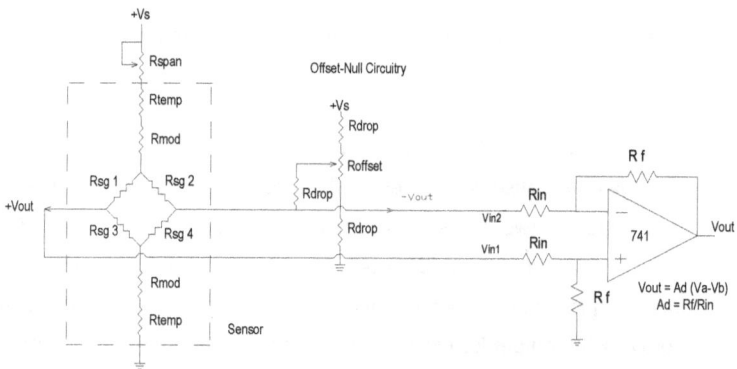

Figure 29 – Bridge circuits may also contain Span and Zero (gain and offset) adjustments as shown.

Bridge gain, or sensitivity, can be varied by including a correctly sized variable resistance, wired as a rheostat, in series between the bridge and the supply voltage. Given the output voltage as being the supply voltage multiplied by the ratio of the output resistance to the total resistance at each side of the bridge, Vout appears as being in direct proportion to the applied voltage source. Or,

$$Vout = \text{Left output voltage} - \text{Right output voltage}$$
$$Vout = Vs\ (R3/(R1+R3)) - Vs\ (R4/(R2+R4))$$
$$\therefore Vout = Vs\ [(R3/(R1+R3)) - (R4/(R2+R4))]$$

As can be seen from the equations, Vout varies in direct proportion to the supply voltage. For any given and constant value of bridge resistances, as the supply voltage increases, the output voltage span increases. As the supply voltage decreases, the output voltage span decreases. Consequently, any variation in the voltage supplied to the bridge produces a proportional change in the bridge output.

Bridges require compensating against two basic types of errors, gain errors and offset errors. Offset errors can usually be biased-out by introducing an external voltage to either output or by introducing a potentiometer into the bridge circuit. An initial adjustment of offset, bias or null before acquiring data from the bridge usually suffices if the components and their leadwires have been properly compensated. Introducing a temperature sensitive resistor into the supply voltage connection between the bridge and the supply minimizes gain errors.

As the ambient temperature of the bridge changes, the components within the bridge, including the strain gauges change their resistance values. The overall effect will appear as a change in the bridge's gain characteristics. In other words, if it is momentarily assumed the stress and strain are held constant and the temperature of the bridge varies, the output from the bridge would respond to the temperature change by varying in a proportional manner. If during the temperature variation the stress/strain of the gauged material were to change, the output would be greater or lesser than the same amount of stress/strain change at a normal operating temperature. This type of temperature sensitivity is often called *thermal drift*, and is measured in Ohms/C°, µin/in/C°, µstrain/C° or Volts/C° depending upon the specific reference of the drift.

To counter-act the effect of the resistance changes, a temperature sensor in series with the bridge supply can detect temperature variations and adjust the voltage supplied to the bridge to reduce or increase the bridge gain. In this way, the bridge sensitivity to an applied stress or strain remains constant, at least over a fixed (and usually fairly narrow) temperature span, typically 10 to 20 degrees Celsius.

For additional bridge circuit material, consult Chapter 2 in the ET 200 manuscript.

Scaling

A final computation can be made if a signal conditioning amplifier or other scaling device is used at the output of the bridge circuit. Since the signal at the bridge output is in the fractional millivolt or microvolt range, to properly display the measured units of stress or strain, mathematical quantities of gain (occasionally called scale factor) and offset will be required. If the desired value to be displayed is Stress, in psi, and assuming the bridge output is nulled at zero applied force, the scale factor would be 2500 psi divided by 213.728 microvolts. The resulting scale factor is 11,697,136.60385. This would be the value of a scale factor entered into a data-acquisition software program.

Scale Factor = Quantity to be displayed / Input voltage to the display amplifier

Figure 30 – A portion of an analog input signal conditioning rack. All modules contain gain and offset (span and zero) adjustments internally, as shown at the top of two modules.

In the case of an analog amplifier, an output voltage of .25 Volts might be used to represent the 2500psi of Stress developed within the strain-gauged sample. Again assuming the bridge output voltage is applied to the amplifier input, this would require an amplifier gain of 11,697.137, and would necessitate an instrumentation grade of amplifier.

Amplifier Gain, Av = Desired amplifier ΔVout span / Bridge ΔVout span

Note the amplifier input and output is indicated as being voltage spans. It is assumed the applied force at the sample will be varying between zero and 5000 pounds, the bridge output will be varying between zero and 213.728 microvolts, and the amplifier output will vary between zero and .25 Volts. It assumed that zero force results in zero bridge output, a null-balanced bridge. Although this is usually close to being the case, it should be assured by including a bridge null circuit if necessary. In any case, the bridge offset is often not known until the strain gauge bridge circuit is connected, and the bridge output is measured. With gains and scale factors as high as 10,000 to 10 million plus, even a slight bridge offset can yield a substantial initial error of the displayed value.

165

Figure 31 – Another older form of strain gauge signal scaling amplifier.

Stresscoat

At this point in the chapter, it should be obvious that strain gauges are applied where a mechanical designer needs to acquire more information about material or component deformation and strength. Upon closer investigation however it may be difficult to establish the specific locations to glue the strain gauge(s) and types of gauges to be laid. For this reason a product called Stresscoat has been developed to determine where the stresses appear and to determine the location to glue the strain gauges. Applied like paint, Stresscoat becomes brittle and exhibits cracks when subjected to stress. Once Stresscoat is applied and allowed to dry the device under test is run through a preliminary test. The resulting cracks allow engineers to determine where the stresses are formed and where to lay the strain gauges.

The following photos demonstrate.

Figure 32 - Stresscoat applied to the hitch of a cart. The cart will next be run through a test and the stresscoated area will be analyzed to determine where the gauges will be laid.

Figure 33 - Curious aliens are often found examining seed carts, not really of course. A technician applies Stresscoat to determine specific locations to lay strain gauges.

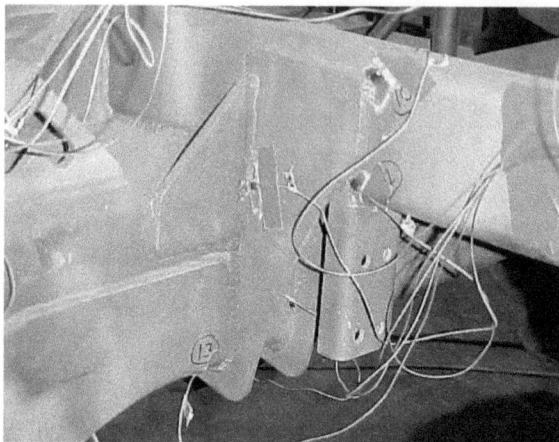

Figure 34 - Another example of a Stresscoat application, several strain gauges also appear in the photo.

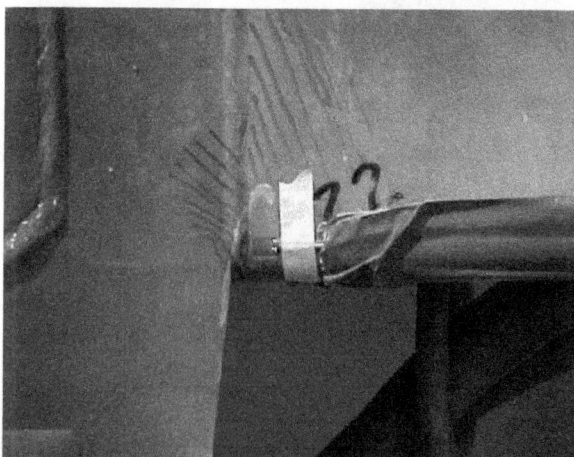

Figure 35 - The cracks resulting from the Stresscoat application has been highlighted with a red marker in this photo. The individual Stresscoat cracks are not obvious in the photo.

Figure 36 - Once the strain gauges are layed the leadwires are run to a junction box where the bridge completion resitors are located. The junction boxes are connected to a portable data acquisition unit.

Figure 37 - A portable data acquisition unit.

Load Cells

Among the more popular forms of strain gauge applications is the *load cell*. A load cell is a force sensor. More specifically, load cells are an accumulation of strain gauges connected to form an electrical (Wheatstone) bridge and mounted

169

on a metal platform that is used for the purpose of determining mechanical load, such as weight or force. Load cells are commercially available in a variety of load ratings and configurations. "Pancake" styles are commonly used to measure forces in the 100's of pounds to hundreds of thousands of pounds. Platform load cells are typically found in grocery scales at the checkout counter and weight between ounces and a few pounds.

Operation

Load cells do _NOT_ measure stress or strain. Although strain gauges are the primary sensing element within each load cell, the purpose of the unit is to output a signal that varies in relation to the applied _force_ and not the applied stress or resulting strain. As will be demonstrated following, determining the output of a load cell is far less complicated than computing the output of a strain gauge bridge.

Figure 38 - Loads cells are available in a variety of sizes and styles depending upon the application. Pictured above are two "Pancake" styles of load cell (left), a load washer (center) and a platform load cell (upper right).

It is suggested that one become reacquainted with the operation of a resistive (Wheatstone) bridge before continuing as the majority of load cells contain a full bridge internally. Specifically, the concepts of differential output signals and _Ratiometricity_ should be fully understood. All bridge circuits are ratiometric; the amount of supply voltage determines the amount of output voltage when the bridge is unbalanced. Since any bridge is merely an application of series resistance voltage divider circuits, it can be proven that a bridges' differential output voltage varies in direct proportion to the amount of supply voltage provided to the bridge.

The following demonstrates the relation between bridge supply voltage and output.

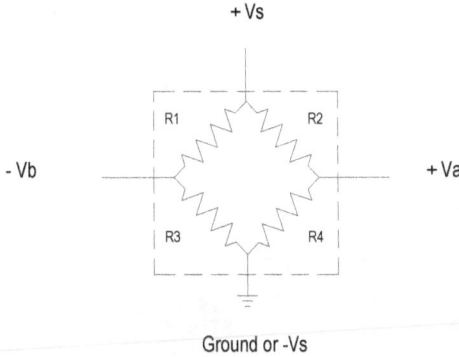

Figure 39 – Load cells are commonly composed of 4 active strain gauges connected in a full-bridge circuit configuration. When a force is applied to the load cell, the positive output (+Va) will increase towards +Vs and the negative output (-Vb) will decrease towards Ground or -Vs. The output signal is the difference between Va and Vb.

$$Vout = Va - Vb$$
$$Va = Vs \left[R4/(R2+R4) \right]$$
$$Vb = Vs \left[R3/(R1+R3) \right]$$
$$Vout = Vs \left\{ \left[R4/(R2+R4) \right] - \left[R3/(R1+R3) \right] \right\}$$

As can be seen from the equation, the bridge output varies in a direct proportion to the applied supply voltage, Vs. In this way varying the supply voltage to the bridge varies the output voltage span and as such, is understood to control or determine the bridge "Gain." Gain is defined as being the distance between output voltage-

171

span end points, and is occasionally referred to and labeled on schematics as bridge sensitivity or Span.

Compensation

As can be seen in the following load cell photo the resistive bridge circuit has been modified to include additional resistance between the supply connections and the strain gauge bridge. This is done for two possible reasons, to compensate the bridge output for changes in ambient temperature (temperature compensation) and secondly, to assure that all load cells of the same model numbers exhibit similar input/output characteristics (modulus determination). Strain gauge and load cell bridge circuits are often modified to compensate for thermal effects and to better match units of similar model numbers for replacement should one fail.

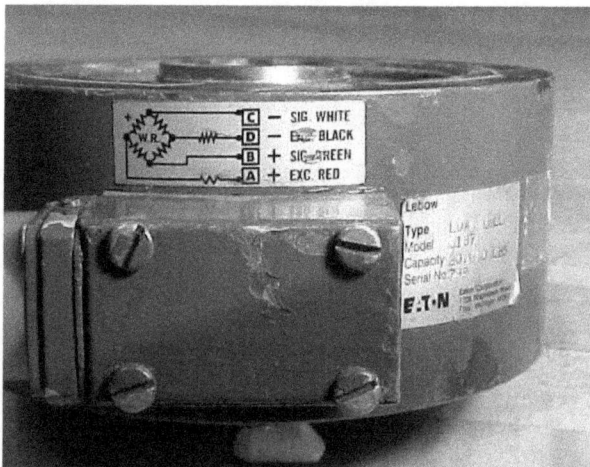

Figure 40 - This 20,000 pound load cell contains a bridge circuit diagram as an aid to making the proper electrical connections. Note the additional gain stabilizing compensating resistors in series wth the supply voltage in the diagram.

Upon inspection of the engineering data sheet that is included with strain gauges one would notice the sensitivity of the strain gauge's gain (gauge) factor (GF) to temperature. A graph of the characteristic is contained in the following figure. The

172

graph demonstrates the linear variation in sensitivity (Gauge Factor) as temperature changes. As temperature increases, the gauge factor, (or gain factor, GF) increases. This means <u>the individual strain gauges that compose the load cell's bridge circuit become more sensitive as the temperature increases</u>. Without some form of compensation the load cell output would tend to increase as a result of a temperature increase even though the applied load remained constant. Obviously, without thermal compensation the results of an experimental test would be disastrous if the product being tested is instrumented and calibrated indoor where the temperature is 70°F and is tested outside where the temperature is 30°F or worse. This is commonly the case with experimental testing of agricultural, military, automotive and construction products.

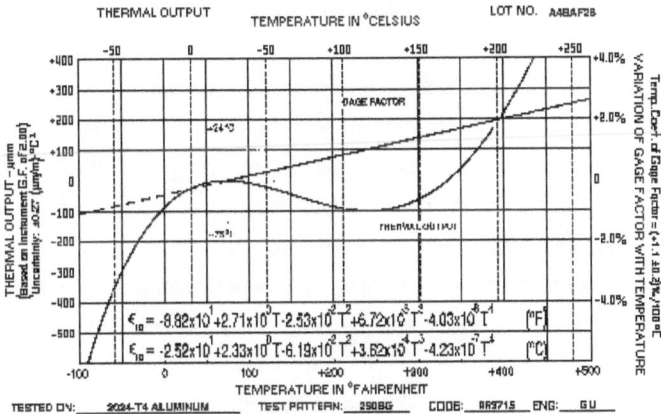

Figure 41 - Thermal effects on strain gauge factor (linear characteristic) and offset (sinusoidal appearing characteristic). Photo courtesy of Vishay Micro Measurements.

Full bridge strain gauge and load cell circuits are inherently temperature compensated against thermally induced *offset* drift. Offset refers to the bridge circuit output voltage at null-balance with no mechanical load applied. A temperature created shift of the strain gauges resistance at zero applied force (no-load) will cause all of the strain gauges in the bridge to change the same amount in the same direction, since the strain gauges are matched. Since the bridge is a pair of voltage divider circuits the effect will be offsetting and the output at no-load will continue to be null-balance, or zero volts at the output.

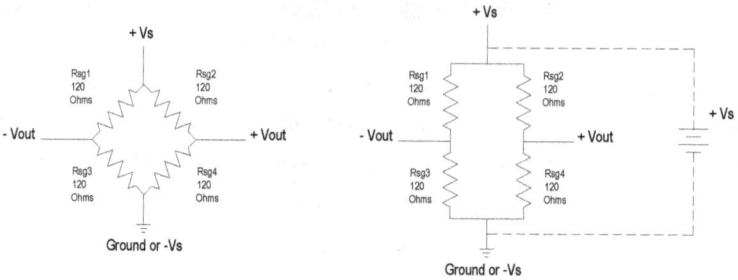

Figure 42 - All Wheatstone bridge circuits (left) can be analyzed and interpretted as series voltage divider circuits (right).

As a precaution against thermally induced _gain_ (gauge factor) error, many load cell manufacturers include temperature compensation components within the load cell. Since the load cell contains strain gauges configured as a full (4-active element) bridge, and since the bridge output voltage is directly proportional to the amount of applied supply voltage (remember ratiometricity?), if the bridge sensitivity should increase with an increase in ambient temperature, the bridge output will increase with an increase in temperature – even if the applied load remains constant. To compensate for the increased sensitivity, temperature-sensing resistors (pure metal resistive temperature detectors, or RTD's) are placed in series with the bridge circuit.

The temperature compensating resistors increase in resistance as the ambient temperature increases. The increase in resistance results in less voltage being applied to the strain gauge bridge circuit. As the voltage applied to the bridge decreases the output voltage from the bridge would normally decrease but since the individual strain gauges have become more sensitive as a result of the temperature increase, the output voltage stabilizes. Overall, an increase in temperature causes the strain gauges to generate more output voltage from the bridge, but since the temperature sensing resistances in series with the bridge have reduced the supply voltage to the strain gauges the output signal voltage remains constant. However, temperature compensation is only valid over very defined limits of temperature extremes. Operation outside of the temperature extremes will result in erroneous data from the load cell.

Resistors are also placed in series with the load cell bridge to match the operating characteristics of similar models. If a load cell should go bad, and they do – mostly from abuse, to minimize the calibration procedure it is highly desirable to replace the bad unit with one of similar operating characteristics. Recognizing this, many load cell manufacturers place "Modulus" resistors in series with the strain gauge bridge. Modulus resistors reduce the sensitivity of the load cell in an attempt to provide all similar units with the same sensitivity to voltage supply and load. Modulus resistors are commonly found on platform load cells where the amount of calibration adjustment associated with the display amplifier may be limited. Some platform load cells contain both temperature compensation and modulus resistances in series with the supply voltage.

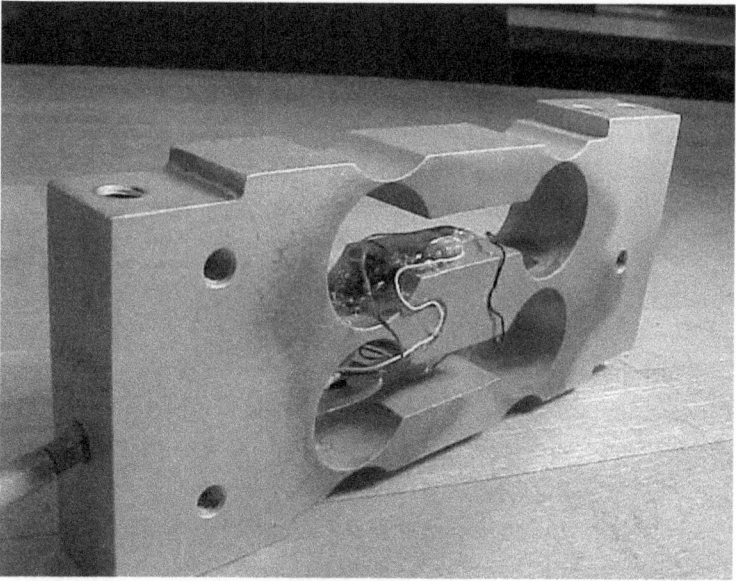

Figure 43 - Internal view of a platform load cell. Note the modulus resistor in the center of the photo.

One final consideration when applying load cells, the manner with which the load force is applied must also be considered critical. Off-center loading may result in an output signal that is less than expected and could result in damage to the load cell. Load cells will typically output a signal in proportion to the vertical component of the applied force, assuming the load cell is mounted in a position to accept vertical forces. If the force is applied at an angle, even a slight angle, the resulting output signal from the load cell will be misleading and will not represent the total applied force, only the sensitive axis component of the applied force.

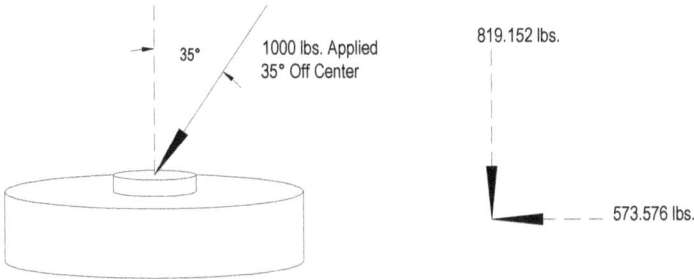

Figure 44 – Since load cells are only sensitive to forces applied upon a specific axis, a 1000 pound load applied 35 degrees from center will generate an output signal equivalent to 819.15 pounds.

Rated Output

Since the load cell contains multiple strain gauges configured into a full-bridge circuit configuration, one need not concern themselves with computations comparable to the stress/strain and strain gauges. To determine the output millivoltage signal of a load cell for a given amount of applied load (force or weight), only the *rated output* and supply voltage values are required.

As an electronic sensor of force, it would be expected that the load cell specifications would contain a statement of gain that involves volts per force. However, since the bridge circuit's output voltage is contingent upon the amount of supply voltage, the load cell's sensitivity coefficient will also contain a reference to the anticipated supply voltage to be used on the load cell bridge circuit. For this reason the "Gain" of a load cell is expressed in terms of overall sensitivity. The load cell's *Rated Output* is the output voltage generated at <u>full rated load</u> (the mechanical load rating of the load cell) divided by the supply voltage.

$$\text{Rated Output (RO)} = \Delta V_{out} \div \text{Supply Voltage}$$

Expressed as a load cell specification, rated output (RO) normally appears in specifications similar to the following. As an example, a load cell with a rated output of 2 millivolts per volt would appear as in the following.

$$RO = 2.0 \text{ mV/V}$$

In this example, the load cell will produce 2.0 millivolts <u>for each volt of applied supply voltage to the load cell, at the full load rating of the load cell</u>.

Figure 45 - A pair of well used 200-pound pancake load cells. The cover has been removed from the unit on the right to expose the strain gauges.

Example Problem:

Assume a load cell is specified at a rated load of 100 pounds, has a rated output of 2.0 mV/V and is to be used with a 5-Volt supply. Determine the output voltage at full rated load, 50 pounds, 20 pounds and 10 pounds.

Solution:

Given the supply is 5V and the RO is 2.0mV/V, the load cell will produce 10mV at 100 pounds.

Since,

$$RO = \Delta Vout \div Supply\ Voltage$$
$$RO = \Delta Vout\ /Supply\ Voltage,\ in\ mV/V$$

Rearranging to solve for the output voltage,
$$\Delta Vout = RO * Supply\ Voltage$$
$$\Delta Vout = (2\ mV/V) * (5V)$$
$$\Delta Vout = 10\ mV\ at\ full\ rated\ load\ (100\#)$$

The output voltage at applied loads other than the full rated load will be proportional to the output at the full rated load. Therefore if 50# is applied, the output will be one-half of the full rated load output, or 5 mV. An applied load of 20# equals,
$$\Delta Vout = 20\#/100\# * RO$$
$$\Delta Vout = 20\#/100\# * (2.0\ mV/V*5V)$$
$$\Delta Vout = 2\ mV$$

And an applied load of 10# equals,
$$\Delta Vout = 10\#/100\# * RO$$
$$\Delta Vout = 10\#/100\# * (2.0\ mV/V*5V)$$
$$\Delta Vout = 1\ mV$$

In short, the load cell is a linear device with an output of approximately zero millivolts (a null-balanced bridge) with no load applied. When a load is applied the output will vary in proportion to the rated output value of the load cell. To approximate the output, first determine the supply voltage and the RO of the load cell. Multiply the RO and supply voltage to determine $\Delta Vout$, or the output voltage change at full rated load. Then, as demonstrated previously, determine the output as a proportion of the full rated output by finding the proportion of the rated load applied. The equation follows.

$$Vout = [(Applied\ load \div Full\ rated\ load) * RO] \pm Offset[1]$$

[1]Note - this equation assumes the load cell output is close to zero millivolts at null-balance, with no load or mechanical force applied. Most of the time the null-balance output of the load cell will be very close to zero millivolts (if not, the load rating of the load cell may have been exceeded and the load cell should be considered defective). Otherwise the offset at no-load must be added to the computation.

Figure 46 - These signal conditioning modules are capable of accepting inputs signals from a variety of sensors including load cells and output both current (4-20 mA) and voltage (0-10V) signals. The calibration span and zero adjustments are obvious at the top of the modules.

Calibration

The process of assuring accurate information from any sensor-based measurement system is referred to as *calibration*. In most experimental measurement applications an amplifier or other form of signal conditioner follows the sensor and "drives" the display or data acquisition device.

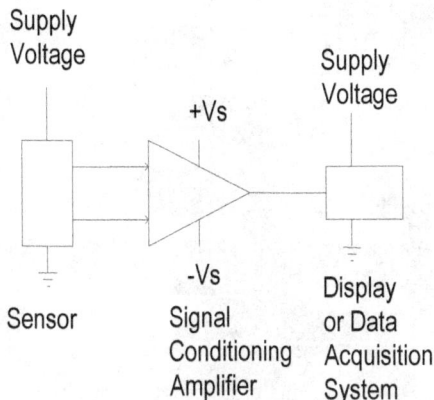

Figure 47 - The signal conditioning amplifier in this diagram is responsible for developing a voltage range that numerically represents the physical variable being measured. Adjustments of gain and offset are normally included to calibrate the output signal range.

Calibration allows the output of the amplifier or other signal conditioner to numerically represent the physical values being measured by the sensor. As an example, assume a load cell is measuring force between zero and 200 pounds. In this example the amplifier output would likely be set to 0-Volts at zero pounds and 2-Volts at 200 lbs. Signal conditioning circuits normally have adjustment provisions to precisely set the output to the desired values representing the variable being measured. Linear sensors are normally connected to signal conditioning amplifiers with gain and offset adjustments for calibration. The gain and offset adjustments are directly related to the mathematical functions of slope and intercept respectively. The gain and offset adjustments are commonly referred to as *span* and *zero* adjustments where gain determines the output signal span and zero determines the initial output offset or minimum output voltage. In the previous example the zero adjust would set the output to zero volts at no applied load and the span (gain) would set the output to 2-Volts with a 200 pound load applied.

Figure 48 - Each of the LED displays in this photo has a unique signal conditioning amplifier. The amplifier modules are located in the rack along the bottom of the photo.

Given that load cells are linear devices; calibration is normally performed at the extremes of the anticipated application forces. Load cells are rarely used throughout the entire range of full rated load however. More commonly, load cells are used between zero applied force and approximately 70% to 80% of full rated load. This is done as a precautionary measure to avoid inadvertently overloading the load cell as replacement costs normally run into the hundreds and in many cases, thousands of dollars.

Load cells rated at small load values, 10-100 pounds, can be calibrated by first applying zero force and then applying the maximum anticipated value of load. With zero force applied, the amplifier, transmitter or data acquisition unit Offset, or Zero adjustment is set for a minimum (usually zero) indication. When the maximum force is applied the amplifier or data acquisition Gain, or Span is set to indicate the value of maximum applied load. With smaller ratings of load cells an actual mass of known weight can be applied to simulate the maximum anticipated load during operation.

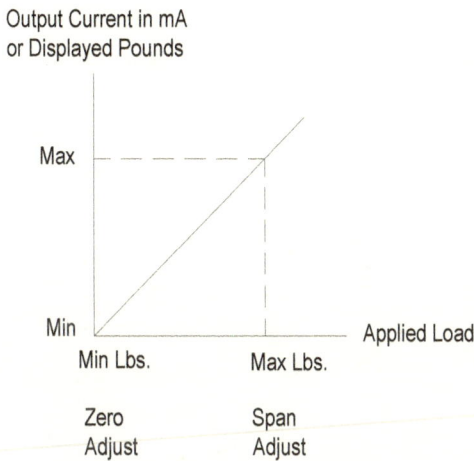

Output Current in mA
or Displayed Pounds

Max

Min

Min Lbs.   Max Lbs.   Applied Load

Zero   Span
Adjust   Adjust

Figure 49 - Graphical representation of the load cell calibration process. Depending upon the application, calibration could be performed upon a process transmitter, digital data acquisition system, analog instrumentation amplifier or digital panel meter.

However, load cells rated into the thousands or tens of thousands of pounds must resort to an alternative calibration technique, as applying a 20,000-pound mass is not only difficult but can also be dangerous.

Shunt Calibration

Most load cells include a *shunt calibration resistor* when purchased. Shunt calibration allows the maximum load to be simulated rather than applied to the load cell. Again, the maximum applied load is anticipated to be 70% to 80% of the full rated load of the load cell and the shunt calibration resistor will simulate this amount when applied. The shunt calibration resistor is applied between two of the four load cell leadwires (see the following figure) with no mechanical load applied to the load cell.

When connected, the parallel combination of the shunt calibration resistor and the particular strain gauge in the load cell bridge will result in a total resistance slightly less then the strain gauge's no-load value. The parallel combination will create an equivalent voltage in the load cell output, simulating the load value specified with the shunt calibration resistor.

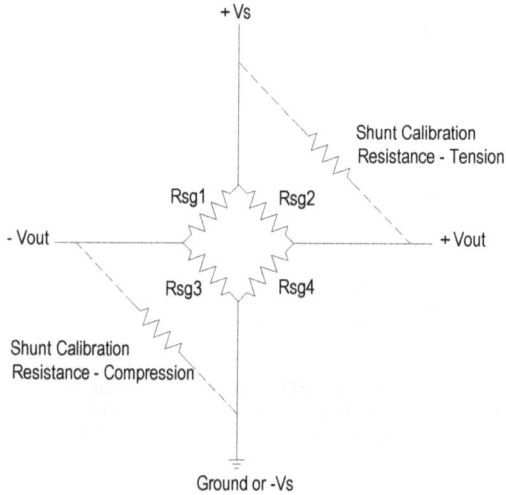

Figure 50 - Shunt calibration resistors are applied to simulate an applied physical load. Calibration resistors can simulate tension or compression loads.

As indicated in the previous figure, two connection possibilities are specified with the shunt calibration resistance, one for tension simulation and the other for compression simulation since most load cells can be used for both tension and compression loads.

Figure 51 - A microprocessor based data acquisition system. All sensors are connected to the four modules at the left. Calibration data values are entered through the keypad.

Scale Factor

Initially during load cell calibration, a minimum value of zero force is applied and an entry is made or an adjustment is performed to display a numerical value of zero. Next, a maximum value of load is applied or simulated and an entry is made or an adjustment is performed to display a numerical value equal to the maximum load. Occasionally during this stage of the process of calibration a value of scale factor is required to determine the maximum value displayed on a data acquisition system or computer monitor. Scale factor is a numerical value of gain that is used to amplify the output of the load cell when the maximum load is applied or simulated. Should a scale factor or gain value be required during this stage of the process a computation will need to be performed. The computation is demonstrated in the following example.

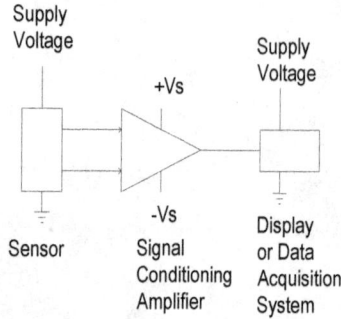

Figure 52 - A block diagram of a typical data acquisition system.

Example Problem:

Assume a 5000-pound load cell is to be connected to a data acquisition system and the system is to be calibrated. Determine the required offset and scale factor to be entered into the system during the calibration process. Assume the Rated Output of the load cell is 2.2 mV/V, the supply voltage is 10 Volts and the shunt calibration resistor simulates 72% of the 5000-pound rated load (3600-pounds).

Solution:

Once the 10 Volt supply is connected to the load cell and the load cell output is connected to the data acquisition system the system is powered-up and allowed to warm-up for a couple of minutes until the displayed value from the load cell is stable. Once the display is stable the calibration process can begin. Initially there should be no physical load applied to the load cell. The data acquisition system display should be zero since no load is applied but will probably be slightly off from zero.

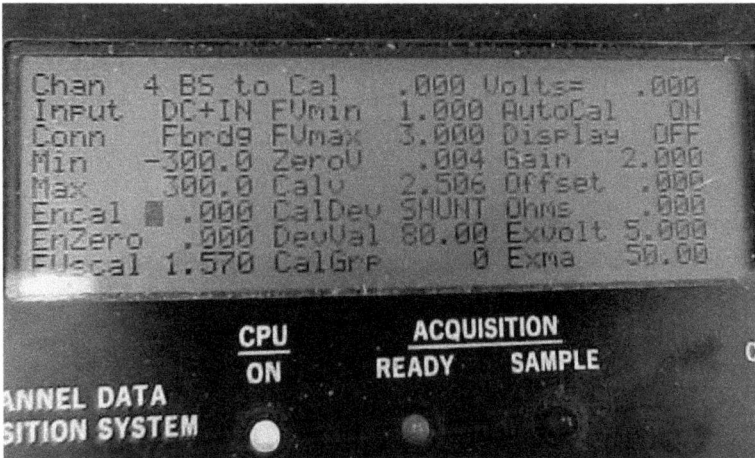

Figure 53 – A data acquisition system display. Note the indicated values, a 100-pound full bridge load cell with a 5 Volt supply and shunt resistor simulating 80 pounds was connected to the data acquisition unit.

The first step in the calibration process is to adjust the display to zero using the "Offset" or "Zero" adjustment provided with the data acquisition system. Occasionally, depending upon the type of system a value equal to the negative of the displayed offset will need to be entered to null the display. Next, the maximum load value or shunt calibration resistor is applied. Most data acquisition systems currently available will provide some form of adjustment labeled "Gain" or "Span." The adjustment should be varied until the displayed value equals the value of the applied load or the value associated with the shunt calibration resistor, 72% of the full rated load of 5000#, or 3600 #in this case.

However, many systems require that a scale factor be entered through a keyboard or other data entry device to display the maximum applied or simulated value. If a scale factor value is solicited the following procedure should be used. First determine the millivoltage output of the load cell at the full rated load. Using the provided specifications, the output voltage from the load cell is determined at 5000#.

$$\Delta Vout = RO * Supply\ Voltage$$
$$\Delta Vout = (2.2\ mV/V) * (10V)$$
$$\Delta Vout = 22\ mV\ at\ full\ rated\ load\ of\ 5000\#$$

Once the output voltage is determined at full rated load, the output voltage at the simulated load can be determined. Since the shunt calibration simulated load is 72% of the full rated load, the output voltage with the shunt calibration resistor applied will be 72% of the output voltage at the rated load, or,

$$Vout = [(Applied\ load \div Full\ rated\ load) * RO] \pm Offset$$

Since the "Applied load ÷ Full rated load" equals 72% and assuming the offset has been set to zero in the previous step,
$$Vout\ at\ max\ calibration = [(72\%) * 22\ mV] \pm 0$$
$$Vout\ at\ max\ (shunt)\ calibration = 15.84\ mV$$

When the calibration resistor is applied the data acquisition display should read 3600#. The scale factor represents the gain of the amplifier in the previous figure. Since gain equals output divided by input, the desired display value of 3600 needs to be divided by the millivoltage from the load cell.

$$Scale\ Factor\ (SF) = 3600/15.84\ mV$$
$$Scale\ Factor\ (SF) = 3600/.\ 01584V$$
$$Scale\ Factor\ (SF) = 227,272.7273$$

Subsequently, a scale factor of 227,272.7273 will yield a displayed value of 3600. At this point both "Ends" of the calibration process should be re-checked and fine-tuned if required.

Figure 54 - A series of eDaq stacked signal conditioning modules. These units offer the flexibility of being calibrated and setup remotely via a laptop computer or a network connection.

A Scale Factor Shortcut

The previous process can be abbreviated if one determines the output voltage at the full rated load, 5000# in this example. Knowing that the display should read 5000 at an output voltage from the load cell of 22 mV (.022V) the required scale factor can be determined by dividing the full rated load (assuming this will be the displayed value) by the load cell output voltage at the full rated load. The following illustrates.

$$SF = \text{Full rated load} \div \text{full rated output voltage}$$

$$SF = 5000 \div .022V$$

$$SF = 227272.7273$$

Homework Problems

Strain Gauges

#1 – Given the included information, determine the following values:

-Given:
Area = 1 in^2
Length = 2 inches
F_{comp} = 5000#
E = 31 E6 $_{PSI/in/in}$ (1020 steel)
μ = .288 $_{in^2/in^2/in/in}$
$R_{Strain\ Gauge}$ = 120 Ohms
$GF_{Strain\ Gauge}$ = 2.03
Vs = 5 Volts (4 element, full bridge)

-Find:
A) Stress
B) Strain (ε_L)
C) New Length
D) Strain (ε_T)
E) New Area
F) ΔRsg
G) Bridge Vout

#2 – Given the included information, determine the following values:

-Given:
Dia = 1 inch
Area = .785 in^2
Length = 12 feet (144 inches)
$F_{Tension}$ = 12000#
E = 31 E6 $_{PSI/in/in}$ (1020 steel)
μ = .288 $_{in^2/in^2/in/in}$
$R_{Strain\ Gauge}$ = 120 Ohms
$GF_{Strain\ Gauge}$ = 2.03
Vs = 5 Volts (4 element, full bridge)

-Find:
A) Stress
B) Strain (ε_L)
C) New Length
D) Strain (ε_T)
E) New Area
F) ΔRsg
G) Bridge Vout

#3 – Given the included information, determine the following values:

-Given:
Dia. = 1 in.
Length = 120 inches
$F_{Tension}$ = 10,000#
E = 10.7 E6 $_{PSI/in/in}$ (Alum)
μ = .350 $_{in^2/in^2/in/in}$
$R_{Strain\ Gauge}$ = 120 Ohms
$GF_{Strain\ Gauge}$ = 2.10
Vs = 5 Volts (4 element, full bridge)

-Find:
A) Stress
B) Strain (ε_L)
C) ΔLength
D) Strain (ε_T)
E) ΔArea
F) ΔRsg
G) Bridge Vout

#4 – Given the included information, determine the following values:

-Given:
Dia. = 1.5 in
Length = 5 feet
ΔF_{comp} = 5K, 10K, 20K lbs.
E = 31 E6 $_{PSI/in/in}$ (1020 steel)
μ = .288 $_{in^2/in^2/in/in}$

-Find:
A) Stresses
B) ΔStrains (ε_L)
C) ΔLengths
D) Strains (ε_T)
E) ΔAreas

$R_{Strain\ Gauge}$ = 350 Ohms   F) ΔRsg's
$GF_{Strain\ Gauge}$ = 2.08   G) Bridge ΔVout
Vs = 5 Volts (4 element, full bridge)

#5 – Given the included information, determine the following values:
Assume two active (strain sensing) gauges mounted on the sample and wired
into a full bridge circuit.

-Given:

    Dia. = .5 in

    Length = 50 feet

    ΔF_{comp} = 1K, 2K, 5K lbs.

    E = 31E6 $_{PSI/in/in}$ (1020 steel)

    μ = .288 $_{in^2/in^2/in/in}$

    $R_{Strain\ Gauge}$ = 350 Ohms

    $GF_{Strain\ Gauge}$ = 2.18

Note: Assume 2 active strain gauges,

    Vs = 5 Volts,

-Find:

A) Stresses

B) ΔStrains (ε_L)

C) ΔLengths

D) Strains (ε_T)

E) ΔAreas

F) ΔRsg's

G) Bridge ΔVout

H) the circuit diagram

I) Label the active gauges in the
circuit diagram

#6 – A strain-gauged column outputs 10.5 millivolts under a load of 1500 pounds,
and –1.2 millivolts when the load is removed. An instrumentation amplifier is to
be used to condition the bridge output into a 0 to 1.5 Volt (0 to 1500 pounds)
signal for display on a digital panel meter. Determine the gain and offset values
required for the amplifier output voltage span.

#7 – For each of the previous strain gauge problems assume an output of .1
millivolts per microstrain is desired. Determine the necessary circuitry and
operating values to develop the desired output. (Hint: For each problem a high
gain amplifier will be required. For each problem, determine the gain from the
strain and bridge Vout computation results.)

Shunt Calibration

#8 – A 350-Ohm bridge with a single active strain gauge (GF = 2.10) is shunt
calibrated with a 100K resistor. Determine the amount of stress represented with
a 40K shunt resistor. Hint: consider using the gauge factor equation.

#9 – A 120-Ohm bridge with a single active strain gauge (GF = 2.10) is shunt
calibrated with a 40K resistor. Determine the amount of stress represented with a
40K shunt resistor. Hint: consider using the gauge factor equation.

Load Cells

#10 – Given the tension/compression load cell specifications for the Eaton model
3169 included in the Stress/strain manuscript material, determine the requested
values for the following conditions.

A)
1000 lbs. rated load cell,
Vs = 5 Volts,
$F_{Tension}$ = 650 lbs.
- Determine the output voltage.

B)
1000 lbs. rated load cell,
- From the diagram determine the supply resistance.
- From the diagram determine the signal resistance.
- Explain the reason for any difference in resistance.

C)
1000 lbs. rated load cell,
Vs = 10 Volts,
$F_{Tension}$ = 650 lbs.
- Determine the output voltage.

D)
1000 lbs. rated load cell,
Vs = 5 Volts,
F_{Comp} = 1200 lbs.
Force is applied 30° from center,
- Determine the output voltage.

E)
- Explain the purpose and operation of the calibration resistor included with the Eaton model 3169.

#11 – A load cell rated at 2000 pounds has a rated output of 1.6 mV/V and is to be used with a 5 Volt supply. Determine the output at rated load and the required scale factor to be used with a shunt calibration resistance representing 75% of the rated load. Assume a displayed value of 2000.

#12 – A load cell rated at 100 pounds has a rated output of 2.2 mV/V and is to be used with a 9 Volt battery. Determine the output voltage at rated load and the required scale factor to be used with a shunt calibration resistance representing 72% of the rated load. Assume a displayed value of 100.

Chapter 5 - Fluid Pressure

The previous chapter introduced pressure within solid materials. This chapter introduces liquid and gas fluid pressure principles and the associated hardware of pressure measurement and control. Mechanisms and components utilized in the testing and automation industries are also investigated for operation and application.

Objectives:

Upon completion of this chapter, you should be able to:

- Explain the operational characteristics of common pressure sensors
- Explain the operation of common pressure regulating and amplifying relays
- Interpret the operational specifications of common pressure sensors
- Address routine pressure sensor signal conditioning concerns
- Suggest calibration techniques for common pressure sensors

Chapter Outline:

Introduction

Arguably, the most commonly measured variable is pressure. Level, flow and temperature systems often utilize a pressure measurement to infer a process value. As a result, pressure measurement ranges cover a wide range of values, from a fraction of a gentle breeze to pressures so great that strong materials are permanently distorted. Pressures and the associated forces they generate can accelerate and deform objects yet cannot be seen and are often difficult to accurately describe. Likewise, it is often best to define pressure by what it can do. For this reason the following description is worth remembering. *The application of pressure, or force, always produces a deflection, a distortion or some change in volume or dimension, no matter how great or small the applied pressure.* It has been said that this is the founding principle of all measurement instrumentation. Certainly, the longer one works with industrial or experimental measurements the more this fact will be demonstrated.

Figure 1 - A flare utilizes draft-range (fractional psi) vacuum pressure sensors to measure and control the burning of methane gas developing underground within landfill operations. The large circular diaphragm in the photo is representative of low-pressure.

Physical Principles

Pressure is defined as a *force applied over an area*. The resulting equation is identical to the definition.

$$P = F \div A = F / A$$

Primarily measured in English units of pounds per square inch (psi) or metric units of Pascals or Newtons/square meter (N/m^2), pressure is also indicated in units of atmospheres, inches of water, millimeters of Mercury, and numerous other dimensions to be explained later. The conversion between N/m^2 (Pascal) and psi is shown following.

$$1 \text{ Pascal} = 1 \text{ } N/m^2 = 0.00014504 \text{ psi}$$

Or,

$$1 \text{ psi} = 6894.8 \text{ Pascal} = 6894.8 \text{ } N/m^2$$

When defined as a force over an area, pressure can be difficult to visualize. Try to imagine compressed air within a tire defined as a force over an area. A mental image of forces working across an area within a tire is difficult to envision. As a result, an algebraic re-arrangement of the previous pressure equation best represents pressure forces and areas.

Since, $P = F / A$

Therefore, $F = P \times A$

As can be seen, the rearranged equation demonstrates the effect of a pressurized gas or liquid when applied to area, a force is created. Again, although fundamental and intuitive this is an essential concept in automatic process control. Self-regulating, negative feedback control systems universally employ the concept of forces generated by pressures applied to areas.

- Example Problem 1:

$P = F \div A$

A vehicle lift in a maintenance garage applies compressed air to elevate vehicles weighing up to 10,000 pounds. Determine the air pressure to support a 10,000-pound vehicle using a 9-inch diameter actuator.

Solution:

Pressure equals force divided by area but the area is given as a diameter. Therefore, the area must first be determined using,

$$A = \pi r^2$$

Where "r" equals the diameter divided by two, or 4.5 inches.

$$A = (3.14 \times 4.5^2)$$
$$A = 63.62 \text{ sq. in.}$$

The required pressure can now be determined by dividing the given force of 10,000 pounds by the area,

$$P = F/A,$$
$$P = 10,000\#/63.62 \text{ in}^2$$
$$P = 157.2 \text{ psi}$$

- Example Problem 2:

$F = P \times A$

An automobile traveling at 60 miles per hour develops a 470 Pascal pressure difference between the interior and the exterior of a 14-inch tall by 25-inch long window. Determine the force in pounds created on the window.

Suggestion:

Before attempting to solve the problem, it might be best to visualize the effect. All moving objects exhibit low pressures along the sides and rear. This effect is referred to as the Venturi effect or Bernoulli's Principle and is discussed in chapter 7. In this case, an auto moving at high velocity (60 mph) experiences a decrease in pressure on the exterior along the vehicle's sides including the windows. Since the windows are closed, the interior remains near atmospheric pressure but the exterior pressure drops slightly below atmospheric causing a slight vacuum. The pressure difference formed by the high pressure on the inside of the vehicle and the lower pressure on the outside places a force upon the window equal to the pressure difference multiplied by the area of the window.

Solution:

Since the force is requested in pounds, a conversion to psi from Pascals is required. Given,

$$1 \text{ Pascal} = 0.00014504 \text{ psi}$$
$$470 \text{ Pascals} = 470 \times .00014504 \text{psi/Pascal}$$
$$470 \text{ Pascals} = .0681688 \text{ psi}$$

Knowing the pressure difference in psi, the force can be determined by multiplying the pressure difference by the area of the window.

$$\text{Area} = 14 \text{ inches} \times 25 \text{ inches}$$
$$\text{Area} = 350 \text{ sq. inches}$$

Finding the force,

$$F = P \times A$$
$$F = .0681688 \text{psi} \times 350 \text{ in}^2$$
$$\boldsymbol{F = 23.86 \text{ pounds}}$$

Since the low pressure is on the exterior, the force would be pushing against the window towards the exterior of the vehicle.

Fluid Pressure - Liquids

Pressure can be observed in the three physical forms of liquid, gas and solid. Pressures observed in solid materials, known as *stresses* and the resulting strains were discussed in chapter 4. Flowing liquid and gas materials, known as *fluids,* are addressed in this chapter and in chapter 7. Standing liquid pressure as within a tank is called *Hydrostatic* pressure and is created when the weight of a volume of liquid is distributed over a supporting area.

Liquid Pressure
Varies With
Depth

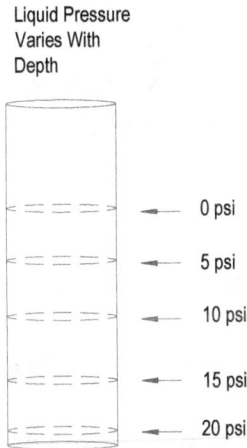

0 psi

5 psi

10 psi

15 psi

20 psi

Figure 3 - As the height of liquid above the measurement location increases, the pressure increases. The pressure above the measurement location is often referred to as a column of pressure or pressure "Head."

When the weight of a liquid volume is distributed over the containing areas of the sides and bottom of a vessel, such as a holding tank, the resulting pressure is determined by the amount of liquid above the point of measurement. This is an example of the basic pressure equation where the resulting pressure equals force (weight of the liquid) over the area of support.

Pressure = Force / Area

Since,

Force = weight

P = weight / Area

The resulting equation can be further manipulated to develop an equation for expressing pressure as a function of liquid height.

Since,

$$Pressure = Weight / Area$$

And from chapter 3 where density equals weight divided by volume,

$$Density = Weight / Volume$$

Therefore,

$$Weight = Volume \times Density$$

Subbing for weight,

$$Pressure = (Volume \times Density) / Area$$

Since,

$$Volume = Area \times Height$$

Then,

$$Pressure = (Area \times Height \times Density) / Area$$

Canceling Areas yields,

$$Pressure = Height \times Density$$

Or,

$$P = H \times D$$

- Example Problem:

$P = H \times D$

A vented water tank contains 4-feet of water. The tank is elevated 5-feet. Determine the pressure at the bottom of the tank and at the outlet 5-feet below the tank.

Solution:

From the diagram it can be seen the water level is 9-feet above the pressure gauge and 4-feet above the bottom of the tank. As a result, the pressures equal:

Since, $P = H \times D$

Pressure at the tank bottom

= 4 Ft. × 62.4 #/Ft³

$P = 249.6$ #/Ft² (÷144 in²/ft²)= **1.73 #/in²**

Pressure at the gauge

= 9 Ft. × 62.4 #/Ft³

$P = 561.6$ #/Ft² (÷144 in²/ft²)= **3.9 #/in²**

Contents vented to atmosphere

4.00

5.00

psig

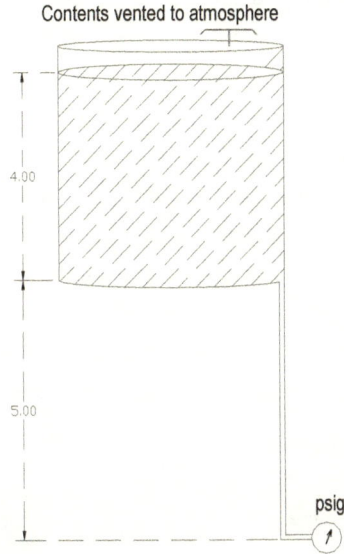

Since liquids are understood to be non-compressible, a pressure applied to a confined liquid transmits the pressure. In a liquid column such as within a tank, the pressure varies linearly with depth and is transmitted virtually undiminished laterally. Hydraulic systems utilize pressure transmission to generate large forces when pump-driven hydraulic fluid pressure is applied to a surface area such as an actuator. Forces generated by hydraulic actuators are yet another example of force equaling pressure multiplied by area.

$$F = P \times A$$

The inability of liquid to compress is also obvious in the weather characteristic of Dew Point. All air contains some amount of moisture; remember this when working with air compressors. When leaving a compressor, air temperature is usually quite high. As the compressed air travels through a system of airlines, the air temperature decreases and the air contracts consuming less space. In the process of contraction the moisture, which cannot be compressed is squeezed out of the air. The dew point temperature is an indication of the temperature required for the moisture to appear, such as outdoors. In the end result, air can be compressed but water cannot. If overlooked in air compressor systems the resulting condensation can cause metals to oxidize and corrode, and the movement of moisture throughout the air system can collect dirt and other particles that have a tendency to reside in or near the small moving parts of pneumatic components. Over time the water carried particles accumulate within components and can cause the components to become sticky in operation.

Figure 4 - The lower portion of this pneumatic relay contains moving components that can become sticky and unpredictable in operation when subjected to dirty air.

Fluid Pressure - Gases

Gas pressure is created when a gas such as air, is compressed and contained within a vessel, like a balloon. In the balloon, pressure is distributed equally upon all three dimensions of containing surfaces. This is not always the case however.

Atmospheric pressure decreases with an increase in altitude. In effect, the earth is at the bottom of an ocean of air, meaning the air pressure is greatest at the bottom of the atmospheric ocean. In this context, gas pressure is similar to liquid pressure. As pressure is measured deeper below the surface of liquid a pressure increase is noted. This effect is obvious with gases when ascending over significant distance in an elevator or small aircraft, a decrease in pressure is noticed. Indeed, if a gas storage vessel were tall enough a variation in gas pressure from top to bottom would also be noticed. In science and for the applications of this text, gas pressure is applied equally upon all containing surfaces and liquid pressure varies according to the depth of the liquid.

Gas Pressure is
Constant
Throughout

Liquid Pressure
Varies With
Depth

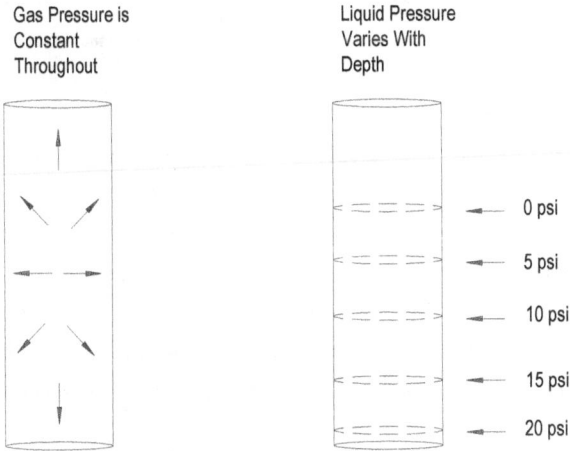

0 psi

5 psi

10 psi

15 psi

20 psi

Figure 5 - A comparison of the effects of gas and liquid pressures. Note - Unless otherwise indicated, all vessels containing liquids are assumed to be vented to atmosphere.

Figure 6 – Due to their spherical shape, gas storage tanks are obvious in chemical plants.

Differential Pressure

All fluid (liquid and gas) pressure measurements are *differential* pressure measurements. In a manner similar to measuring voltage where two meter probes are required, fluid (gas and liquid) pressure will always be measured as the difference between two specific individual pressures, hence the term *Differential* pressure. Differential pressure sensors require that two distinct pressures appear at a pressure sensor's inputs. As will be seen, two pressures may also need to be considered when performing a pressure computation.

Figure 7 - High and low differential pressure ports are obvious in the lower right portion of this transmitter.

A differential pressure instrument has two pressure connection ports. The input ports are commonly labeled High and Low, or Input and Reference or something representative of the two required pressures. Differential measurement instruments output a signal representative of the difference between the two applied input pressures.

Figure 8 - This type of flowmeter (orifice plate) generates a differential pressure in relation to the flow rate of fluid material passing through the pipe. A differential pressure sensor and/or transmitter are required to develop the flowmeter output signal.

However, true differential pressure measurements and sensors compose only one of three categories of pressure measurement and sensor types. *Absolute* and *Gauge* pressure measurement sensors compose the remaining two. In both cases, absolute and gauge measurement instruments are performing a differential measurement but the reference pressures are fixed at uniquely specific values. Differential pressure sensors can be adapted to perform Absolute and Gauge measurements but are more commonly used where the subtraction of two non-specific or unique pressures results in information about the state or condition of a process. With absolute and gauge pressure sensors the reference pressure port may not always be obvious but in one form or another is always available since again, two pressures are always required to sense pressure. Flowmeters and liquids under pressure often require the use of a true differential pressure sensor or differential pressure transmitter.

Absolute Pressure

Absolute pressure measurement devices utilize a vacuum as the reference pressure and are understood to measure a pressure applied to the High-pressure port in relation to the total absence of any reference pressure on the Low-pressure port. Absolute pressure measurements are common when working with pressures near or below atmospheric pressure but are also common in applications where atmospheric pressure variations should not be allowed to influence the measurement result. Absolute pressure sensors are constructed as a differential pressure sensor but has the reference chamber sealed and evacuated, replaced with a total vacuum or as close to an absolute vacuum as is practically possible.

Figure 9 - A Barometer is an example of an absolute pressure measurement component.

A barometer measuring atmospheric pressure is an example of an absolute pressure instrument. Atmospheric pressure is understood to be approximately (it changes continuously) 14.7 psia, or 14.7 pounds per square inch above a total vacuum. Expressed oppositely, a total vacuum of 0 psia is the equivalent of 14.7 psi below atmosphere. Pressures below atmosphere are considered a vacuum, or *negative pressure*. Pressures above atmosphere are considered *positive pressure*.

$$0 \text{ PSIA} = -14.7 \text{ PSI}$$

Industrial absolute pressure measurements are labeled and displayed on pressure gauges and instruments in units of PSIA, or pounds per square inch absolute.

Gauge Pressure

Gauge pressure measurement devices utilize an atmospheric pressure as the reference input. As an example of a Gauge pressure measurement, consider the pressure in your car tires. The pressure indicated on the gauge is the difference between the tire pressure itself, and the atmospheric pressure surrounding the pressure gauge.

Figure 10 - Although each of the pressure sensors in this photo are differential pressure sensors, the sensor in the lower left is a gauge-type pressure sensor.

When atmosphere is used as the reference pressure, the measurement is referred to as a *Gauge* pressure measurement. Sensors measuring pressures above atmosphere are called gauge pressure sensors. Gauge pressure measurements and sensors assume atmospheric pressure is 0 psi. A total vacuum is represented as −14.7 psi gauge, or 14.7 psi below atmospheric pressure.

$$0 \text{ PSIG} = +14.7 \text{ PSIA}$$

Industrial gauge-type pressure measurements are labeled and displayed on pressure gauges and instruments in units of PSI or PSIG, expressed as pounds per square inch gauge.

Gauge pressure measuring instruments often appear to contain only a single pressure input port. However, upon close inspection a means of sensing atmospheric pressure is usually obvious.

Figure 11 - The reference input on this gauge pressure sensor is obvious to the right of the base of the input port.

Absolute Pressure Conversions

Given that atmospheric pressure is assumed to be 0 psig and 14.7 psi above a total vacuum, or 14.7 psia, conversion between gauge and absolute quantities requires the addition or subtraction of 14.7 psi.

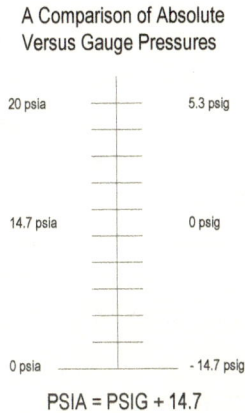

A Comparison of Absolute
Versus Gauge Pressures

20 psia	5.3 psig
14.7 psia	0 psig
0 psia	- 14.7 psig

PSIA = PSIG + 14.7

Assuming atmosphere is 14.7 psi absolute (psia), mathematically,

Differential Pressure,

$$PSID = P_{Hi} - P_{Lo}$$

Atmospheric Pressure,

$$P_{atmos} = 14.7 \text{ psia}$$

Absolute Pressure,

$$PSIA = PSI\,(G) + 14.7$$

Gauge Pressure,

$$PSI \text{ or } PSIG = PSIA - 14.7$$

Negative Pressure

Pressure less than atmosphere is called a *vacuum,* or *negative pressure.* Vacuum pressure is normally measured with low-pressure instruments yielding indications in ranges of inches of water (in H_2O), inches water column (in. W.C.)

or inches of mercury (in Hg). An absolute vacuum is 0psia or −14.7psig and is understood to "pull" a column of approximately 30-inches of mercury when measured with a manometer.

Manometers

Most low pressure (within 20 psi of atmosphere), measurement applications specify pressure in terms of a liquid column as if a vacuum referenced manometer were used in making the process measurement. As an example, consider atmospheric pressure, which is normally expressed as 30 inches of mercury (30" Hg). Many pressure transmitters also utilize this concept, preferring to specify pressure ranges in inches of water (in. w. c., inches water column). To determine the numerical value associated with the specification, multiply the distance (height) times the density of the indicated liquid.

Given,

$$Pressure = Force \div Area$$

And,

$$Force = Weight$$
$$Pressure = Weight \div Area$$

Since,

$$Weight = Volume * Density$$

Therefore,

$$Pressure = (Volume* Density) \div Area$$

Since,

$$Volume = Area * Height$$

Substituting,

$$Pressure = (Area*Height*Density) \div Area$$

Canceling areas yields,

$$Pressure = Height * Density$$

Or,

$$P = Ht. * D$$

As an application example, take the atmospheric pressure stated above of thirty inches of Mercury (30" Hg). Since,

$$Pressure = Ht. * D.$$

$$Pressure = 30 \text{ inches} * D_{Hg}$$

$$Pressure = 30 \text{ inches} * 848.6 \text{ lbs/ft}^3 * (1/1728 \text{ in}^3/\text{ft}^3, \text{ conversion to in}^3)$$

$$Pressure = 30 \text{ inches} * .491 \text{ lbs/ in}^3$$

$$Pressure = 14.73 \text{ PSIA}$$

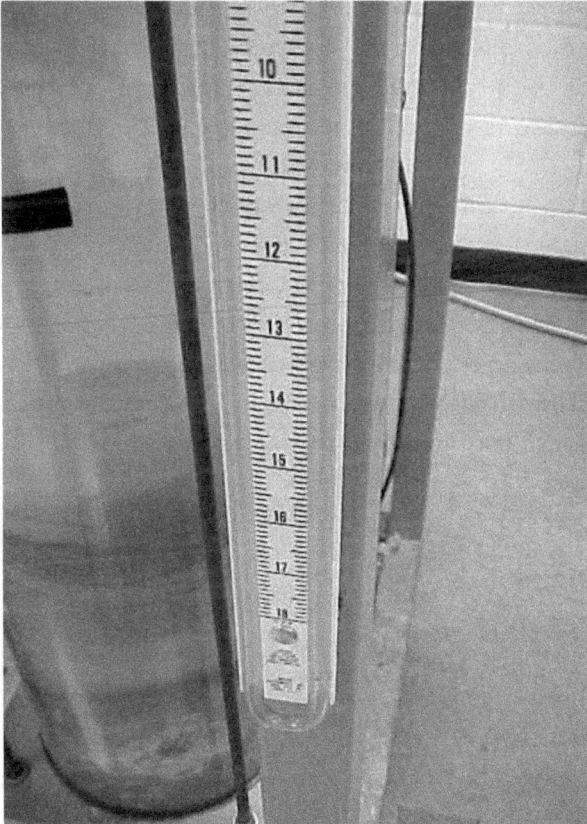

Figure 12 - A U-tube manometer uses the displacement of liquid, usually Mercury or water, to measure positive and negative pressures.

Note that atmospheric pressure is expressed as an absolute quantity in PSIA. Unless otherwise indicated all pressure measurements are assumed gauge quantities however, atmospheric pressure is absolute. Therefore,

30" Hg = 14.73 PSIA.

Some process measurements such as flow through an orifice plate or a Venturi tube, or liquid level under pressure utilize a true differential pressure measurement where the reference pressure is neither atmosphere nor a vacuum. Gauge pressure applications far outnumber either absolute or differential. Consequently, PSIA and PSID are used to signify the appropriate type of measurement application, and PSI (or occasionally PSIG) is utilized for all others.

Instrument Mechanics – Pressure Mechanisms

Torque Balance Mechanism – Applied Forces
The following material applies principles utilized in the vast majority of pneumatic and hybrid electro-pneumatic components. Although one will rarely be expected to disassemble and make repairs to the interior of the components, the concepts presented assist in preparing the student for a study of mechanisms and provide a foundation from which to study the physical principles in subsequent chapters.

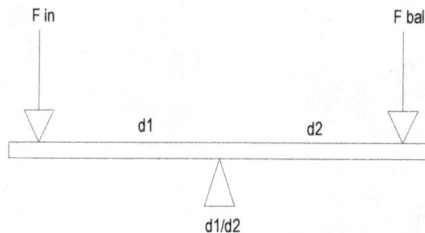

Figure 13 - An input force is applied to the mechanism; the balancing force is also called the output force.

Earlier chapters presented the lever with input and balancing forces, and demonstrated the inverse ratio between the forces and distances.

$$Tccw = Tcw$$
$$Fin\ d1 = Fbal\ d2$$
$$d1/d2 = Fbal\ /\ Fin$$

In this manner mechanical advantage is provided, where large forces can be generated with relatively small applied forces.

Example Problem 1

Torque Balance Mechanism – Applied Forces

The application of additional forces would appear to greatly complicate the analysis of the mechanism. In effect, only two additional forces (torques, or moments) have been applied. Mechanisms such as these are used where more than one or two input variables influence a single outcome, such as valve position in a batch mixing process. When assembled in this particular configuration, the mechanism is capable of performing another very common mathematical function. The torque-balance operational analysis follows.

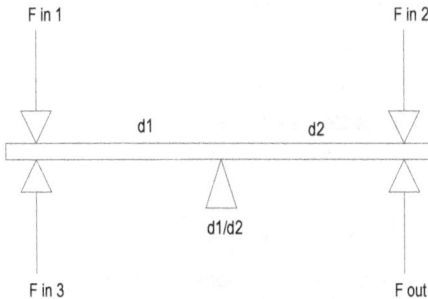

Assuming the mechanism is at balance, torque in each direction is equal. Therefore,

$$T_{CW} = T_{CCW}$$

Note: Each direction (clockwise and counter-clockwise) contains two torque-generating components.

Since,

$$T = F*d$$

$$T_{CW} = (F_{IN\,2} *d_2) + (F_{IN\,3} *d_1)$$

And,

$$T_{CCW} = (F_{IN\,1} *d1) + (F_{out} *d_2)$$

Since,

$$T_{CW} = T_{CCW}$$

$$(F_{IN\,2} *d_2) + (F_{IN\,3} *d_1) = (F_{IN\,1} *d1) + (F_{out} *d_2)$$

Solving for F_{out},

$$(F_{IN\,2} *d_2) + (F_{IN\,3} *d_1) - (F_{IN\,1} *d1) = (F_{out} *d_2)$$

$$(F_{IN\,2} *d_2)/d_2 + (F_{IN\,3} *d_1)/d_2 - (F_{IN\,1} *d1)/d_2 = (F_{out} *d_2)/d_2$$

$$F_{IN\,2} + (F_{IN\,3} - F_{IN\,1}) *d_1/d_2 = F_{out}$$

Or,

$$F_{out} = d_1/d_2 \ (F_{IN\,3} - F_{IN\,1}) + F_{IN\,2}$$

The Fout equation appears similar to the general equation for a linear function, Y=MX+B. This implies that the torque balance mechanism with multiple applied forces is capable of performing in the capacity of an analog computer, or performing simple linear scaling equations of gain and offset. The ratio of the lever arm lengths (d_1/d_2) determines the gain (slope), and the amount of input force at $F_{in\,2}$ determines the offset (Y-axis intercept).

The opposing locations of the input forces provide the added benefit of differential inputs (note the $F_{IN\,3} - F_{IN\,1}$ terms in the output equation). Differential inputs are useful when measuring liquid levels under pressure, fluids flowing through constriction flowmeters or when subtraction of two quantities needs to be performed.

Example Problem 2

Torque Balance Mechanism – Applied Pressure

Torque balance mechanisms form the basis behind the operation of the vast majority of transducers and transmitters in process automation. To assist in understanding the operation of the mechanisms and in the performance of homework exercises, the following example is provided. It is suggested that the student become familiar with the mathematical manipulation of the equations and the operation of the basic mechanism.

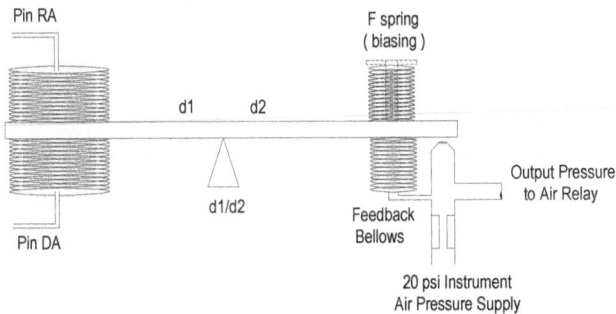

Figure 14 - A differential pressure transmitter.

The differential pressure transmitter in the previous diagram contains two input pressure bellows, a lever (moment arm, d1 and d2), fulcrum (d1/d2), a variable force generating spring, and an orifice/baffle assembly. The moment arm lever, or beam is operating in the capacity of the baffle.

Qualitative Assessment

The application of a compressed fluid (liquid or gas) to a surface area results in a generated force. Observing the following transmitter diagram, the surface areas

215

are provided in the form of bellows (or diaphragms, Bourdon tubes, etc.) at the two inputs ($P_{IN\,RA}$, $P_{IN\,DA}$) and at the *feedback* locations. Assume for the moment that an input pressure is applied to only the $P_{IN\,DA}$ input bellows. As the input pressure increases, the moment arm lever (d1, d2) attempts to rotate in a clockwise direction. As the beam rotates, the right side of the beam is drawn closer to the orifice, increasing the pressure within the orifice. As the orifice pressure increases, the output pressure increases. The output pressure changes in proportion to the input pressure increase (see following diagram) and the pressure within the feedback bellows increases. The increasing pressure within the feedback bellows places a restoring or balancing torque back upon the input (hence the term, feedback), causing the transmitter to achieve equilibrium. The required moment arm lengths (d_1, d_2) and spring force (Fbias) are determined from the gain and offset computations following in the Quantitative Assessment section.

Orifice/baffle (also known as flapper/nozzle) arrangements are commonly used where physical displacement of bellows, Bourdon tubes, diaphragm and other types of pneumatic components are required to create a corresponding output pressure.

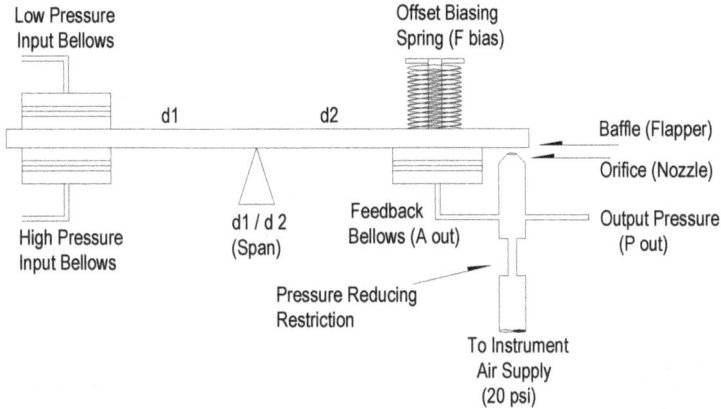

Figure 15 - Pressure transmitters convert a process pressure span to a standard output pressure span. The output pressure span is commonly 3-15 psi. In this diagram the orifice (also called a nozzle) and the baffle (often called a flapper) are responsible for increasing the output pressure when the baffle is torqued in a clockwise direction. The resulting nozzle pressure is provided to the output port and to the feedback bellows to balance all of the applied torques. The output pressure span is normally applied to a controller.

Quantitative Assessment

As with the previous examples, an assessment begins with recognition of the torque-balance nature of the instrument's operation. Knowing this, an initial statement can be made about the instrument at balance. The sum of the clockwise-applied torques equals the sum of the counter-clockwise torques.

$$\Sigma T_{CW} = \Sigma T_{CCW}$$

Since,

$$T = F*d$$

Using subscripts from the following diagram,

$$T_{CW} = (F_{spring} *d_2) + (F_{IN\ DA} *d_1)$$

And,

$$T_{CCW} = (F_{IN\,RA} *d_1) + (F_{feedback} *d_2)$$

Since at balance,

$$T_{CW} = T_{CCW}$$

$$(F_{spring} *d_2) + (F_{IN\,DA} *d_1) = (F_{IN\,RA} *d_1) + (F_{feedback} *d_2)$$

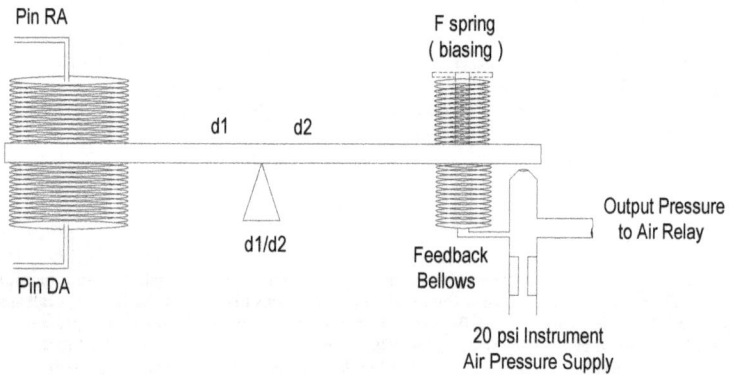

However, instead of determining F_{out} it is desired to know the output pressure, P_{out}. If all other variables are given, by rearranging the previous equation to solve for the feedback force the output pressure can be determined. At this location in the previous equation, a substitution is made for all of the pressure-generated forces.

Since,

$$P = F/A$$

$$F = P*A$$

Therefore,

$$(F_{spring} *d_2) + (F_{IN\,DA} *d_1) = (F_{IN\,RA} *d_1) + (F_{feedback} *d_2)$$

Becomes,

$$(F_{spring} *d_2) + [(P_{IN\,DA} *A_{IN\,DA})*d_1] = [(P_{IN\,RA} *A_{IN\,RA})*d_1] + [(P_{feedback} *A_{feedback})*d_2]$$

Since the feedback pressure and the output pressure are the same,

$$P_{feedback} = P_{out}$$

Therefore,

$$(F_{spring} *d_2) + [(P_{IN\ DA}*A_{IN\ DA})*d_1] = [(P_{IN\ RA}*A_{IN\ RA})*d_1] + [(P_{out}*A_{out})*d_2]$$

Isolating the P_{out} term,

$$(F_{spring} *d_2) + [(P_{IN\ DA}*A_{IN\ DA})*d_1] - [(P_{IN\ RA}*A_{IN\ RA})*d_1] = [(P_{out}*A_{out})*d_2]$$

Dividing by d_2 yields,

$$(F_{spring} *d_2)/ d_2 + [(P_{IN\ DA}*A_{IN\ DA})*d_1]/d_2 - [(P_{IN\ RA}*A_{IN\ RA})*d_1]/d_2 = (P_{out}\ A_{out})$$

Dividing by A_{out} yields,

$$F_{spring}/A_{out} + [(P_{IN\ DA}*A_{IN\ DA})*(d_1/d_2)]/ A_{out} - [(P_{IN\ RA}*A_{IN\ RA})*d_1/d_2]/ A_{out} = P_{out}$$

Rearranging,

$$P_{out} = \{(d_1/d_2) * [(P_{IN\ DA}*A_{IN\ DA}) - (P_{IN\ RA}*A_{IN\ RA})]/ A_{out} \} + F_{spring}/A_{out}$$

Assuming the input and feedback bellows areas are equal yields,

$$P_{out} = (d_1/d_2)(P_{IN\ DA} - P_{IN\ RA}) + F_{spring}/A_{out}$$

Or, the output pressure equals the input pressure difference (differential) multiplied by the transmitter gain, and biased by the spring force divided by the area of the feedback (or output) bellows. All terms and the resulting output will be in psi.

At this point, the most common question usually raised is "Shouldn't the A_{out} quantity have cancelled in the derivation?" If one truly understands the operation of the transmitter, the purpose behind the denominator in the bias term (A_{out}), and in the instrument will be obvious. As an aid to understanding the purpose behind A_{out}, imagine the unit operating without an input to either of the bellows on the left and with zero applied spring force. In this situation the output pressure would be zero. Now imagine gradually increasing the variable spring force, or screwing the spring downward, compressing the spring and applying the spring's resulting force (and torque) in a clockwise manner on the beam. Without an input pressure applied on the left side of the beam, the balancing pressure would be developed within the output (feedback) bellows as a result of the force applied by the spring and the area of the feedback bellows. For this reason, the area of the feedback bellows must be contained within the output pressure equation.

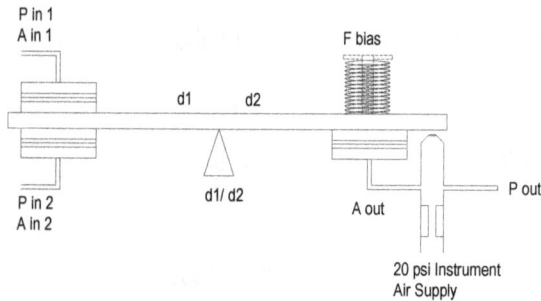

Figure 16 - A pressure transmitter. In this diagram, the bellows have been substituted with pressure diaphragm capsules at the inputs and at the feedback locations.

Note also from the equation that changing the area of the feedback bellows in relation to the area of the input bellows can aid or oppose the d1/d2 gain.

$$P_{out} = (d_1/d_2)[(P_{IN\ DA}*A_{IN\ DA})-(P_{IN\ RA}*A_{IN\ RA})]/A_{out} + (F_{spring}/A_{out})$$

Again, assuming the input and feedback bellows areas are equal yields,

$$P_{out} = (d_1/d_2)(P_{IN\ DA} - P_{IN\ RA}) + (F_{spring}/A_{out})$$

It is suggested that these examples be used as reference items for any subsequent homework problems.

Pressure Instrument Hardware

Only a handful of basic components are applied in every pneumatic instrument. The Bourdon tube, the bellows and the diaphragm are three primary components when generating force from an applied pressure is required. Although variations of each are available, each applies the input fluid (liquid or gas) pressure to an area to generate the required force, utilizing the $F = P * A$ equation. (See the following figures). The result of the generated force will be the displacement of a mechanism. Springs and levers (gears are only rarely used, mostly in pressure gauges) are most commonly used to link mechanisms depending upon the instrument type, its application and whether the instrument is of a torque-balance or force-balance variety. The resulting displacement of a mechanism will eventually output a pressure signal, position an indicator on a recording instrument, be converted into an electronic signal or provide some other automation or control function.

Figure 17 – The three types of Bourdon tubes, clockwise from upper left C-type, helix and spiral.

Figure 18 – A common bellows used to convert pressure into force.

Figure 19 – A stacked Diaphragm for conversion of pressure to force.

Bourdon, Bellows and Diaphragm

Within pressure measurement and control apparatus pneumatic components are required to either generate a force from an applied pressure or use an applied force to create pressure. The Bourdon tube, the bellows and the diaphragm are the three primary components when generating a force from a gas or liquid pressure is needed. Although variations of each are available, all apply the input fluid (liquid or gas) pressure to an area to generate a required force, thereby utilizing the F = P * A equation. (A review of the Pressure Fundamentals and Transmitter Selection material following this chapter is suggested, specifically figures 12-15 within the material). The result of the generated force will be the displacement of a mechanism. Displacement will always result in the application of a force that opposes and balances the initial motion. Springs are most commonly used to balance pneumatic mechanisms. The resulting displacement will eventually be converted into an electrical current (usually 4 to 20 milliamps), create an output pressure signal, position an indicator on a recording instrument, or provide some other measurement or control related function.

Figure 20 - Most pneumatic instruments are a hybrid of electronic and mechanical components. This Fisher Controls type 546 current to pressure (I/P) transducer, or servo valve, uses magnetics and a 20 psi air supply to convert 4-20 mA into 3-15 psi. The 546 is a common sight in power plants, chemical factories and petroleum refineries.

Figure 21 – The feedback bellows of a differential pressure transmitter. A pneumatic airline applies air pressure to a bellows; the bellows applies the resulting force to the lower end of a lever (moment arm) that pivots around the thumbwheel. The resulting transmitted force is used to output a 3-15 psi air signal.

Once the pressure is converted into a force, and the force has been translated into motion any number of functions can be performed. The mechanical motion can be used to indicate the applied input pressure, transmit an analog of the input pressure, position the pen of a recording instrument, or position a valve to control the input pressure. Each of these functions is performed routinely in industrial measurement and controls currently. Again, if one were to investigate the evolution of industrial automation, the present age of computer integration began with analog computers of a mechanical variety similar to those described.

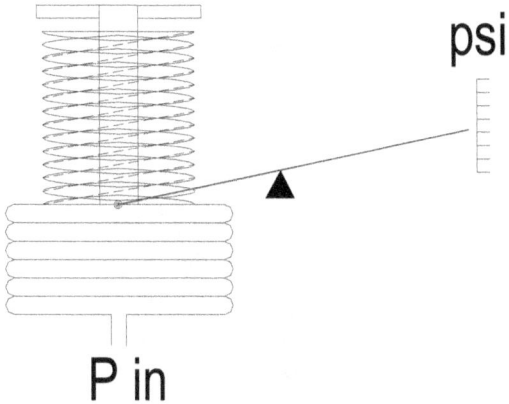

Figure 22 - A simple but common concept in industrial instrumentation, a force-balance system to measure pressure. The force generated by applying an input pressure to the bellows compresses the spring located above the bellows. As the spring compresses, the generated spring force increases and balances the force created by the bellows input pressure. The resulting pressure indication is displayed on the "psi" scale.

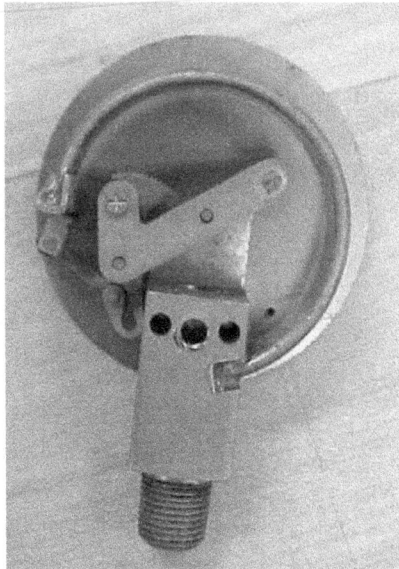

Figure 23 - A common pressure gauge with the cover removed. The C-type bourdon tube expands as pressure is applied due to the outer area of the Bourdon tube being greater than the inner area; therefore the outer force is greater than the inner force. The resulting indication is provided through the use of a pinion-sector gear arrangement.

Pressure Standards

In order to accurately measure and control pressure, pressure measurement standards are required for purposes of comparison. Measurement standards are available in varying degrees of accuracy and expense. At the top of the calibration pyramid, primary standards are used to calibrate secondary standards. Secondary standards are available regionally for calibration of tertiary and working standards used in laboratories to calibrate working instruments. All standards should be traceable to the National Institute for Standards and Technology (NIST, formerly the National Bureau of Standards, or NBS). Without standards and the comparison (calibration) process, accurate measurement of any type would not be possible.

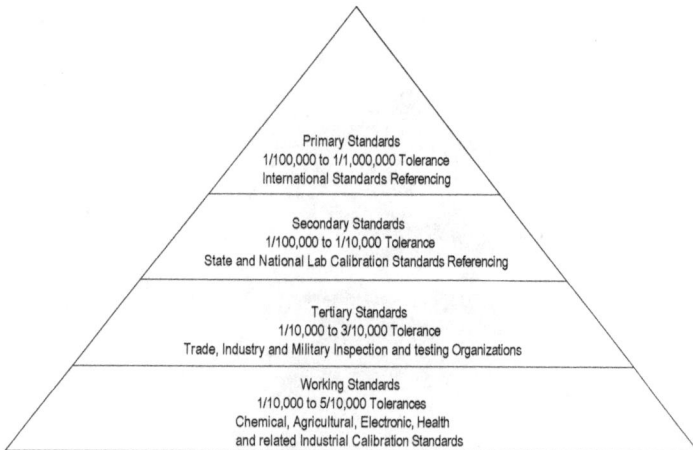

Primary Standards
1/100,000 to 1/1,000,000 Tolerance
International Standards Referencing

Secondary Standards
1/100,000 to 1/10,000 Tolerance
State and National Lab Calibration Standards Referencing

Tertiary Standards
1/10,000 to 3/10,000 Tolerance
Trade, Industry and Military Inspection and testing Organizations

Working Standards
1/10,000 to 5/10,000 Tolerances
Chemical, Agricultural, Electronic, Health
and related Industrial Calibration Standards

Figure 24 - The hierarchy of measurement standards.

Calibration, or more specifically *error quantification and correction* are also necessary with electronic component measurement apparatus. Measurement standard resistors, inductors and capacitors are commercially available for such purposes. Industry standard components such as cathode ray tubes and deflection coils, metal frame or chassis parts and other assembled parts are commonly available in electronic assembly and manufacturing facilities. Industry standards may not be NIST traceable but precisely meet manufactured specifications and are used solely for the purpose of in-plant comparison. In any case, if accurate measurement and control processes are desired, standards are required.

When working with apparatus for measurement and control of physical variables, standards often assume the form of a basic physical principle or concept. A common pressure measurement calibration standard is a dead weight tester. The dead weight tester operates according to the $P = F/A$ equation where accurately known weights are applied to a precisely known hydraulic piston area. The resulting compression pressure is used as an accurate input to the device under test.

Figure 25 - A dead weight tester is used to calibrate pressure gauges and sensors. Precisely known weights are taken from the storage tray (front center of photo) and placed upon a known area to

Another common low-pressure (around atmosphere by approximately ±10 psi) measurement standard is the manometer. The manometer is available in several forms. U-type, well type, slant tube and others, review the Pressure Fundamentals and Transmitter Selection material following the manuscript text, (handout figures 6-9). All operate according to the pressure = height * density (P=Ht*D) equation. A simple U-tube manometer is easily made by bending a transparent and flexible length of hose into a "U" shape. A liquid (usually water, mercury or alcohol) of known weight density is allowed to partially fill the transparent, U-shaped hose. Fluid pressure is applied to one leg of the manometer with the other leg being left open to atmosphere or applied to some other accurately known reference pressure.

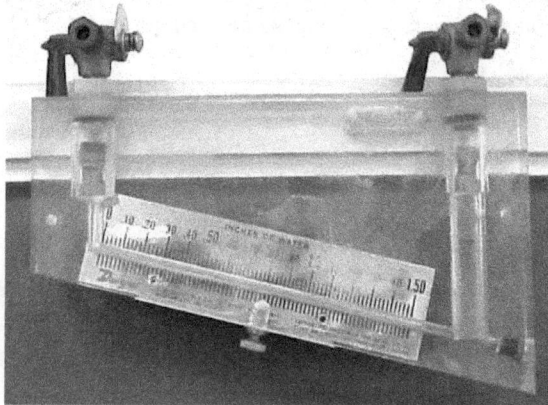

Figure 26 - An inclined-tube or slant-tube manometer allows for accurate reading of small pressure changes. These types of instruments are commonly found in control centers and weather stations where ambient pressure changes can effect process conditions.

Figure 27 – The backside of a well-type manometer. The high pressure is applied to the circular well at the bottom and the reference pressure is applied to the top.

Figure 28 – The front side of the well-type manometer contains a scale for measuring column length in inches (left side), and a scale for measuring pressure (right) when using Mercury as the measurement fluid.

As the pressure under test enters the input leg, the liquid will be displaced towards the reference pressure, (if the test pressure is greater than the reference), or away from the reference leg (if the pressure is less than the reference). After the liquid settles to equilibrium, the liquid displacement (or difference of height between the two liquid columns) is representative of the applied test pressure.

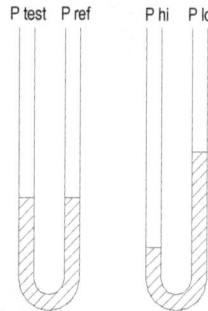

P test   P ref   P hi   P lo

Figure 29 - A U-Tube manometer before and after the application of test pressures. Note the designations for the input pressures. The difference in the liquid level heights in the vessel at the right are used measure the difference between the two applied pressures. By multiplying the height difference by the density of the manometer fluid, the difference in the applied pressures is determined.

The manometer is a pressure balance instrument. The sum of the pressures acting above the bottom or bend of the manometer on each side is equal. To measure the unknown or test pressure, locate the leg with the lowest liquid level and use the low liquid level as a reference line for both legs. The leg with the lowest liquid level will contain the highest applied input pressure. The amount of the pressure difference will be equal to the height of the liquid level difference between the legs, multiplied by the weight density of the manometer liquid used. If a reference pressure other than atmosphere is used, it must be added or subtracted from the height * density product.

Again, once the liquid pressure difference is known, adding or subtracting the computed liquid pressure to the reference pressure can easily determine the unknown test pressure. In short, add the computed manometer liquid pressure to the reference pressure if the liquid moves toward the reference. Subtract the computed pressure from the reference pressure if the liquid moves away from the reference input.

The procedure for determining an unknown pressure appears complicated but is reasonably straightforward. Through observation, one can easily determine the quantity of unknown test pressure by applying the pressure balance concept and computing the pressures in each leg above the reference line. The pressure in each leg down to the reference line must be equal or the manometer will not be in equilibrium. Knowing the reference pressure, liquid density and distance of the difference in the two columns are the only requirements for accurately measuring a test pressure with a manometer.

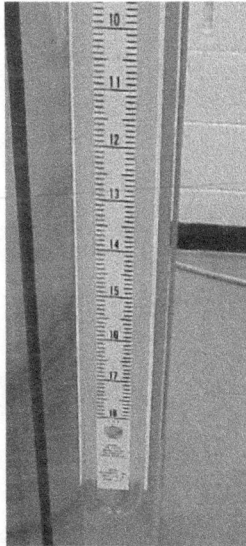

Figure 30 - The bottom portion of a U-tube manometer. At almost 5 feet tall the U-tube is too long to adequately capture in a single photo.

Mercury filled well-type manometers are heavy, awkward and are currently considered unsafe to carry into most plants for calibration purposes. In cases where the manometer cannot be taken to the site, the apparatus to be calibrated is replaced in the field and brought into the lab. Accurate calibration is possible if the unit under tested is linear and if span and zero adjustments are available. If calibration adjustments are not available such as with most pressure gauges, the amount of error is recognized (as offset or gain error), quantified and labeled on the device. In many manufacturing plants, calibration tests are performed routinely on incoming sensors and related measurement apparatus before being placed into operation.

Pressure Components

Transmitters

Although electronic components appear to dominate the industrial automation market, many pneumatic (compressed air) components are still in use today. Many applications require pneumatics due to the lack of available electric power, for safety, or for force generating concerns. Pneumatics in some form will always be required whenever a large force needs to be developed within a confined area.

Figure 31 – A common differential pressure transmitter connected with a single input as a gauge pressure transmitter.

In industrial process measurement and control systems transmitters convert a process quantity into a "standard" signal span. Common standard signal spans are 4 to 20 milliamps and 3 to 15 psi. The transmitter is normally connected to the output of a sensor. During typical operation in a pneumatic (3 to 15 psi output pressure) transmitter, an applied and increasing process pressure enters a bourdon, bellows or diaphragm at the input causing expansion of the particular pressure input device. As the bourdon, bellows or diaphragm expands against a mechanical linkage, the connecting linkage transmits the increasing input force (from the pressure increase) to a device directly opposing and balancing the input increase, such as a spring. The resulting spring motion allows a proportional amount of supply pressure to the instrument output.

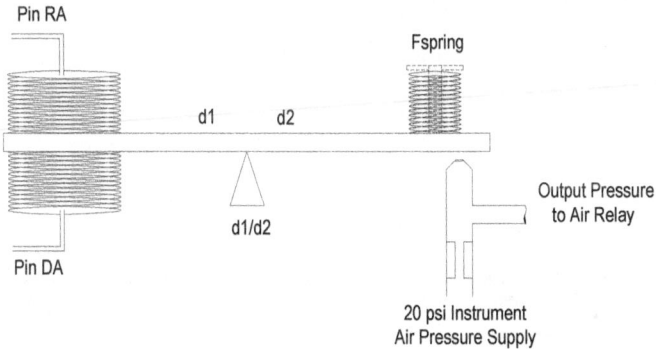

Figure 32 - This diagram is representative of the operation of pressure transmitters with pneumatic outputs. The direct-acting (DA) input has a non-inverting relation with the output (input↑, output↑) and the reverse-acting (RA) input has an inverting relation with the output (input↑, output↓).

Typical output devices vary between poppet valves, orifice/baffle arrangements (also called flapper/nozzles), air relays, and other similar valve-like components capable of converting linkage motion position displacement into a proportional air pressure span. It should be remembered that mechanical components will come to rest in equilibrium (at balance), and the output pressure described in the example will always be derived from an instrument air supply, usually 20psi (1.4 kg/cm^2). Although not indicated in the figure, balancing of all components is achieved by applying all, or a portion of the resulting output, back against the input. Thereby resulting in an output that is assured of having a direct relation with the input. In essence, each component within a control system contains a closed-loop control system internally.

Torque-balance and force-balance are the most commonly applied mechanical operations within transducers, transmitters, actuators and pneumatic relays.

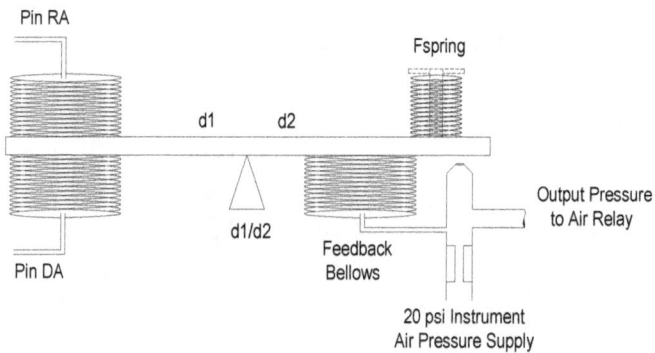

Figure 33 - A pneumatic pressure transmitter with internal feedback to assure the output is a precise analog of the input. Sensitivity is determined by the d1/d2 ratio and output offset by the spring force.

Torque-balance components are easily recognized by the levers required to transmit and distribute the forces throughout the instrument. If a lever or fulcrum is obvious within the instrument or if any mechanical component appears to be moving through a partial rotation, torque balance is utilized. The gain of a torque-balance instrument is determined by the moment arm lengths of D1 and D2, the fulcrum location. Output offset is determined by the amount of bias applied to the moment arm, usually through a bias spring force. Torque balance is commonly used in transmitters, transducers (P/I's) and pneumatic controllers. Force balance is apparent when the pressure input device, usually a diaphragm or bellows, is

234

working directly against a spring without the use of a lever or fulcrum. A single linear axis of motion is also normally associated with force balance components. Air relays, (such as amplifying, regulating or computing components), valve positioners and actuators are common force balance components.

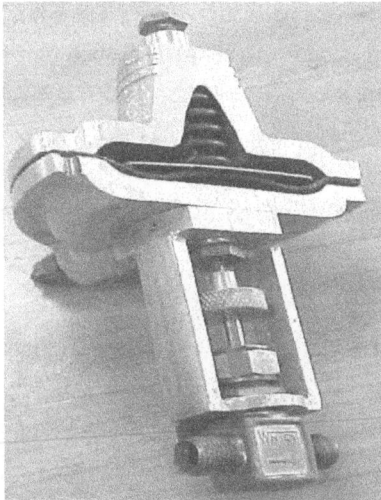

Figure 34 – An inexpensive actuator is cut-away to expose the internal diaphragm and force-balancing spring. The Whitey valve is attached below the actuator. Actuators are force-balance devices.

Although few students of instrumentation today will repair or maintain transmitters, transducers or actuators at the component level, an appreciation of their internal construction and operation is beneficial since contained within each is a process undergoing automatic control. In addition, most manufacturing and experimental applications employ an obvious form of torque, or/and force balance.

Transmitter Calibration

Process transmitters (also called secondary elements) convert the output signal from a sensor (also called a primary element) into a standard electronic signal span that is compatible with controllers, recorders, display devices and other automation hardware and components.

Currently, the most common transmitter output standard is the 4-20 milliamp current loop signal. The current loop is used to transmit signals over long distances without concern for signal loss (called attenuation) due to wire resistance. The current loop forms a series circuit between the transmitter, the controller, the DC power supply (usually 12-45 VDC) and any other components requiring information about the process being measured. Since current is constant throughout a series circuit, the current loop signal can be transmitted at a constant value throughout the signal carrying wires independent of distance.

Figure 35 - A pressure transmitter with a local display providing process pressure.

Pressure transmitter calibration involves setting the minimum and maximum output current values of 4 mA and 20 mA to specific minimum and maximum pressures. Two calibration adjustments labeled Zero and Span are available on traditional transmitters. As demonstrated in the following figure, the Zero adjustment is set to 4 mA when the pressure is at the desired minimum measured value. Subsequently, the Span adjustment is set to 20 mA when the measured pressure is at the desired maximum value. Each adjustment

236

corresponds to the effects of varying the slope (Span) and Y-axis intercept (Zero) of a linear Y=mX+b graphical relationship. The Y-axis intercept is often referred to as *offset* or *output offset* since the offset quantity represents the amount the output span is offset from an output of zero. It is suggested that one study the following figure in detail as the provided concepts apply not only to level transmitters, but also to calibration of all process transmitters.

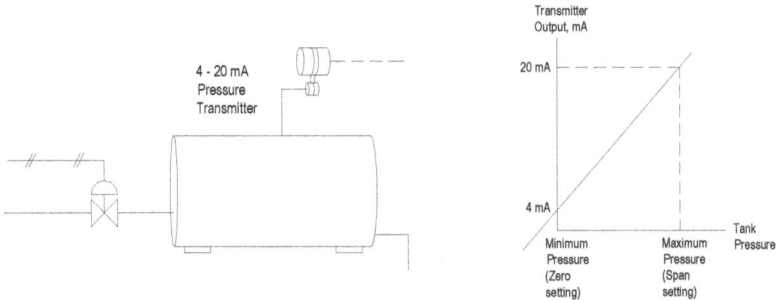

Figure 36 - Transmitter calibration requires setting the Zero and Span adjustments at corresponding minimum and maximum pressures. The technique directly relates to linear, Y=mX+b relationships (right).

Output Signal

When the minimum and maximum calibration pressures are known, the transmitter output current can be easily related to the measured pressure using percentages. Linear, Y=mX+b mathematics can also be used but will require a bit more effort. As an example, consider a calibration example where 4 to 20 mA corresponds to 0 psi to 100 psi. Assume the transmitter output is measured as 8 mA and it is desired to know the associated input pressure. First, determine the percentage of the measured current within the 4-20 mA span by subtracting the minimum value (4 mA) from the measured value (8 mA) and dividing the difference by the current span (16 mA). Multiply the result by 100%. Second, multiply the resulting percentage by the span of the measured level, and then add the minimum level value. (Note: span is defined as the difference between end points; the pressure span is 100 psi minus 0 psi, or 100 psi.)

237

First,

% mA = [(Measured mA − Minimum mA)\(Maximum mA − Minimum mA)] * 100%

% mA = [(8 mA − 4 mA) ÷ (20 mA − 4 mA)] * 100%

% mA = [(4 mA) ÷ (16 mA)] * 100%

% mA = (.25) * 100% = 25%

Therefore, 8 mA is 25% of the 4-20 mA output span. This corresponds to a pressure that is 25% of the calibrated span of 0 to 100 psi.

Second,

Pressure in psi = [25% * (100 psi − 0psi)] + 0psi

Pressure in psi = [25% * (100 psi)] + 0psi

Pressure in psi = (25 psi) + 0psi

Pressure in psi = 25 psi at 8 mA of transmitter current

Mathematically, the input pressure determination process is an application of the Y=mX+b relationship between transmitter output and pressure input. To demonstrate, assemble a linear graph with pressure on the input (X-axis) and transmitter current on the output (Y-axis). The resulting graph should appear as in the following diagram.

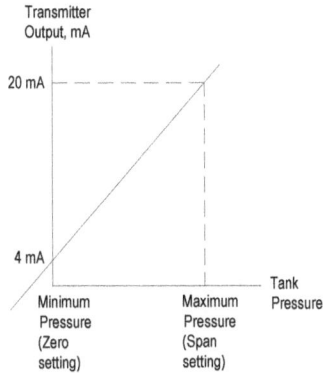

Figure 37 – Pressure transmitter calibration is an application of linear Y=mX+b concepts.

To determine the input pressure corresponding to a given transmitter current, first recognize that it is desired to determine an input quantity (pressure) given an output quantity (current). This can be observed in the graph of the preceding figure. The linear graph is described by Y=mX+b or using the applied quantities, output current (Y) equals transmitter gain (slope, m) multiplied by the input level (X) plus the output offset (intercept, b). Written mathematically the relationship appears as follows.

$$Y=mX+b$$

Or,

$$\text{Current in mA} = (\text{Gain} * \text{Level}) + \text{Offset}$$

The transmitter gain is then determined from the input and output quantities.

$$\text{Gain} = \text{Slope} = \Delta Y/\Delta X = (20\ mA - 4\ mA)/(100\ psi - 0\ psi)$$
$$\text{Gain} = \text{Slope} = (16\ mA\ /\ 100\ psi)$$
$$\text{Gain} = .16\ mA/psi$$

The transmitter offset is determined from the input and output quantities. The Y=mX+b equation is rearranged to solve for the Y-axis intercept, b.

239

$$\text{Offset, or Y-axis intercept, } b = Y - mX$$
$$\text{Offset} = \text{Current} - (\text{Gain} * \text{pressure})$$

Choosing corresponding input and output values from the graphed X-axis (100 psi) and Y-axis (20 mA) quantities and inserting the previously computed gain yields,

$$\text{Offset} = 20 \text{ mA} - (.16 \text{ mA/psi} * 100 \text{ psi})$$
$$\text{Offset} = 20 \text{ mA} - (16 \text{ mA})$$
$$\text{Offset} = 4 \text{ mA}$$

The computed offset value of 4 mA can be verified by observing the previous figure and noting where the plot crosses the Y-axis, since the Y-axis intercept is defined the Y-axis value when X = 0.

Once the gain and offset are known they can be substituted into the rearranged Y=mX+b equation along with the output current value. The resulting equation for determining the input given the output becomes X=(Y-b)/m. Using the previous example of 8 mA yields the following.

$$X = (Y - b)/m$$
$$\text{Input} = (\text{Output} - \text{Offset}) \div \text{Gain}$$
$$\text{Input Pressure} = [(8 \text{ mA} - 4 \text{ mA})] \div (.16 \text{ mA/inch})$$
$$\text{Input Pressure} = (4 \text{ mA}) \div (.16 \text{ mA/inch})$$
$$\text{Input Pressure} = 25 \text{ psi}$$

In the field, the percentage approach is much more rapid. Often the desired value of input or output need only be approximated to satisfy the investigation. Either of the previous techniques can be used to solve the problems at the end of the chapter although both should be understood.

Figure 38 - A common 0-60 psi pressure regulator.

Pressure Regulators

Among the more common components found in industrial automation applications are air relays (also called volume boosters), biasing relays and pressure regulators. All are introduced here due to the similarity of operation. Pressure regulators maintain constant output pressure under conditions of changing airflow demand. However, an air relay is a proportional *pneumatic amplifier* capable of pressure amplification or reduction. Air relays are also known as volume boosters due to the increased volume of air they are capable of providing. Biasing relays add or subtract a constant amount of air pressure to an applied input signal span.

Pressure regulators utilize a force-balance operating concept to maintain a constant output air pressure under conditions of changing airflow demand. The input force is the result of a spring located in the upper chamber of the regulator component. The output force is the result of applying the output pressure upon a gasket-like output diaphragm. The application of the output pressure to the diaphragm area results in a force that is applied upwards against the input spring force. The following diagram illustrates.

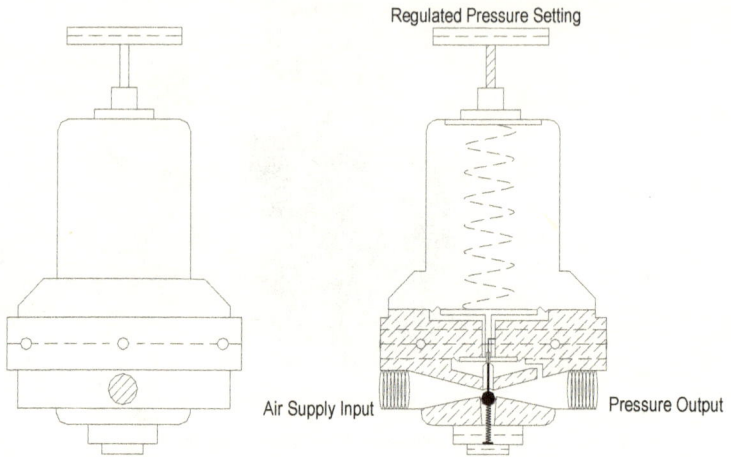

Figure 39 - External and internal views of a common pressure regulator.

As can be seen in the previous figure, the internal construction of a common pressure regulator contains relatively few components. A spring, diaphragm, and pilot valve compose the majority of the function determining mechanisms. In operation, the pressure setting knob is rotated clockwise compressing the spring in the upper chamber. As the spring compresses, the pilot valve is moved downwards causing the supply air to pass around the valve and appear in the output port.

Regulation, the ability to maintain constant pressure, is achieved by applying the output pressure up against the input spring. Upon close inspection in the following figure an internal feedback port can be observed in the output chamber of the regulator. The feedback port allows the resulting output air pressure to appear under the diaphragm.

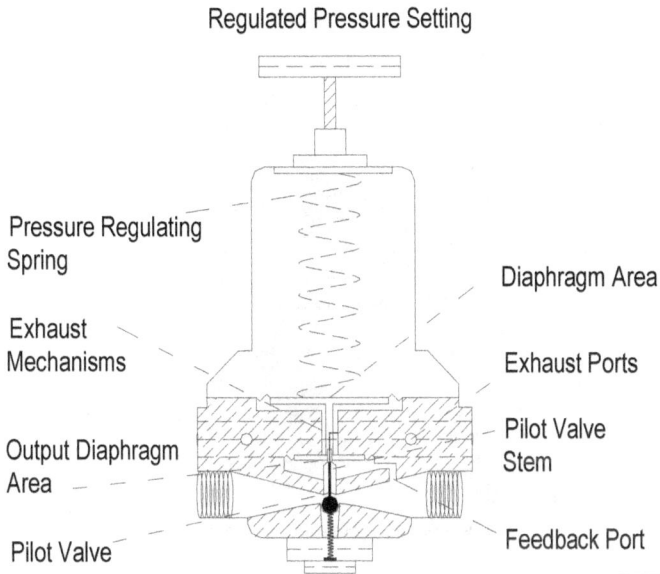

Regulated Pressure Setting

Pressure Regulating
Spring

Diaphragm Area

Exhaust
Mechanisms

Exhaust Ports

Output Diaphragm
Area

Pilot Valve
Stem

Pilot Valve

Feedback Port

Figure 40 - The essential components of a common air pressure regulator.

The diaphragm "Floats" inside the regulator, it has no fixed or rigid connections to any of the other internal components but rests on top of the pilot valve stem. As the fed-back output air pressure is applied to the lower portion of the diaphragm, an upward force results. Any difference between the downward regulating spring force and the output pressure generated upward force will cause the pilot valve to reposition until the upward and downward forces are balanced. Should the output pressure be larger than dictated by the spring force, the lifting motion of the output pressure on the diaphragm will uncover the top portion of the pilot valve stem. Lifting the diaphragm above the pilot valve stem will open a path for the excessive output air pressure to escape via the exhaust ports. Once the output air pressure has discharged and decreased, the diaphragm lowers to rest atop the stem and the exhaust path is again closed.

Figure 41 - A cutaway view of a pressure regulator.

Should the load (airflow demand) on the regulator increase, the output pressure and corresponding upward diaphragm force will momentarily decrease resulting in a slight downward motion of the diaphragm and valve stem. The downward movement of the valve stem will open the pilot valve causing the output airflow and pressure to increase. The increased output pressure will cause the upward force to again balance the downward spring force. Pressure regulation is afforded in this manner.

A mathematical analysis of the pressure regulator forces can be demonstrated by assuming an initial condition of force-balance.

At balance,

$$F\downarrow = F\uparrow$$

The downward force is created by the regulating spring and the output pressure applied to the diaphragm area creates the upward force.

Since,

$$F\downarrow = P*A$$

Substituting for output pressure,

$$F\downarrow = Pout*Aout$$

The output pressure is a result of the regulating spring force and the diaphragm area.

$$Pout = F\downarrow \div Aout$$

Therefore,

$$Pout = F/Aout$$

And for the regulating spring force,

$$Fspring = Pout * Aout$$

Example Problem:

Determine the spring force required to output 20-psi if an output diaphragm area of 1 square inch is used.

Solution:

Using the previous equation for spring force,

$$Fsp = Pout * Aout$$
$$Fsp = 20psi * 1in^2$$
$$Fsp = 20\#$$

Example Problem:

Determine the variation in spring force required to accommodate a pressure change of 15 to 20 psi. Assume an output diaphragm area of .5 square inch is used.

Solution:

The problem requests the change in spring force for a 5-psi pressure increase. If a spring rate (#/in) and thread information (threads per inch) were provided, one could determine the number of clockwise adjustment rotations required to change the pressure. However, only the change in spring force is requested and the previous equation is again applied.

$$Fsp = \Delta Pout * Aout$$
$$Fsp = 5psi * 1in^2$$
$$Fsp = 5\#$$

Designers of computing and regulating relays attempt to minimize motion within the components to maximize life expectancy and accuracy. As a result, a rather stiff spring with a substantial spring rate (approximately 100#/in) would be used. In which case only a partial rotation of the regulating adjustment would most likely be required.

Example Problem:
A 120#/inch spring used in a pressure regulator is compressed .05 inches during a half-rotation adjustment of the regulated pressure setting. Using a 1.5in² output diaphragm area, determine the *change* in the regulated pressure output.

Solution:
If you've been paying attention to the reading, you probably saw this coming. To determine the amount of output pressure change the amount of downward spring force must first be determined.

Since spring rate (SR) equals force (F) divided by distance (D),

$$SR = F/D$$
$$F = SR * D$$
$$F = 120 \#/in * .05in$$
$$F = 6.0\#$$

Knowing the applied spring force and the diaphragm area, the resulting pressure change can be found. Since the pressure change is the result of the change in spring force applied to the output diaphragm area,

Since,

$$P = F/A$$

And,

$$\Delta P = \Delta F/A_{out}$$
$$\Delta P = 6.0\#/1.5 in^2$$
$$\Delta P = 4\ psi$$

Amplifying Relays

Understanding the operation of pneumatic (compressed air) components first requires that you forget any terminology associated with electronic components. In electronics, a relay is a discrete (two-state), electro-magnetically actuated switch capable of multiple circuit connections. In pneumatics, a relay is a proportional device capable of amplification (output span is greater than the input span) or reduction (output span is less than the input span). Fortunately, all pressure regulating and computing components operate according to a single basic principle, force-balance.

Figure 42 – A 1:6 amplifying air relay or "volume booster" is an example of force-balance components.

As mentioned previously, air relays, biasing relays, reversing relays and pressure regulators operate according to a *force-balance* principle where an applied input force is equaled a generated output force. In operation, the input pressure is applied to an input area resulting in a downward input force. The input force is ultimately opposed and balanced by an output-pressure generated upward force. The output force is the result of the output pressure being applied against an output diaphragm area.

Relays consist of few internal components. The internal components include an input diaphragm, output diaphragm, pilot valve and an occasional spring for output-offset biasing. As an example, refer to the amplifying relay of the following figure.

Input Pressure Port

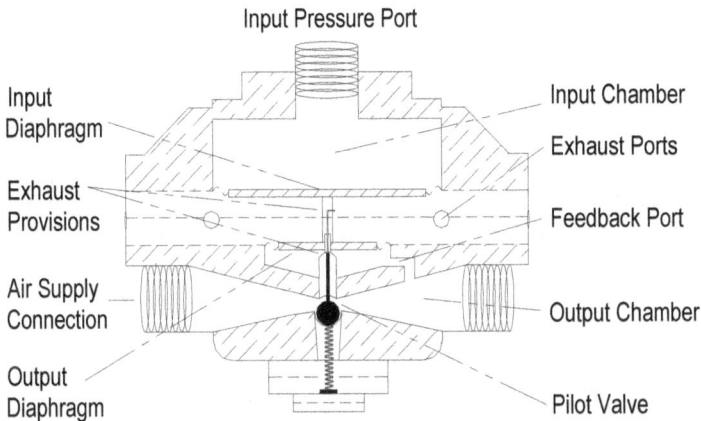

Figure 43 - A simplified diagram of a volume booster computing relay, or air relay.

In the diagram, the input air pressure (also called "Loading" pressure) to be amplified is applied into the upper portion (input chamber) of the unit. Once introduced into the input chamber, the input pressure is applied to an input diaphragm area at the bottom of the input chamber. Since pressure applied to an area equals force, the applied input pressure results in a downward force. The downward force causes the pilot valve to open allowing the supply pressure to pass around the pilot valve and appear in the output chamber. The resulting output pressure appears at the output connection but also passes through a small internal port and is applied to the bottom of the output diaphragm area. Once applied to the bottom of the output diaphragm area, the resulting output pressure becomes an upward force closing the pilot valve. In this manner the resulting output pressure is compared against the input pressure and results in a mechanical form of negative feedback. The fed-back output pressure assures that only a specific output pressure value is allowed to appear in the output port.

Should the output pressure be too large, the resulting output pressure will cause the pilot valve to close, reducing the output pressure. Should the output pressure not be large enough, the downward input force will overdrive the output pressure's feedback force opening the pilot valve and causing an output pressure increase.

Eventually a balance between the pressure created forces at the output and input diaphragm areas will result in a fixed relation between the input and output pressures. The fixed relation is the gain (or reduction factor) of the air relay and is determined by the ratio of the input and output diaphragm areas. The following mathematics demonstrates.

At balance,

$$Fin\downarrow = Fout\uparrow$$

Since,

$$F = P*A$$

Substituting,

$$Pin*Ain = Pout*Aout$$

Rearranging to solve for gain, K,

$$Gain, K = Pout \div Pin = Ain \div Aout$$

Therefore,

$$\therefore Gain, K = Pout/Pin = Ain/Aout$$

The previous equation demonstrates that the gain of an air relay is determined by the ratio of the input and output diaphragm areas. Note that an air relay gain of greater than unity requires an output diaphragm area to be smaller than the input diaphragm area.

Close inspection of pneumatic components will note exhaust ports at various locations around the center of the unit. Exhaust ports are required to discharge excess output air pressure that would otherwise be trapped in the output chamber. Should the input pressure change from a high to a low condition the pilot valve closes and the output air pressure should follow the input by decreasing. If the output is connected to a non-vented component such as a pressure gauge, air-powered actuator, bellows or similar device, the relay must discharge the excess air pressure to atmosphere since the load is incapable. Discharge of excess output air pressure is performed by momentarily allowing the output diaphragm to lift above the top of the pilot valve stem. As the diaphragm area lifts above the pilot valve stem, a passage to atmosphere via an exhaust port is provided and the excess output pressure is discharged until the forces equalize and the output diaphragm area settles on the pilot valve stem once again. The following diagram attempts to illustrate.

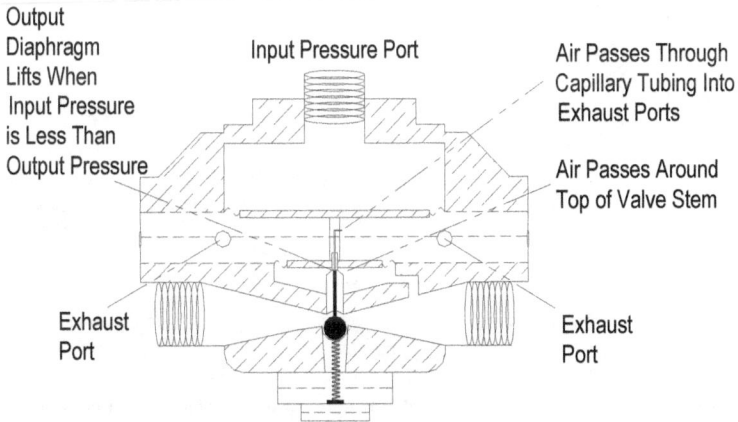

Output Diaphragm Lifts When Input Pressure is Less Than Output Pressure

Input Pressure Port

Air Passes Through Capillary Tubing Into Exhaust Ports

Air Passes Around Top of Valve Stem

Exhaust Port

Exhaust Port

Figure 44 - Excess output air pressure is discharged to atmosphere through an internal capillary and exhaust ports.

It's often said that pneumatics is not a precise science. This attitude is probably the result of inexpensive pressure gauges having the reputation of being notoriously inaccurate. In addition, many pneumatic applications utilize working pressures as "Ball-park" values, meaning the intended input and output pressures need not be exact, just close enough for proper operation. Unlike electronic systems that are designed for specific operational conditions, limits and extremes, a pneumatic system may be designed to develop and output pressure span of 100-psi when a maximum of only 50 or 60 psi may be required and used. One of the reasons for this is due to air relays being commercially available with limited gains of only 1:1 (an airflow amplifier, the output ports are larger in diameter than the input ports but the output and input pressures are equal), 1:2, 1:4 and 1:6. Note that these ratios designate input to output pressure ratios, Individuals working with pneumatic systems commonly refer to amplifying air relays as a "One to two" or "One to four" or "One to six." Air relays are cascaded (connected in series succession) to result in higher gains.

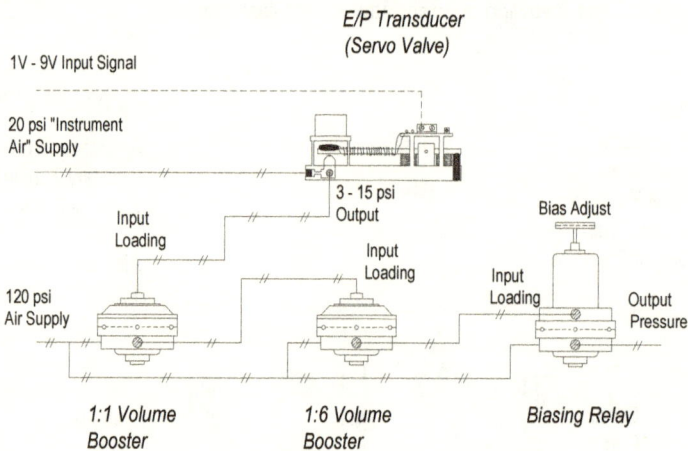

Figure 45 - An electronic signal is converted to a pneumatic signal and mathematically conditioned using amplifying and biasing relays.

As an example, the output of a 1:4 can be applied to a 1:6 for an overall gain of 24. In this case a 2-psi pressure change applied to the 1:4 results in an output pressure change from the 1:6 of 48 psi. However, over the year's pneumatics have become quite precise with some computing and regulating components having the ability to compensate for atmospheric pressure and ambient temperature variations. Yet in many industrial applications pneumatics are still perceived as inaccurate components used primarily to generate large forces.

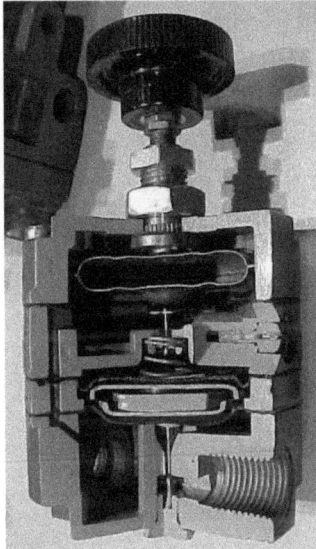

Figure 46 - The internal capsules in this sensitive pressure regulator compensate for variations in atmospheric pressure.

Amplifying air relays commonly convert 3-psi to 15-psi pressure spans from an E/P or I/P transducer into higher pressure ranges for final control purposes and are commonly used in place of valves. Reducing relays are available with "Gains" of 6:1, 4:1 and 2:1, and are commonly used to convert large pressure ranges of hundreds of psi into 3-psi to 15-psi pressure ranges for compatibility with commercially available control components. In either case, amplifying or reducing relays are usually used when a proportional pressure is needed for final control element applications such as disc-brake pressure on web tension machines.

Figure 47 - Pneumatic regulators, volume boosters, biasing and reversing relays have similar internal construction. The input is applied at the top and results in an opening or closing of a pilot valve in the lower section which allows supply air to appear at the output and to be applied back against the input as feedback to assure an accurate output pressure.

Biasing Relays

Whenever a pressure range is to be increased or decreased by a constant amount of pressure, a biasing relay is required. Biasing relays, or bias boosters, utilize a force-balance operating principle as described previously with air relays to add or subtract a constant amount of pressure to the input pressure range. Biasing relays are typically rated at +20 to –20 psi. As demonstrated in the following diagram, the input pressure range is applied to an input diaphragm area to create a downward input force. The resulting input force is applied in conjunction with a biasing spring force to open a pilot valve that allows a portion of the air supply pressure to appear in the output chamber. The output pressure is applied to the bottom of an output diaphragm resulting in an upward balancing force equal to the algebraic sum of the input pressure created force and the spring force.

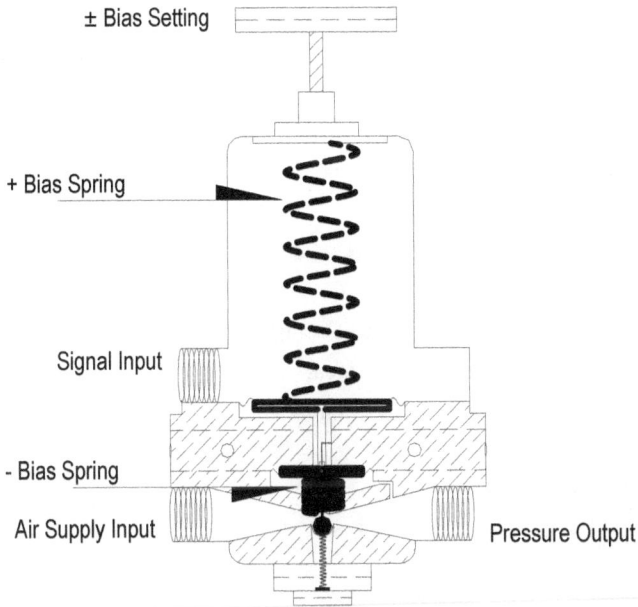

Figure 48 - A cut-away view of a biasing relay.

Note that the biasing force is provided by two individual springs labeled +Bias Spring and –Bias Spring in the previous diagram. Clockwise rotation of the bias setting knob compresses the upper spring opening the pilot valve and establishing a positive output bias pressure. The applied input pressure range will be elevated (increased) by a fixed amount. Rotating the bias setting knob in a counter-clockwise direction allows the –Bias Spring to hold the pilot valve closed until the applied input range achieves a large enough pressure to overcome the – Bias Spring force opening the pilot valve and increasing the output pressure. The result will be a negative bias or suppressed output pressure range. Negative bias has the same effect as subtracting a constant pressure from the input pressure range. As an example, a 3-psi to 15-psi input pressure range would become 0-psi to 12-psi if the bias adjust is set an upward force equivalent of 3-psi.

Note that biasing relays can amplify in conjunction with providing biasing provisions. As with air relays, manipulation of the input and output diaphragm areas facilitates amplification. The following demonstrates.

At balance, the sum of the downward and upward forces in computing relays is equal,

$$Fin\downarrow = Fout\uparrow$$

The input forces are created by the bias spring force and the input pressure range applied to the input diaphragm area,

$$Fsp\downarrow + Fin\downarrow = Fout\uparrow$$

Since,

$$F = P*A$$

Substituting,

$$Fsp + Pin*Ain = Pout*Aout$$

Rearranging to solve for Pout,

$$Pout = Pin (Ain/Aout) + Fsp/Aout$$

Note the similarity to the Y=mX+b equation where,

- Fsp/Aout = output offset, bias (Y-axis intercept value)
- Pin = Input pressure
- Ain/Aout = Gain, (slope)
- Pout = Output pressure

Note also that the influence of the bias spring is contingent upon the size of the output diaphragm area.

To determine gain, the Fsp term is set to zero and the equation is arranged to solve for Pout/Pin.

Since,

$$Pout = Pin\ (Ain/Aout) + Fsp/Aout$$
$$Pout = Pin\ (Ain/Aout) + 0$$

And,

$$Pout/Pin = Ain/Aout$$
$$\therefore Gain,\ K = Pout/Pin = Ain/Aout$$

Solving the equation for Fsp yields,

$$Fsp = Pout*Aout - Pin*Ain$$

Example problem:

A unity gain (1:1) biasing relay is to convert a 3-psi to 15-psi input pressure range into an 8-psi to 20-psi output pressure range. If the biasing relay utilizes a .5 square inch output diaphragm area, determine the required input diaphragm area, the required spring force and the spring force direction.

Solution:

Since gain equals the output span divided by the input span, the gain equals –

$$\Delta Pout/\Delta Pin = Ain/Aout$$
$$12psi/12psi = Ain/Aout = 1$$

The unity gain requirement is verified. Since the ratio of the input to the output area is 1, the input area must equal the given output area.

$$Ain/Aout = 1$$
$$Ain = Aout = .5\ sq.\ in.$$

To determine the required spring force, the given values must be substituted into the spring force equation.

Since,

$$Fsp = Pout*Aout - Pin*Ain$$

Subbing areas and corresponding input and output pressure values,

$$Fsp = (20psi)*(.5\ in^2) - (15psi)*(.5\ in^2)$$
$$Fsp = 10\# - 7.5\#$$
$$Fsp = 2.5\#$$

Since the spring force computed as a positive quantity, the spring force will be the same direction as originally assumed in the equation. The resulting spring force will be pushing down to create a positive output bias. If the spring force had computed as a negative quantity the spring force direction would be upward, indicative of a negative bias pressure.

Figure 49 - A common -18 to +2 psi biasing relay.

Example Problem:

Assume the biasing relay is to perform an amplification function as well. Determine the required diaphragm areas, bias spring force and direction to convert 3-15 psi into 0-72psi. Assume the biasing relay has an input diaphragm area of 1 square inch.

Solution:

Since gain equals the output span divided by the input span, the gain equals –

$$\Delta Pout/\Delta Pin = Ain/Aout$$
$$72psi/12psi = Ain/Aout = 6$$

A 1:6 gain is required of the biasing relay. Since the input diaphragm area is given as 1 in^2 and,

$$Gain, K = \Delta Pout/\Delta Pin = Ain/Aout$$
$$Gain, K = 72psi/12psi = Ain/Aout$$

Substituting the given input area,

$$72psi/12psi = (1 \text{ in}^2)/Aout$$

Solving for Aout,

$$Aout = 1 \text{ in}^2/6$$
$$Aout = .166 \text{ in}^2$$

To determine the spring force, assume the spring direction is to be applied downward aiding the input pressure created force. The resulting balance equation would appear as follows.

At balance,

$$F\downarrow = F\uparrow$$
$$Fsp\downarrow + Fin\downarrow = Fout\uparrow$$

Subbing for pressures,

$$Fsp\downarrow + PinAin\downarrow = PoutAout\uparrow$$
$$Fsp = PoutAout - PinAin$$
$$Fsp = (72psi)(.166in^2) - (15psi)(1in^2)$$
$$Fsp = 12\# - 15\#$$
$$Fsp = -3\#$$

The actual spring force is upwards indicating a negative bias or an equivalent subtraction of 3 psi ($3\#/1in^2$) from the input pressure range or 18psi ($3\#/.166in^2$) from the output pressure range.

Transducers

In simplest terms, a transducer is an energy converter. Common examples of transducers include the conversion of mechanical to electrical energy with electrical generators, and conversion of electrical energy to mechanical energy as with motors. Yet, transducers assume a variety of forms especially in the industrial automation and experimental testing fields.

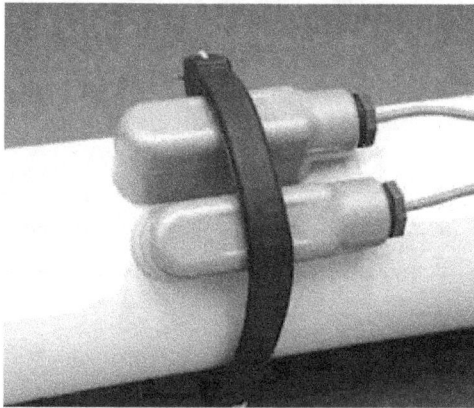

Figure 50 - Ultrasonic transducers determine flow within closed systems without pentrating the pipe or contacting the flowing material.

In measurement and control components transducers convert process signals through the use of a sensor or, convert the output signal of a controller from an electronic signal into a pneumatic or mechanical signal for positioning a valve or other final control element. Computer-based systems commonly employ transducers to convert pressure into electrical current (P/I) or electrical current into pressure (I/P). Transducers converting a process variable into an electrical signal are called sensors. Sensors are usually used in conjunction with a transmitter to convert the sensor output into a signal that is recognized by commonly available industrial measurement and control apparatus, usually 4 milliamps to 20 milliamps.

In the controls field however the term "transducer" usually refers to a component responsible for converting an electrical signal, usually from a controller, into another physical form such as pneumatic (compressed air). In short, an industrial transducer is usually a current to pressure servo valve or I/P, (pronounced "I to P").

Figure 51 - The Fisher Controls type 546 I/P (with the cover removed) is a standard in the controls industry and is commonly found attached directly to a valve actuator.

I/P's are a combination of electrical and mechanical technologies and commonly use the electrical 4-20 mA input signal to generate a magnetic field. The electro-magnetic field works in conjunction with a permanent magnet to position a baffle in close proximity to an orifice (see the following figures). The resulting pressure change is applied to an internal air relay for amplification before being sent to the output pressure port. Most, but not all, industrial I/P's feed-back the resulting output signal for comparison against the input. As with other pneumatic components, internal feedback assures an accurate output. The applied 4-20 mA input signal usually results in a 3-15 psi output pressure span.

Figure 52 - Common servo valves (E/P or I/P transducers) use magnetics to position an orifice/baffle. The resulting output pressure is created within the orifice and often requires a volume booster to send the output pressure over long distances or through high diameter airlines without a time lag.

Permanent Magnet

S

N

Connection Terminals

Span Adjust

Electro-magnetic Coil

Orifice/Baffle

Air Supply Port

Output Port

Mechanical Zero Adjust

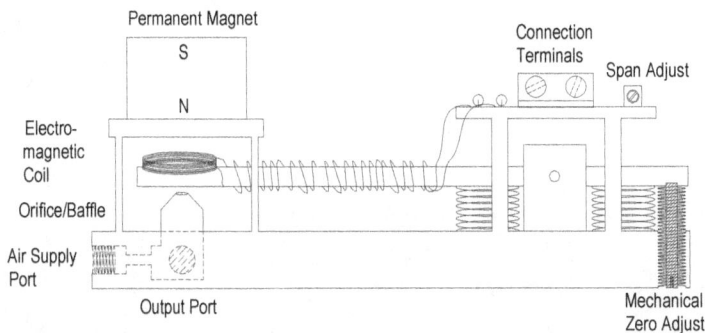

Figure 53 - A simplified diagram of the Fairchild E/P transducer in the previous diagram.

Shown in the previous two figures is a Fairchild T5100 series of voltage to pressure (E/P) transducer. The T5100 model in the figures is a slight variation on the common industrial transducer in that it converts voltage instead of current to pressure, contains no internal feedback and the output pressure is developed within the unit's orifice. E/P's are commonly found in laboratories and industrial applications where the leadwire run is short and air volume demands are minimal. Since the T5100 does not contain an internal air relay the amount of air delivery through the orifice is quite small. This can result in an undesired time lag between the time a signal change is imposed at the input and the time the output pressure charges the output airline volume to the requested pressure. For this reason, a 1:1 volume booster (air relay) is commonly used with the E/P and is located in close proximity to the output connection to minimize the time lag. Most common industrial transducers include an integral air relay and employ internal feedback to assure input/output accuracy.

Figure 54 - A group of Moore Products I/P transducers, or servo valves. These units output 3-15 psi for an input of 4-20 mA.

I/P transducer calibration is accomplished by locating and setting the span and zero adjustments under appropriate input conditions. For direct-action (input increases, output increases), the input is set to 4 mA and the zero adjustment is set to 3-psi. The input is then set to 20 mA and the span adjustment is set to 15-psi. For reverse-action (input increases, output decreases), the input is set to 4 mA and the zero adjustment is set to 15-psi. The input is then set to 20 mA and the span adjustment is set to 3-psi. The following diagram illustrates.

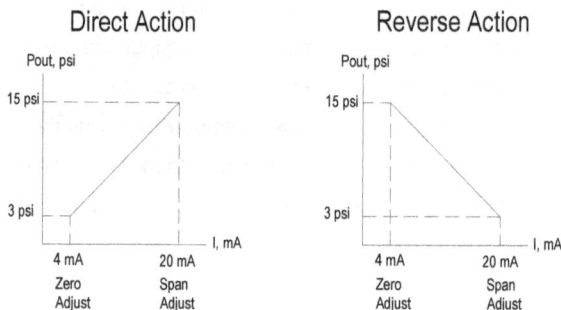

Figure 55 - Transducer calibration is accomplished by setting the zero at the minimum input and setting the span at maximum input.

Fluid Pressure Application Notes

The following application information is taken from automation manufacturer's advertising, maintenance materials and personal experience. The manufacturer's materials are located before the homework problems at the end of this section. The materials are provided to assist the student in understanding pressure sensor and transmitter specifications, to provide an overview of typical applications, and to demonstrate the technologist's responsibilities and perspective. It is suggested that the student closely examine the following information.

Application Notes:

1 – Pressure gauge line selection must take into consideration the pressure and corrosion of the measured media. Flexible copper tubing and compression fittings are suggested for most low-pressure installations, particularly for very low pressure and high vacuum applications where leaks cannot be tolerated in the lines.

2 – The volume of pressure lines (internal area and length) will influence the response speed of pressure measurement and control systems. Improper matching of the pressure line to the system will result in objectionably slow response. Line volume is especially critical for systems with controllers, low-pressure and vacuum ranges. The following table details suggested line sizes.

Pressure	Length	Standard Pipe	Copper Water Tube, (type K)
0 to 12 in H_2O	Up to 10 feet	1/4 inch	3/8 inch
0 to 12 in H_2O	10 to 25 feet	1/2 inch	5/8 inch
0 to 12 in H_2O	25 to 60 feet	3/4 inch	1 inch
12" H_2O to 15 psi	Up to 25 feet	1/4 inch	3/8 inch
12" H_2O to 15 psi	25 to 100 feet	1/2 inch	5/8 inch
15 to 50 psi	Up to 50 feet	1/4 inch	3/8 inch
15 to 50 psi	Over 200 feet	1/4 and 1/2 inch*	3/8 and 5/8 inch*
Above 50 psi	Up to 200 feet	1/4 inch	3/8 inch
Above 50 psi up	Over 200 feet	1/4 and 1/2 inch*	3/8 and 5/8 inch*

*Use smaller line for the first 50 feet nearest gauge and large line for the balance of the length. If pressure requires, use extra-heavy pipe of internal diameter similar to standard pipe in table.

3 – When assembling a manifold for gauge, recorder and controller calibration or maintenance, the installation of a gauge cock (valve) and tee-fitting in the line is recommended to permit removal for testing or replacement without shutting down the system. The tee may also be used to facilitate filling or draining the line when required as in the following diagram.

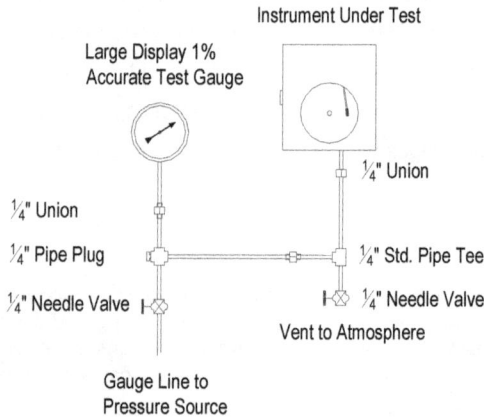

Figure 56 - A manifold arrangement for instrument testing and calibration.

4 – Although occasionally unavoidable, pulsating line pressures are objectionable because:
- Recorders produce an unreadable display,
- Violent pulsations may result in premature rupture of pressure elements,
- Calibration can be negatively influenced,
- Pulsations causes excessive wear on mechanical elements,
- Loop tuning becomes unpredictable and difficult.

Pulsations cannot be completely eliminated but may be minimized by adding a variable restriction (resistance) in series with the line and capacity (capacitance) connected to the pressure line. Just as the installation of a resistance and capacitance in electronic circuits forms a filter to remove unwanted frequencies or voltage variations, the installation of restriction and capacity forms a *damper* to minimize pulsations in pneumatic systems.

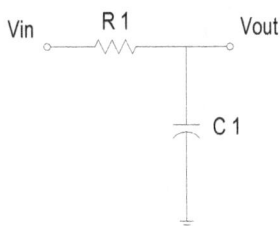

Figure 57 - Lo-pass filters remove unwanted frequencies in electric circuits by placing a restriction in series and a capacitor in parallel with the output.

The simplest pulsation damper is a variable restriction installed in the gauge line several feet from the pressure gauge or instrument. Usually, the volume of the pressure line is enough to form the required capacitance for the damper. If the resulting pulsations are not adequately damped or sufficient capacity does not exist in the pressure line, a "Snubber" can be assembled and installed. The snubber is composed of a variable restriction (resistance) and small capacity of larger diameter pipe to form the pneumatic equivalent to an electrical RC low-pass filter. The following diagram illustrates. For more violent pulsations, commercial dampeners such as glycerin-filled instruments are available.

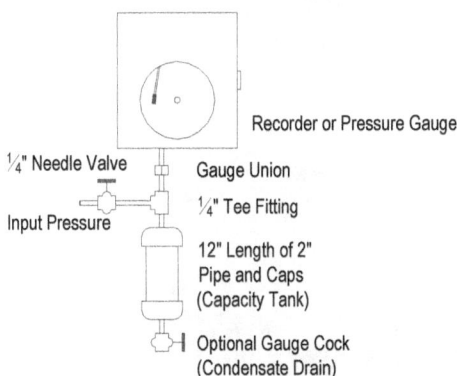

Figure 58 – In a manner similar to the electric circuit lo-pass filter, a damper or "Snubber" minimizes pressure pulsations in instrument lines.

The needle valve should be closed slowly until the instrument indicates a constant pressure yet responds to significant average pressure changes. Generally, the valve can be adjusted to close and the opened ¼ to ½ turn.

Note – After installing ancillary pressure line components always check for proper and accurate operation of the associated measurement and control apparatus. The insertion of needle valves to reduce pulsations, especially with differential pressure instruments, should not be performed haphazardly. Special precautions should be made to ensure the values being displayed or recorded are correct.

5 – Current generation pressure instruments can usually accommodate small amounts of pressure above their rated operating values with minimal effect upon calibration and operation. However, prolonged exposure to *overrange pressure*, or *overpressure*, conditions risks permanent damage to the pressure instrument. A simple means of overrange protection for low-pressure air and gas gauges can be assembled in the field is shown in the following figure. The height of the sealing fluid, L, above the bottom end of the bubble pipe should be about 10% greater than the head in sealing liquid equivalent to the maximum range of the gauge, according to the following formula:

L, inches = (1.1) (gauge range, inches of H_2O) ÷ (Sp Gr of seal liquid)

A temperature correction for the fluid specific gravity may be necessary. A sealing liquid which will not freeze, evaporate or be contaminated by gas should be used. Do not use a viscous liquid, because the pressure element may be damaged before the overpressure bubbles can escape through the liquid. After the gauge and overpressure seal are installed, close the needle valve and pour sufficient sealing liquid through the filling plug to produce the correct value of "L." Open the needle valve just enough to provide a satisfactory speed of response to normal pressure variations.

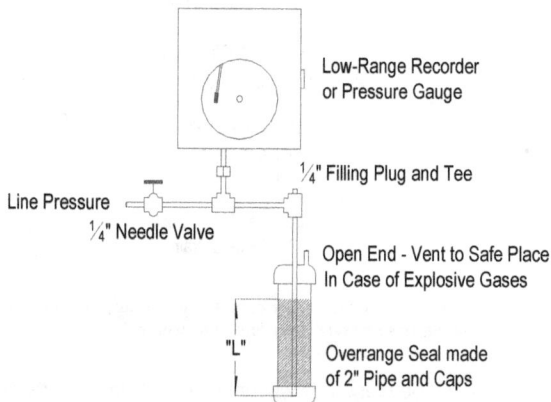

Figure 59 - Overpressure protection for low-range instruments.

Example: To protect an instrument for pressures above 5-psi using Mercury (Hg) as the sealing liquid, first determine the depth in inches of water of the liquid column at the maximum pressure. This is performed by arranging the Pressure=Height*Density equation to solve for Height (Ht).

Since,

$$P = Ht * D_{H2O}$$
$$Ht = P/D_{H2O}$$
$$Ht = 5psi/.0361\#/in^3$$
$$Ht = 138.5 \text{ inches}$$

Subbing Ht into the previous equation,
$$L = (1.1) * (138.5 \text{ inches}) \div 13.6$$
$$L = 11.2 \text{ inches}$$

The overrange seal should be filled with Mercury to 11.2 inches. During operation, the sealing liquid will be displaced from the center pipe into the larger surrounding volume. As the pressure approaches an overrange condition at 5-psi, the Mercury is forced out of the center pipe and into the outer volume of the pipe container. Line pressures of approximately 5.5-psi will result in bubbling through the sealing liquid, into the vent and surrounding atmosphere as the line pressure overcomes the backpressure created by the standing height of Mercury. The following diagram illustrates.

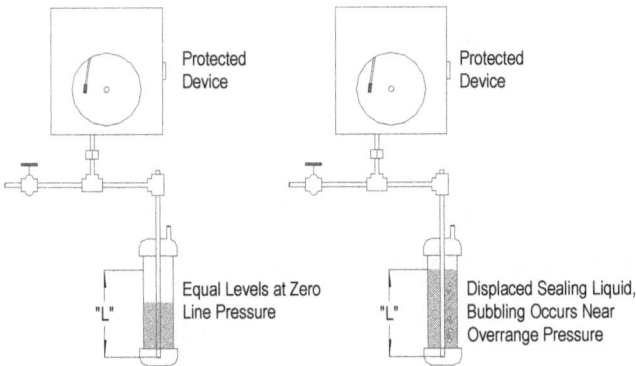

Figure 60 - Overrange protection using a sealing liquid. Operation of the protection component is similar to a manometer.

269

6 – When gauge lines for the measurement of air and gas are installed, provision must be included for the draining and venting of any condensate. Slope instrument airlines at 1 inch per foot of run to drain condensate towards the process connection. When liquids are being measured, equal attention must be given to draining and filling the lines.

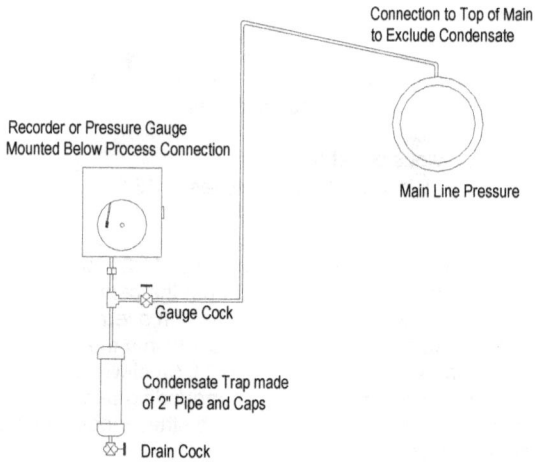

Connection to Top of Main
to Exclude Condensate

Recorder or Pressure Gauge
Mounted Below Process Connection

Main Line Pressure

Gauge Cock

Condensate Trap made
of 2" Pipe and Caps

Drain Cock

Figure 61 - Connection considerations for mounting pressure instruments below the process.

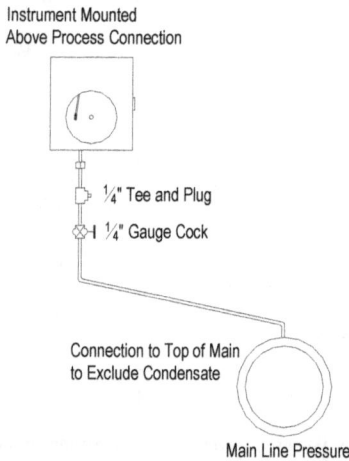

Instrument Mounted
Above Process Connection

¼" Tee and Plug

¼" Gauge Cock

Connection to Top of Main
to Exclude Condensate

Main Line Pressure

Figure 62 - Connection considerations for mounting instruments above the process.

7 – To assure accurate measurement, consider installing barometers in locations where low-pressures or vacuum pressures are measured. Elastic pressure instruments employing Bourdon tubes, bellows, and diaphragms measure the difference between the internal and external pressure. When measuring low-range pressures with uncompensated instruments considerable errors may result from variations in ambient pressure. Gauges installed in rooms ventilated by exhaust and circulating fans may result in significant errors. Opening and closing the case doors of low-range (close to atmosphere) instruments may cause a momentary reading variation that will be especially apparent on recording instruments. The following diagram illustrates the effect of ambient pressure variations.

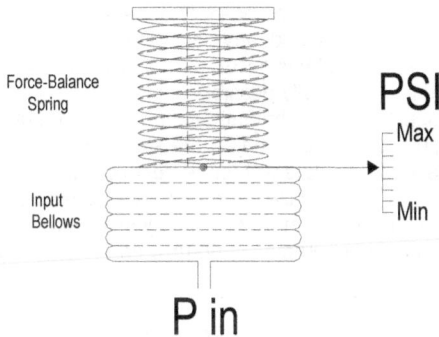

Figure 63 - As the ambient pressure surrounding the input bellows changes, the resulting input force and indicated pressure will also change.

8 – Unless otherwise specified, do not subject absolute pressure sensors and especially mechanical pressure gauges to more than 10 psi above atmospheric pressure; atmospheric pressure is 14.7 psia.

9 – The temperature of fluids entering pressure gauges and measurement instruments should never exceed 150° F. When instruments are measuring steam or other hot condensable vapors, mounting the instrument below the steam line will provide temperature protection by the condensate that forms in the line. The line should be sloped upwards (see previous diagrams) towards the process line or vessel so that gases and vapors may eliminate themselves.

For gauges mounted above the steam or hot condensable vapor connection, a coil pipe seal should be used. Before steam is applied to the coil, enough water should be applied through the filling plug to ensure that the pipe is full. The gauge lines should not be smaller than ½ inch standard pipe size to provide free counter-flow of steam and condensate. Avoid low points in the line that can trap condensate and create head (pressure) errors.

10 – To minimize the possibility of equipment damage and injury, always open and close hand valves and pressure adjustment mechanisms slowly. One cannot always accurately assess the physical condition of industrial hardware by visual inspection. To reduce the possibilities of valve stem breakage, internal component damage or violent process upsets (often called "Bumps"), always proceed cautiously when making field adjustments.

11 – The cooling of system air is inevitable between exiting the compressor and moving to a load. As the air cools in temperature it contracts and moisture is squeezed out from contraction. Moisture can cause pressure components to corrode and even worse, can move dirt and other undesired particulates through the air system. The particulates are certain to appear in the narrow passages within air relays and regulators, and may become stuck holding the pilot valve or exhaust capillary open. Should this occur, air will often be heard "Bleeding" from the exhaust ports. If it is practically possible, a rapid exercising (changing) of the input or output pressure (remove the output connection and rapidly cover and uncover the output pressure line, or adjust the setting knob in and out several times) usually frees the particulate without removal or replacement of the component in question. To minimize the opportunity of this occurrence, always provide accommodations for moisture and dirt by installing air dryers, oilers and filters, and performing the required maintenance on the ancillary systems at regular intervals.

12 – Measurement instruments should be installed in locations to be free from dust, dirt, corrosive fumes, vibrations and extremes of temperature.

Author's note – Application material is continuously under development. Feel free to submit additions, recommendations, corrections and clarifications.

Homework Problems

Pressure Mechanisms

#1 – The following transmitter utilizes diaphragm capsules in place of bellows but operates in an identical manner to the units described previously. Assume an input pressure is applied to $P_{IN\,2}$ of .5 to 2.5 psig, $P_{IN\,1}$ is vented to atmosphere and an output pressure of 3 psi to 15 psi is desired. Assuming all diaphragm areas are equal at 1 square inch, determine the fulcrum setting (d1/d2 ratio) and the spring force to achieve the output pressure span.

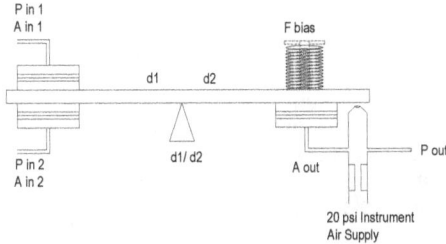

#2 – A differential pressure source of approximately zero to 100 inches of water is to be connected to the following transmitter for conversion into a proportional 3-15 psi signal. Assuming all diaphragm areas are equal at 1 square inch, determine the connection locations for the inputs (high and low differential pressures), fulcrum setting (d1/d2 ratio) and the spring force to achieve the output pressure span.

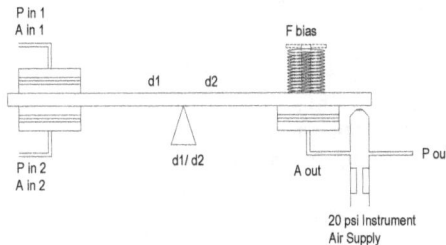

#3 – It is desired to use the following transmitter to reverse a 15-3 psi signal span. As an example, assume a liquid level process transmitter outputs a 15-psi to 3-psi signal as the volume increases. Assuming all diaphragm areas are equal at 1 square inch, determine the connection location for the input signal, fulcrum

273

setting (d1/d2 ratio) and the spring force to achieve the output pressure span (3-15 psi).

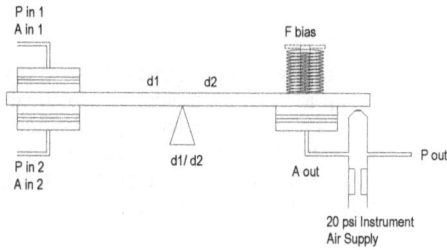

#4 – A single-ended input pressure span of 100 to 500 psig is to be connected to the following transmitter for conversion into a proportional 3-15 psi signal. Assuming the input diaphragm areas are .1 square inch and the output area is 2 square inches, determine the connection location for the input, the fulcrum setting (d1/d2 ratio) and the spring force to achieve the output pressure span.

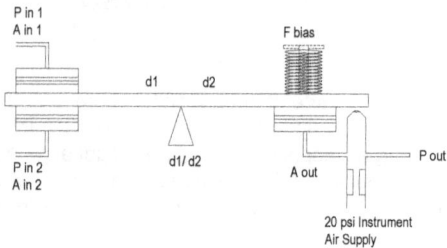

5 – Given the following diagram, assume all diaphragm areas are equal at 1 square inch and a input pressure span of 10 to 50 psi, referenced to atmosphere is applied. Determine the input connections, the d1/d2 setting (gain) and the spring force to establish an out put pressure span of 3 to 15 psi direct acting with the input span.

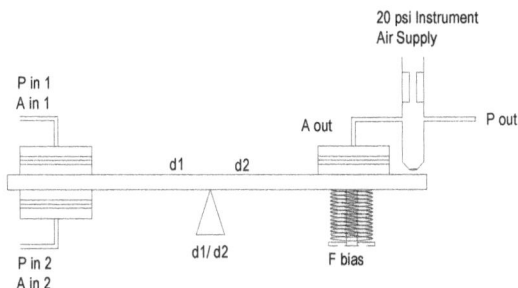

20 psi Instrument
Air Supply

P in 1
A in 1

A out

P out

d1   d2

d1/ d2

P in 2
A in 2

F bias

6 – The transmitter of the previous problem is to be converted into a reverse-acting unit. Using the same input pressure span of 10-50 psi, show the new input connections and determine the gain (d1/d2) and spring force to establish an output of 15 to 3 psi, reverse-acting.

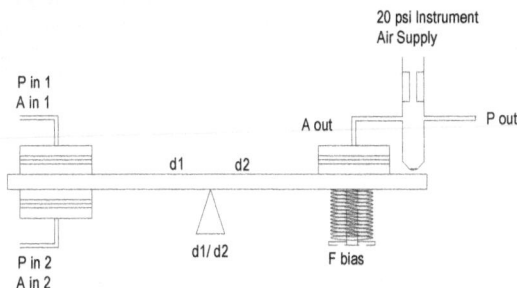

20 psi Instrument
Air Supply

P in 1
A in 1

A out

P out

d1   d2

d1/ d2

P in 2
A in 2

F bias

7 – Some genius designed the following transmitter as a new generation in pressure transmitters. The Purchasing Department in your plant, not being able to pass up a bargain, has acquired a couple of these units and now you must find an application for them. Determine the output pressure equation and from the equation, determine a possible application for the transmitter. Are there any application limitations?

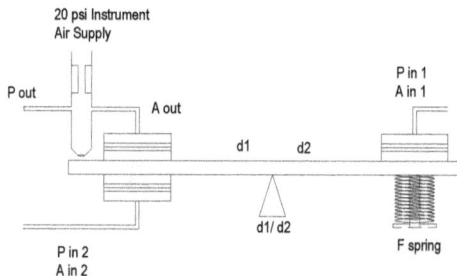

8 – Given the following transmitter, assume all diaphragm areas are equal at 1 square inch. An input pressure span of 20 to 150 inches of water is applied to the direct acting input to generate a 3-15 psi output pressure span. Determine the d1/d2 and spring force values if the distance between the biasing spring and the center of the output diaphragm is 1 inch. If beneficial, assume the beam length is eight inches.

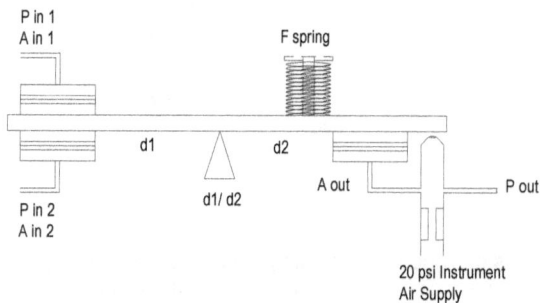

Regulators

9 – A regulator uses a .5 in^2 output diaphragm to establish a 20-psi output pressure. Determine the required amount of regulating spring force to be applied to the upper portion of the output diaphragm.

10 – The regulator of the previous problem is adjusted to apply an additional 10 pounds of spring force. Determine the output pressure change.

11 – A load change causes an output pressure decrease of 1.2 psi. Using a diaphragm area of .8in^2 determine the resulting spring force change.

Air Relays

12 – An input pressure span of 3-15 psi is to be amplified to 6-30 psi. If the input diaphragm of the air relay to be used is 1in^2, determine the required gain ratio and corresponding output diaphragm area.

13 – An input pressure span of 10-100 psi is to be reduced to 3-15 psi. If the input diaphragm of the air relay to be used is 1in^2, determine the required gain ratio and corresponding output diaphragm area. (Hint: Recall that pneumatics is not a precise science and that only components of 2:1, 4:1 and 6:1 are commercially available. Choose the component that results in getting as close to the desired output range as possible.)

14 – Determine the output pressure range in the previous problem using a 6:1 reducing air relay.

15 – A biasing relay is used to elevate a 3-15 psi pressure span into 10-22 psi. If an output diaphragm area of .75in^2 is used, determine the required biasing spring force.

16 – A biasing relay is used to suppress a 3-15 psi pressure span into 0-12 psi. If an output diaphragm area of .6in^2 is used, determine the required biasing spring force.

17 – Determine the output pressure spans of each of the following components if the biasing relay is set to subtract 18 psi for a transducer input of 1V to 9V.

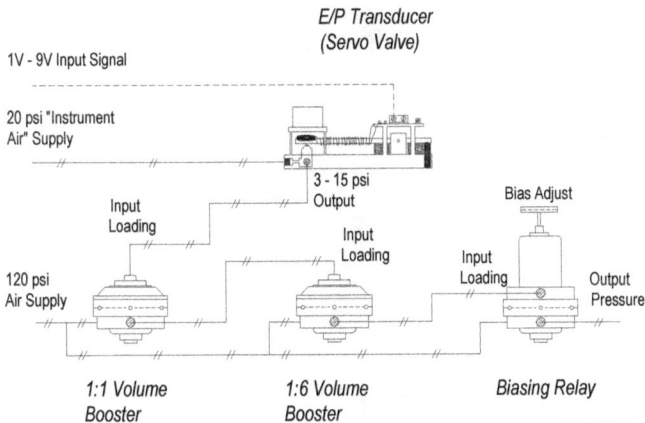

E/P Transducer
(Servo Valve)

1V - 9V Input Signal

20 psi "Instrument Air" Supply

3 - 15 psi Output

Bias Adjust

Input Loading

Input Loading

Input Loading

120 psi Air Supply

Output Pressure

1:1 Volume Booster

1:6 Volume Booster

Biasing Relay

18 – The 1:6 volume booster and biasing relay in the previous diagram are to be replaced with a single unit capable of both amplifying and biasing. However, the biasing specifications indicate that the unit can only subtract a maximum of 15 psi. Using a 3-15 psi signal from the 1:1 and assuming a desired output pressure of 0-72 psi, determine if the new unit will function by finding the required relay gain, the input diaphragm area and the required bias spring force. Assume the output diaphragm area is 1in^2.

Chapter 6 – Level, Density and Buoyancy

This chapter discusses the various purposes and mechanisms for level measurement. Density, specific gravity, hydrostatic pressure and buoyancy are introduced. Hydrometers are examined and torque-balance transmitter theory is re-visited. The chapter is under construction but current topics in order of appearance include,

Objectives:

Upon completion of this chapter, you should be able to:
- Explain the operational characteristics of common level sensors
- Explain level sensor applications involving density and buoyancy
- Interpret the operational specifications of common level sensors
- Apply routine and unique level sensor signal conditioning concerns
- Suggest calibration techniques for common level sensors

Introduction

Level measurement is a common process variable with a variety of sensing and transmitting apparatus available. In many processes product quality is highly contingent upon accurately measuring and controlling liquid and solid material level. Measuring and controlling level is especially important in continuous, semi-continuous and batch processes where vessel level can influence the pressure or material flow in and out of a container.

Figure 1 - As can be seen from the number of tanks in this photo, level is also a common non-industrial measurement.

Level is measured for three primary reasons - to determine *position*, to determine material *volume* or to determine material *weight*. Position measurement determines the level to be at or below a fixed point. Position determination disregards considerations of density, pressure, temperature or other influencing factors and focuses solely on the presence or absence of material at a given height within a container. Volume determination uses a level measurement to determine the amount of three-dimensional consumption of space a material exhibits when stored. Results of cubic feet, gallons, quarts, liters, bushels and barrels are common from volumetric level measurement. Weight determination uses a level measurement to assess the weight of stored material. Tons, pounds, kilograms and ounces result from using level measurement for weight determination.

Figure 2 - Level measurement for position determination provides an opportunity to refill this seed cart befores the storage tanks run empty.

The techniques used to determine material level can be organized into several categories. One category of level measurement could be assembled around the basic sensing techniques. In which case, most of the available sensing components would be grouped according to the following.

-Level determination by measuring hydrostatic or pneumatic pressure

-Level determination by measuring differential pressure

-Level determination by measuring the position or movement of a float

-Level determination by strain gauge related measurements

-Level determination by radiation absorption measurement

-Level determination by heat transfer

-Level determination by acoustic, magnetic or light reflection

Figure 3 - Level measurement can provide interesting challenges. How would you inventory the volume or weight of material in each of these piles? (It's actually done by measuring material flow.) Photo courtesy of River Stone Group, Moline, IL.

Level measurement may be further categorized by placing all level related measurements into two categories. Level systems that utilize the surface itself to yield a measurement are categorized as *Direct Methods* of level measurement. Examples of Direct Methods include the sight glass, the float, heat transfer and electrical conductivity. All require the material level itself to effect a measurement. Other types of level measurement imply or infer liquid level through manipulation or measurement of a related physical variable. These types of measurements are referred to as *Indirect Methods* or *Inferential Methods* of level measurement. Examples of Inferential Methods are hydrostatic pressure, strain gauge or load cell, radiation instruments and differential pressure measurement.

An awareness of Direct and Inferential Methods of level measurement is invaluable during measurement system design and troubleshooting. As always, when specifying or diagnosing a control system or system components, the entire manufacturing process must be understood to the greatest extent possible. If not, the end-result could be expensive and disastrous. In the case of level measurement, the processes, vessels and materials must be considered in conjunction with the desired measurement results. A symmetrical process vessel may be able to utilize a simple float mechanism when sensing liquid, but would yield less than desirable results if measuring crushed rock.

Physical Principles

A study of level measurement begins with an investigation of the physical characteristics materials exhibit when contained. Liquids conform to the shape of the containing vessel however solids usually do not. When amassed, solids conform according to an angle of repose. Angle of repose describes the maximum angle that dry (solid) material can be stacked before sliding. Assuming agitation, vibration or other material conditioning is not applied; dry matter will accumulate in a storage vessel according to the specific material's angle of repose. Knowing this limits the types of level sensing apparatus that can be used to detect material level, volume or weight.

Density

An understanding of weight density (D) is essential to a study of level measurement. Weight density, or simply density, is a comparison of a material's weight to its volume. Measured in pounds per cubic foot, weight density is the reason materials separate and become buoyant (float) when mixed. Of similar importance are the variations of the density equation as indicated following.

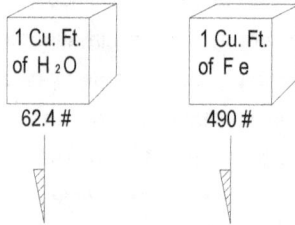

Figure 4 - Iron is about 8 times heavier (more dense) than water.

$$Density = Weight \div Volume$$

Or,

$$D = W/V$$

Rearranging,

$$Weight = Volume * Density$$

Or,

$$Wt. = V * D$$

And,

$$Volume = Weight \div Density$$

Or,

$$Vol. = Wt./ D$$

The following table provides a listing of common materials and their associated densities.

Material	Density		Specific Gravity (Relative to H_2O)
	Grams/Cm^3	#/Ft^3	
Water (H_2O)	1.000	62.4	1.000
Sea Water	1.026	64.0	1.026
Carbon "Tet"	1.594	99.2	1.59
Glycerin	1.261	78.6	1.261
Ethyl Alcohol	.789	49.3	.79
Mercury	13.65	848.6	13.60

Gasoline	.68	42.4	.68
Ice	.92	57.0	.92
Aluminum	2.70	169.0	2.70
Iron	7.85	490.0	7.85

Figure 5 - Common material densities.

Example Problem #1:

A container measuring 3 feet deep, 2 feet wide and 5 feet tall will be storing ethyl alcohol. Determine the weight of the stored material.

Solution:

Knowing the weight density of ethyl alcohol, if the volume can be determined the weight can be found by applying the Volume * Density equation.

$$Volume = Width * Depth * Height$$
$$Volume = 2ft * 3ft * 5ft$$
$$Volume = 30 \text{ cubic feet}$$

Since the density of ethyl alcohol is 49.3 #/Ft³,

$$Wt. = Vol. * D$$
$$Wt. = 30Ft^3 * 49.3\#/Ft^3$$
$$Wt. = 1479\#$$

Example Problem #2:

Determine the weight of 20 gallons of gasoline.

Solution:

Given the density of gasoline is 42.4 pounds per cubic foot, and a cubic foot contains 7.48 gallons, the weight of 20 gallons can be determined by first determining the number of cubic feet in 20 gallons.

Since,

$$7.48 \text{ gallons} = 1 \text{ Ft}^3$$
$$20 \text{ gallons} = (20/7.48) * 1 \text{ Ft}^3$$
$$20 \text{ gallons} = 2.6738 \text{ Ft}^3$$

Once the volume is determined in cubic feet, the weight can be found by applying the Volume * Density equation.

$$\text{Wt.} = \text{Vol.} * D$$
$$\text{Wt.} = 2.6738 \text{Ft}^3 * 42.4 \#/\text{Ft}^3$$
$$\text{Wt.} = 113.37\#$$

Example Problem #3:

An aluminum ingot is found to weigh 20,280 pounds. If the ingot is 3 feet wide and 2 feet tall, determine the length.

Solution:

Knowing the density and the weight of the aluminum ingot, the volume can be determined using the Weight * Density equation.

$$\text{Vol.} = \text{Wt.}/ D$$
$$\text{Vol.} = 20,280\# \div 169 \#/\text{Ft}^3$$
$$\text{Vol.} = 120 \text{ Ft}^3$$

Since volume equals area * length, or width * height * length, the length can be determined by dividing the volume by provided area dimensions.

Since,

$$\text{Volume} = \text{Width} * \text{Height} * \text{Length}$$

$$\text{Length} = \text{Volume} \div (\text{Width} * \text{Height})$$
$$\text{Length} = 120 \text{ Ft}^3 \div (3\text{Ft} * 2\text{Ft})$$
$$\text{Length} = 20 \text{ Ft}$$

Hydrostatic Head

Anyone that has been swimming to any depth has an appreciation of hydrostatic pressure concepts. Simply put, hydrostatic pressure is the result of a liquid's weight distributed over an area, which physically and mathematically defines pressure.

$$P = F/A, \text{ psi or kg/cm}^2$$

Where,

P = Pressure

F = Force (or weight)

A = Area

Hydrostatic pressure is specifically defined as pressure created by a liquid column and as such, is often referred to as a "Head of pressure," or simply "Head." The pressure "Head" expression is the result of using a measured pressure above a point in a vessel to interpret liquid level, volume or weight. "Head" is a reference to the liquid above the point of measurement. Given this, "Head" has become a reference to determining and expressing pressure as a column of liquid, or more specifically a height and implied density such as inches water (H_2O) or Mercury (Hg). Listen closely to how television weathermen express atmospheric pressure; it's usually similar to "30 inches of Mercury," a reference to a column of liquid or, a height and density. Pressure is defined and commonly computed by dividing force by area but as is demonstrated following, pressure can also be computed by multiplying height by density.

Since,

$$\text{Pressure} = \text{Force} / \text{Area}$$

Given that Weight is a Force,

$$\text{Pressure} = \text{Weight} / \text{Area, in psi or kg/cm}^2$$

Rearranging the Density equation to solve for Weight,
Since,

$$\text{Density} = \text{Weight} / \text{Volume}$$
$$\text{Weight} = \text{Volume} * \text{Density}$$

Substituting the Weight equation into the pressure equation for force,

$$P = F / A$$
$$P = \text{Weight} / \text{Area}$$
$$P = (\text{Volume} * \text{Density}) / \text{Area}$$

Since,

$$\text{Volume} = \text{Area} * \text{Height,}$$

Substituting for Volume,

$$\text{Pressure} = (\text{Area} * \text{Height}) * \text{Density} / \text{Area}$$

Canceling areas yields,

$$P = Ht * D, \text{ in psi or kg/cm}$$

This equation directly represents the pressure "Head" reference and the weatherman's reference to "30 inches of Mercury." Assuming uniform and constant material density, it can be seen from the derivation the pressure above a certain measurement location is directly proportional to the height of the liquid, and the density of the stored material. The pressure equation also applies to some dry material as long as the material seeks a uniform level and maintains constant density.

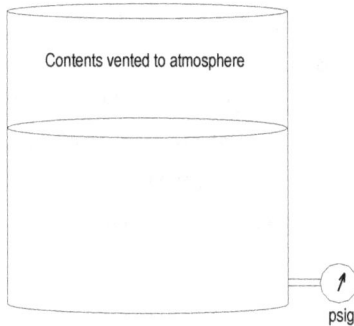

Figure 6 - A simple and effective means of inferential liquid level measurement.

Example Problem:

A pressure gauge mounted at the bottom of a ventilated tank is used to measure the level of methyl alcohol. The gauge reads approximately 12 psi. Determine the level of alcohol in the tank.

Solution:

To determine the head of liquid above the pressure gauge the P=Ht*D equation should be rearranged to solve for height.

Since,

$$P = Ht * D$$
$$Ht = P/D$$

Substituting the given values and converting density from $\#/Ft^3$ to $\#/in^3$ yields,

$$Ht = 12psi \div (62.4\#/Ft^3 \div 1728in^3/Ft^3)$$
$$Ht = 12psi \div (.0361\#/in^3)$$
$$Ht = 332.4 \text{ inches} = 27.7 \text{ Feet of alcohol}$$

Buoyancy

Anyone that has attempted to hold a beach ball underwater appreciates the effects of an upward force called buoyancy. Simply put, buoyancy explains why some objects float and others don't. According to Archimedes, <u>a body partially or entirely submerged will experience an upward buoyant force equal to the *weight of the fluid displaced.*</u> When the italicized words of the previous description are placed into an equation, buoyancy becomes a force equal to the weight of a quantity of fluid (note - fluid represents a liquid or gas). A better description of buoyancy resides in an investigation of the weight density (D) equation.

Since,

$$\text{Density} = \text{Weight} \div \text{Volume}$$

Solving for Weight,

$$\text{Weight} = \text{Volume} * \text{Density}$$

Since weight is a <u>force</u> measured in pounds or kilograms, substituting Force for Weight yields,

$$\text{Force, } F_B = \text{Volume} * \text{Density}$$

Specifically,

$$\text{Buoyant Force, } F_B = \text{Volume} * \text{Density}$$

Or,

$$F_B = V * D,$$

Measured in pounds or kilograms

Where,

F_B = the upward force applied to the submerged object

Volume, V = the volume submerged in the liquid or gas

Density, D = the weight density of the liquid or gas

Therefore, buoyant force equals the volume submerged multiplied by the density of the fluid the body is within. According to this definition, all objects contained within a liquid or gas experience an upward, buoyant force. Helium balloons, human beings, objects that float in liquids and objects that sink – all objects placed into any liquid or gas experience an upward force. The upward force varies in proportion to the amount of the object in the fluid and the density of the fluid.

The concept of buoyancy can be reduced to an examination of the effects of combined or mixed material densities. Heavier (more dense) materials sink; lighter (less dense) materials rise to the surface. Helium rises in air because helium is less dense than air. Petroleum products like oil and gasoline float on water because water is heavier (more dense) than petroleum. In the final analysis, materials of lesser density are always forced upwards when mixed with materials of greater density, and downwards when mixed with materials of a lesser density.

Example Problem:
Assume an object of 1 Ft³ volume and weighing 100 pounds in air is placed into water. Will the object sink or float? Determine the object's apparent weight in water.

Solution:
Given that the density of water is 62.4 #/ft³ and the object's density is 100 #/ft³, the object will sink due to its greater density. To determine the apparent weight of the object in water, the forces acting on the object must be determined. Using the following figure to assist yields a weight in air of 100#. The object's weight does not diminish when placed in water although the object appears to weight less. The apparent loss of weight is due to the upward buoyant force applied by the density of the water to the object.

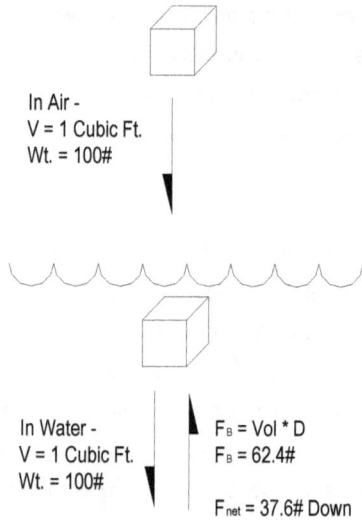

In Air -
V = 1 Cubic Ft.
Wt. = 100#

In Water -
V = 1 Cubic Ft.
Wt. = 100#

F_B = Vol * D
F_B = 62.4#

F_{net} = 37.6# Down

Figure 7 - Objects appear to lose weight in water when in fact, the weight remains the same and the buoyant force acting on the object increases due to the increase in density when moving the object from air into water.

The resulting net (total) force, F_{Net}, on the object is determined by subtracting the buoyant force from the object's weight. The net force can also be used to determine the direction the object will move if not suspended, or the amount of force required by the suspension system.

Given,

$$\downarrow F_{WT} = 100\#$$
$$Vol = 1Ft^3$$
$$D_{Fluid} = 62.4\ \#/Ft^3$$

Finding the buoyant force, F_B,

$$\uparrow F_B = Vol * D$$
$$\uparrow F_B = 1\ Ft^3 * 62.4\ \#/Ft^3$$
$$\uparrow F_B = 62.4\#$$

"Summing" the individual forces results in the net force, F_B,

$$F_{Net} = \downarrow F_{WT} - \uparrow F_B$$
$$F_{Net} = 100\# - 62.4\#$$
$$F_{Net} = 37.6\# \downarrow$$

As can be seen from the resulting force, the object will sink if not suspended due to the larger downward force created by the object's weight. This explains why objects *appear* to weigh less in water than in air.

Figure 8 - Close inspection of the float in this photograph yields a minimum specific gravity (weight density) required of the liquid for the object to remain buoyant. With permission, courtesy K-Tek.

Volume Determination – Level Measurement
Volume determination utilizes a level measurement to determine the volume of liquid within a vessel or container. Volume determination assumes the storage

vessel is of uniform cross-sectional area. By measuring the material height, and knowing the containing vessel's area, material volume can be determined using the following equation.

$$V = A*H$$

Or,

$$V = A * L$$

Units of in^3, Ft^3, Gallons, etc

Where,

    V = Volume

    A = Area

    H = the measured material height or level (L)

Example Problem:

A level transmitter outputs a 4-20 mA signal over level span of 2 to 30 feet. The tank is 10 feet in diameter. Determine the liquid volume in gallons if the transmitter output is 14 mA.

Solution:

Using V = A*H, the tank level needs to be determined from the transmitter output of 14 mA. The current span (difference between end points) of 16 mA corresponds to a level span (again, difference between end points) of 28 feet. 14 mA is ten-sixteenths (5/8) of the distance between 4 mA and 20 mA, and represents 5/8 of the level span between 2 and 30 feet. Five-eighths of 28 feet is 17.5 feet above the minimum level of 2 feet. Therefore, 14 mA corresponds to a tank level of 17.5, or 19.5 feet above the bottom of the tank.

Since,

$$Area = \pi \, r^2 = \pi d^2/4$$

A tank diameter of 10 feet equates to an area of 78.54 square feet. With a tank level of 17.5 the tank volume becomes,

$$V = A*H$$

$$V = 78.54 \text{ ft}^2 * 17.5 \text{ ft}$$
$$\text{Volume} = 1374.45 \text{ ft}^3$$

Since 1 ft^3 = 7.48 gallons,
$$\text{Volume} = 1374.45 \text{ ft}^3 * 7.48 \text{ gallons/ ft}^3$$
$$\text{Volume} = 10280.886 \text{ gallons}$$

Minimum level is rarely an empty tank. An empty tank can result in questionable transmitter output and unpredictable control system operation. Likewise, liquid level calibration usually begins with a small amount of material level in the tank.

The previous example demonstrates that stored material volume can be determined from a measurement of level when the tank is symmetrical and temperature-density effects are ignored. If temperature induced changes in density are a significant concern, the temperature or density will need to be determined (measured) and factored into the previous volume equation.

The example problem also demonstrates how level can be continuously monitored while yielding and displaying a quantity of volume. Inferential values are a common measurement result and can be easily implemented using scale factor and offset functions within available software-based technology. The calibration and display scaling concepts are demonstrated in the lab experiment at the end of the chapter.

Volume Determination – Pressure Measurement
Volume can also be determined from hydrostatic pressure where the containing vessel's area is multiplied by the quotient of the measured pressure and the liquid density.

Since,
$$\text{Volume} = \text{Area} * \text{Height}$$

And,

$$\text{Height} = \text{Pressure/Density}$$

Substituting,

$$V = A*(P/D)$$

Units of in^3, Ft^3, Gallons, etc

Where,

V = Volume

P = Hydrostatic Head Pressure

A = Area

D = Weight Density for the existing pressure conditions

Example Problem:

Head pressure is measured as 10 mA from the bottom of an 8-foot diameter water tank. The 4-20 mA transmitter is calibrated over a 0 psi to 20 psi range of pressure. Determine the volume in gallons of water in the tank.

Solution:

A diameter of 8 feet equated to an area of 50.26 ft^2. Transmitter current of 10 mA is 6/16, or 3/8, or 37.5 % of the 4 mA to 20 mA range. 37.5% of the 0 psi to 20 psi calibrated range of the pressure transmitter is 7.5 psi.

Using the volume equation and subbing for area, pressure and the weight density of water,

$$V = A*(P/D)$$

$$\text{Volume} = (50.26 \text{ ft}^2) * (7.5 \text{ psi} \div 62.4 \text{ \#/ft}^3)$$

However, the units of pressure (psi) and density (#/ft^3) are not compatible and a conversion must be applied. Inches must be converted to feet or vice versa. To convert psi into pounds per square foot (psf), multiply the pressure by 144 in^2/ft^2. To convert weight density in units of #/ft^3 into units of #/in^3, divide the density by 1728 in^3/ft^3. (Remember that weight density is weight per unit volume. Since volume is three dimensional, conversion between in^3 and ft^3 involves a 12*12*12 computation of 1728 in^3/ft^3.)

Continuing the volume computation and converting psi into psf,

$$V = A*(P/D)$$

Volume = (50.26 ft^2) * (7.5 psi *144 in^2/ft^2 ÷62.4 #/ft^3)

Volume = (50.26 ft^2) * (17.31ft)

Volume = 869.89 ft^3

Converting to gallons,

Volume = 869.89 ft^3 * 7.48 gal's/ ft^3

Volume = 6506.78 gal's

This problem is another example of inferential measurement, where one quantity is measured and used to provide information about a different yet related quantity. This type of measurement is possible because of the proportional relationship that exists between the two quantities of volume and pressure. Similar relationships exist in all areas of instrumentation and process control and often provide a less expensive and more effective means of process measurement and control.

Weight Determination – Pressure Measurement
If the weight of the liquid (and some solid material) is measured without concern for temperature-induced changes in density, then this is a determination not of level but of an equivalent level. The weight or force of the contained material on the bottom of the container or supporting members is equal to,

Since,

$$Force = Area * Pressure$$

And since,

$$Weight = Force$$

Substituting,

$$\text{Weight} = \text{Area} * \text{Pressure}$$

Since,

$$\text{Pressure} = \text{Height} * \text{Density}$$

Substituting,

$$\text{Weight} = \text{Area} * \text{Height} * \text{Density}$$

Therefore,

$$W = A*H*D$$

Or,

$$W = A*P$$

Units of pounds, kilograms, etc.

Where,

  W = Weight of the material and the container

  P = Hydrostatic Head Pressure

  A = Cross-sectional Area

  H = Material Height (Level)

Example Problem:

A pressure transmitter is calibrated to output 4-20 mA over a pressure span of 1 psi to 5 psi when used in conjunction with a 12' diameter circular tank. Given a transmitter output of 10.5 mA, determine the material weight in the tank.

Solution:

Since the transmitter is calibrated for 4-20 mA over 1 psi to 5 psi, a current of 10.5 mA is 6.5/16 or 40.625% of the full output range. 40.625% of the 1 psi to 5-psi range is 2.625 psi. The area of the storage vessel is given as a 12' diameter circular tank, or 113.1 ft². Determining the material weight from the equation yields,

$$W = A*P$$
$$W = 113.1 \text{ ft}^2 * 2.625 \text{ \#/in}^2$$

Including the in² to ft² conversion factor,

$$W = 113.1 \text{ ft}^2 * 2.625 \text{ \#/in}^2 * 144 \text{ in}^2/\text{ft}^2$$
$$W = 42750.793 \text{ \#}$$

Weight Determination – Level Measurement

Assuming the material density remains constant, weight can be inferred from the level measurement using the following equation.

$$W = A*H*D$$

Units of pounds, kilograms, etc.

Where,

> W = Weight of the material and the container
>
> A = Cross-sectional Area
>
> H = Material Height in storage vessel
>
> D = Weight Density

Example Problem:

A pressure transmitter is calibrated to output 4-20 mA over a pressure span of 0" to 150" w.c. (Water column) when used in conjunction with a 10' by 16' rectangular tank. Given a transmitter output of 16 mA, determine the material weight in the tank.

Solution:

Since the transmitter is calibrated for 4-20 mA over 0" to 150" of H_2O, a current of 16 mA is 12/16 or 75% of the full output range. 75% of 0" to 150" is 112.5" of water. The area of the storage vessel is given as 10' by 16' or 160 square feet. Determining the material weight from the equation yields,

$$W = A*(H*D)$$
$$W = (160 \text{ ft}^2 * 112.5" * 62.4 \text{ \#/ft}^3)$$

Before continuing the computation the units of inches and feet must be made compatible. Dividing the 112.5" level by 12 in/ft yields the following.

$$W = (160 \text{ ft}^2 * 112.5" \div 12 \text{ in/ft} * 62.4 \text{ \#/ft}^3)$$
$$W = (160 \text{ ft}^2 * 9.375 \text{ ft} * 62.4 \text{ \#/ft}^3)$$
$$W = 93,600 \text{ pounds}$$

As can be seen in the preceding, in addition to level, weight or volume measurement techniques, weight density and buoyancy topics appear in a discussion of level measurement. Floats are a common means of detecting material level. At some point, one should ask why some objects float and others sink. The answer appears in the subject of density. As will be demonstrated later, even objects that sink are capable of measuring level.

Level Transmitter Calibration
Process transmitters (also called secondary elements) convert the output signal from a sensor (also called a primary element) into a standard electronic signal span that is compatible with controllers, recorders, display devices and other automation hardware and components.

Currently, the most common transmitter output standard is the 4-20 milliamp current loop signal. The current loop is used to transmit signals over long distances without concern for signal loss (called attenuation) due to wire resistance. The current loop forms a series circuit between the transmitter, the controller, the DC power supply (usually 12-45 VDC) and any other components requiring information about the process being measured. Since current is constant throughout a series circuit, the current loop signal can be transmitted at a constant value throughout the signal carrying wires independent of distance.

Figure 9 - A gauge-type (single input connection) level transmitter with a local display providing tank level. This transmitter is a Highway Addressable Remote Transducer (HART) type. HART transmitters can be calibrated from any location where the output signal can be accessed.

Level transmitter calibration involves setting the minimum and maximum output current values (of say, 4 mA and 20 mA) to specific minimum and maximum tank levels. Two calibration adjustments labeled Zero and Span are available on traditional transmitters. As demonstrated in the following figure, the Zero adjustment is set to 4 mA when the tank material is at the desired minimum measured value of level. Subsequently, the Span adjustment is set to 20 mA when the tank material is at the desired maximum measured value of level. Each adjustment corresponds to the effects of varying the slope (Span) and Y-axis intercept (Zero) of a linear Y=mX+b graphical relationship. The Y-axis intercept is often referred to as *offset* or *output offset* since it represents the amount the output span is offset from zero. It is suggested that one study the following figure in detail as the provided concepts apply not only to level transmitters, but also to calibration of all process transmitters.

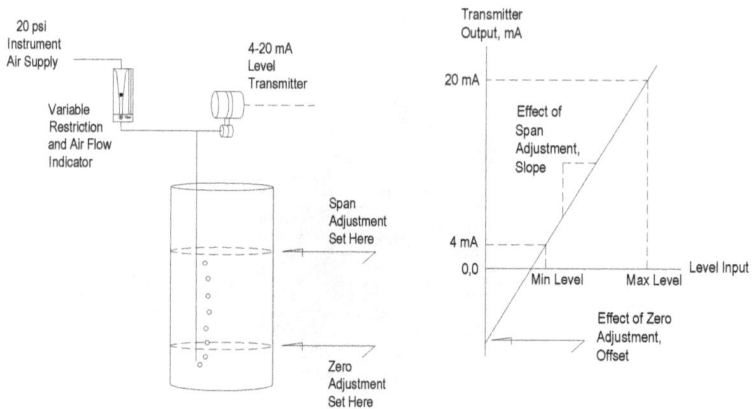

Figure 10 - Traditional transmitter calibration requires setting the Zero and Span adjustments at corresponding tank levels (left). The technique directly relates to linear, Y=mX+b relationships (right).

More recently, HART (Highway Addressable Remote Transducer) transmitters are being employed in process measurement and control facilities. HART transmitters perform calibration by digitally communicating with a handheld programmer or HART-modem over the two signal (and DC power) carrying leadwires. A handheld HART communicator or computer-based HART modem is required to establish the communication link to the transmitter. HART-based calibration eliminates the traditional span and zero adjustments and allows for direct entry of the levels being measured. As an example, assume a HART transmitter is to output a 4-20 mA signal between water levels of 30 inches and 150 inches of water. Using the HART communicator, the minimum level of 30 inches is entered as the 4 mA output value and the maximum level of 150 inches is entered as the 20 mA value. The transmitter will then output 4 mA at minimum level (30 inches), 20 mA at maximum level (150 inches), 12 mA at 90 inches, 8 mA at 60 inches, 16 mA at 120 inches and so forth. The HART communicator can also be used to diagnose possible transmitter problems should the transmitter be suspected of improper operation.

HART instruments are capable of providing digital communication to microprocessor-based "Smart" analog process control components. Developed by Rosemount during the 1980's, HART was later offered as an open standard and is currently organized and promoted by the HART Communication Foundation that includes over 100 member companies. Originally intended to allow convenient calibration and operational adjustments of analog process transmitters; HART was the first bi-directional digital communication technique for process transmitters that did not interfere with the analog 4-20 mA output signal. The result was an instrument that would allow the process to continue operating during transmitter communication. As the current de-facto standard for data communication in "Smart" analog field instruments, HART is found in all chemical processing, power generating and related process industries. Well over 5 million HART nodes have been installed currently. Among the many HART products currently available are analog process transmitters, digital-only process transmitters, multi-variable process transmitters, process receivers (valves), local (field located) controllers, modems, interfaces, gateways, isolators, calibrators and software packages.

Output Interpretation
When the specific minimum and maximum calibration levels are known, the transmitter output current can be easily related to the measurement level using percentages. Linear, Y=mX+b mathematics can also be used but will require a bit more effort. As an example, consider a calibration example where 4 to 20 mA corresponds to 20 inches to 60 inches of water level. Assume the transmitter output is measured as 12 mA and it is desired to know the associated tank level. First, determine the percentage of the measured current within the 4-20 mA span by subtracting the minimum value (4 mA) from the measured value (12 mA) and dividing the difference by the current span (16 mA). Multiply the result by 100%. Second, multiply the resulting percentage by the span of the measured level, and then add the minimum level value. (Note: span is defined as the difference between end points; the level span is 60" minus 20", or 40 inches.)

First,

% mA = [(Measured mA − Minimum mA)\(Maximum mA − Minimum mA)] * 100%

% mA = [(12 mA − 4 mA) ÷ (20 mA − 4 mA)] * 100%

% mA = [(8 mA) ÷ (16 mA)] * 100%

% mA = (.5) * 100% = 50%

Therefore, 12 mA is 50% of the 4-20 mA output span. This corresponds to a level that is 50% of the calibrated span of 20 to 60 inches of water.

Second,

Level in inches = [50% * (60" - 20")] + 20"

Level in inches = [50% * (40")] + 20"

Level in inches = (20") + 20"

Level in inches = 40" at 12 mA of transmitter current

Mathematically, the tank level determination process is an application of the Y=mX+b relationship between transmitter output and level input. To demonstrate, assemble a linear graph with tank level on the input (X-axis) and transmitter current on the output (Y-axis). The resulting graph should appear as in the following diagram.

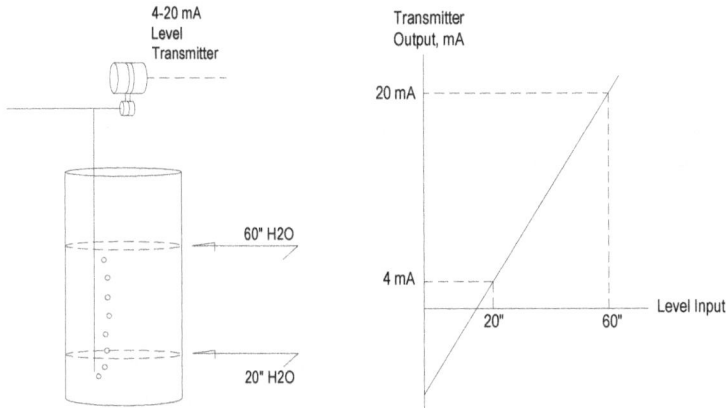

Figure 11 – Level transmitter calibration is yet another application of linear Y=mX+b concepts.

To determine the tank level corresponding to a given transmitter current, first recognize that it is desired to determine an input quantity (level) given an output quantity (current). This can be observed in the graph of the preceding figure. The linear graph is described by Y=mX+b or using the graphical quantities, output current (Y) equals transmitter gain (slope, m) multiplied by the input level (X) plus the output offset (intercept, b). Written mathematically the relationship appears as follows.

$$Y=mX+b$$

Or,

$$\text{Current in mA} = (\text{Gain} * \text{Level}) + \text{Offset}$$

The transmitter gain is then determined from the input and output quantities.

$$\text{Gain} = \text{Slope} = \Delta Y/\Delta X = (20 \text{ mA} - 4 \text{ mA})/(60" - 20")$$
$$\text{Gain} = \text{Slope} = (16 \text{ mA} / 40")$$
$$\text{Gain} = .4 \text{ mA/inch of water}$$

The transmitter offset is determined from the input and output quantities. The Y=mX+b equation is rearranged to solve for the Y-axis intercept, b.

Offset, or Y-axis intercept, b = Y-mX

Offset = Current – (Gain * Level)

Choosing corresponding input and output values from the graphed X-axis (60")
and Y-axis (20 mA) quantities and inserting the previously computed gain yields,

Offset = 20 mA – (.4 mA/inch * 60 inches)

Offset = 20 mA – (24 mA)

Offset = -4 mA

Once the gain and offset are known they can be substituted into the rearranged
Y=mX+b equation along with the output current value. The resulting equation to
allow for determining the input given the output becomes X=(Y-b)/m. Using the
previous example of 12 mA yields the following.

X=(Y-b)/m

Input = (Output – Offset) ÷ Gain

Input Level = [(12 mA – (-4 mA)] ÷ (.4 mA/inch)

Input Level = (16 mA) ÷ (.4 mA/inch)

Input Level = 40 inches

In the field, the percentage approach is much more rapid. Often the desired value
of input or output need only be approximated to satisfy the investigation. Either of
the previous techniques can be used to solve the problems at the end of the
chapter although both should be understood.

Level Measurement Hardware

As with other process variables, level measurement hardware applies the
physical principles associated with density and buoyancy. As will be seen with
the following instruments, the better one understands the physical principles
associated with level, the better one understands the measurement apparatus.

Level Switches

Switches, specifically level switches are often used to detect position by determining the presence or absence of the sensed material level. Optical, thermal, magnetic, mechanical float, electrical conductivity, acoustic resonance and radio frequency propagation are among the more common operational switching techniques applied to level detection. In each of the detection and switching techniques the appearance of material level will cause an operational change within the detector. This in turn causes the level switch's output to change states and in this manner the position of the material level is detected to be either at, or below a predetermined point. The resulting change in switching states can be utilized to open or close a solenoid valve, turn on or off a pump, sound an alarm or take some other form of control action.

Figure 12 - An interesting variation on the level switch concept; these probes vibrate at a constant audible frequency similar to tuning forks. The frequency changes when the probes are immersed in dry or liquid material. The change in frequency causes the switch output to change states.

Most industrial level switches provide the user with the ability to change the switching action. A direct-acting (DA) level switch closes as the level increases while reverse-acting switches open upon a level increase. The following figure is a good example.

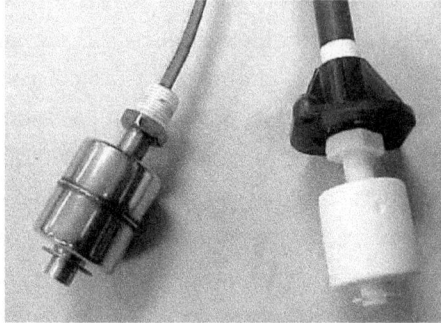

Figure 13 - Two level switches, also called float switches, used to detect position of material level. These units contain magnets within the floats and magnetic reed switches within the centered shafts. Inverting the float reverses the RA/DA switching functions

A final word about switches, in the two preceding paragraphs the author has attempted to avoid using the words "Sensing" and "Sensor" when describing a 2-state, "at or below" level measurement function. Detection is a far more appropriate term when describing process switching functions. Although sensing switches are sensors in a broad description, switches are usually classified differently than process sensors due to their 2-state output. It is generally assumed that "Sensors" refer to a proportional measurement device while "Switches" refer to a discrete, discontinuous, 2-state, on/off detection device. Pressure, flow, temperature, proximity, velocity and numerous other industrial switches are commonly applied throughout the control industry and should not be operationally confused with continuous, proportional sensors and transmitters.

Figure 14 - Level switches, or float switches are available with multiple actuating points for control, alarming and related functions.

Gage Glass

Among the earliest techniques for continuous measurement of level employed an open-end manometer called the Sight Glass, or Gage Glass. Utilizing a pressure-balance concept similar to a manometer, the Sight Glass provides observation of the contained material level through a parallel, side-mounted transparent leg. Observing the following diagram, it can be seen that pressure balance between the vessel and the gage glass can be obtained at any points in the fluid which are at equal distances above or below some reference. If the liquids in the vessel and in the gage glass are subjected to the same external pressure and have the same weight density (specific gravity), the level will be the same in both vessels.

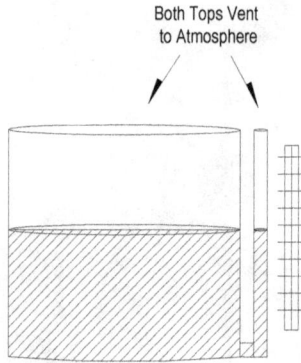

Figure 15 - The Sight Glass or Gauge Glass utilizes a pressure-balance technique to display liquid level through a side-standing transparent sight mechanism. The associated scale can display position in inches, volume in gallons or weight in pounds.

Unfortunately, tall tanks will require a tall gage glass. A means of reducing the height of the column in the gage glass is to isolate the liquid in the vessel with a seal and replace the liquid contained in the gage glass leg with a material of a greater density.

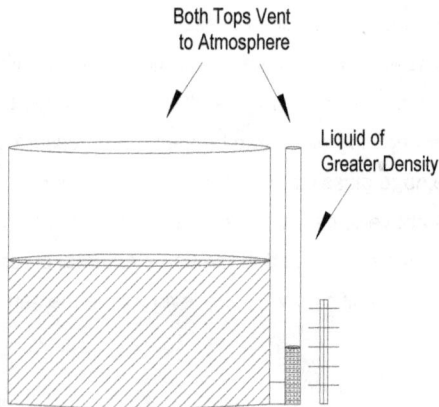

Figure 16 - A liquid having a greater specific gravity than that in the storage vessel creates a condition of pressure balance with a column that does not rise to the tank level. The tank level must be inferred from calibration.

Occasionally, the stored liquid is maintained under pressure - where the pressure above the liquid surface is greater than atmosphere. Carbonated beverages, juices, paints, food products and petroleum-derived materials susceptible to evaporation are commonly stored under pressure or at pressures less than atmosphere. If the pressure above the liquid is maintained constant during level variations the height of the liquid in the gage leg will be offset by a constant value and may easily be biased-out with the display or transmitting device. However, if the surface pressure varies with variations in level, the resulting gage indication will be erroneous. To compensate for the variations in surface pressure a differential form of indication is utilized as in the following diagram.

As demonstrated in the following figure, a differential pressure form of gage glass could be used with the low pressure input connected to the top of the storage vessel and the high pressure input connected to the bottom. In this manner the displayed level in the gage glass will accurately represent the material level in the tank.

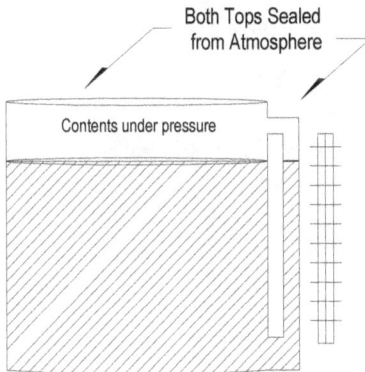

Both Tops Sealed
from Atmosphere

Contents under pressure

Figure 17 - When stored liquids are subjected to varying surface pressures a differential configuration is employed in the gage leg. A differential pressure transmitter could also be applied in place of the gage glass.

The gage glass could also be replaced with a differential pressure transmitter as in the following diagram.

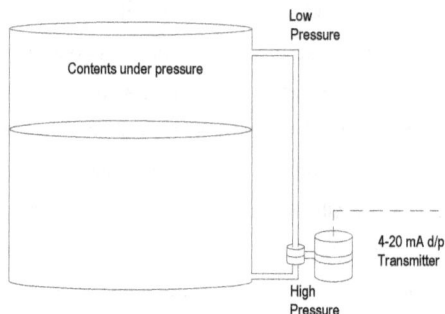

Figure 18 - A differential pressure transmitter is used to measure level of liquids under pressure.

Liquid material contained in a vessel ventilated to atmosphere can use a gauge instrument since gauge instruments also use atmosphere as the reference pressure. However, sealed and unventilated vessels must utilize other types of measurement and transmission means such as differential pressure instruments. A differential pressure transmitter can be connected into a vessel to measure the reference pressure at the top of the liquid, and the liquid's hydrostatic head at the bottom. As the following demonstrates, the transmitted signal will be the difference between the total pressure at the bottom, reference and head, minus the reference at the top of the liquid. The transmitted signal will be an analog of the liquid only.

$$\text{Transmitter Output Current, } I_{out} = K * (P_{HI} - P_{LO}) \pm \text{Offset}$$

Since,

$$P_{HI} = P_{Level} + P_{Reference}$$
$$P_{LO} = P_{Reference}$$

Transmitter Output Current = $K*(P_{Level} + P_{Reference} - P_{Reference}) \pm$ Offset

Therefore,

Transmitter Output Current = P_{Level} = Ht*D

Where,

I_{out} = Transmitter output current range of 4-20 mA

K = Transmitter gain (Span) setting

P_{HI} = the combined input pressure created by the liquid and the pressure above the liquid

P_{LO} = the pressure above the liquid

$\pm$ Offset = the offset associated with the transmitter output, usually 4 mA

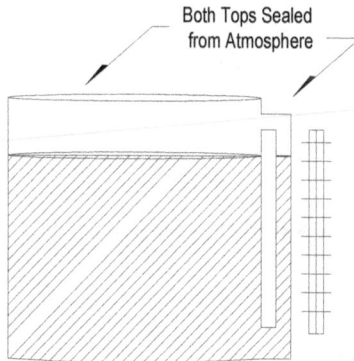

Figure 19 - When liquids are contained under pressure a differential pressure form of measurement is used.

Hydrostatic Pressure Sensing

Among the more common level measurement techniques is the use of hydrostatic pressure to infer tank level. Hydrostatic pressure is created in a gradient below the surface of any contained liquid. Likewise for a material of given density, the deeper the measurement location - the higher the measured pressure. This can be proven by manipulation and substitution of the basic pressure equation.

Since,
$$P = Force / Area, \text{ in psi or } kg/cm^2$$

Since weight is a downward force, pressure can be created by weight being supported over an area.
Therefore,
$$P = Weight / Area, \text{ in psi or } kg/cm^2$$

Since,
$$Density = Weight / Volume$$
And,
$$Weight = Volume * Density$$

Substituting,
$$P = Weight / Area$$
$$P = (Volume * Density) / Area$$

Since,
$$Volume = Area * Height,$$

Substituting for Volume,
$$Pressure = (Area * Height) * Density / Area$$

Canceling areas,
$$\therefore P = Ht * D, \text{ in psi or } kg/cm$$

Assuming uniform and constant material density, it can be seen from the derivation the pressure above a given measurement location is directly proportional to the height of the liquid being stored, and the material density. This may also hold true for some dry material as long as the material seeks a uniform level and maintains a constant density, although weight (load cells) is a far more common measurement technique for dry material level.

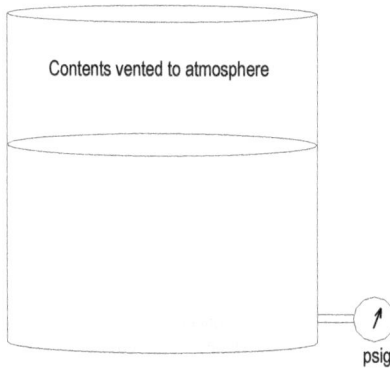

Figure 20 - A simple and effective means of inferential liquid level measurement.

When using a pressure indicating or transmitting instrument to measure level as in the previous diagram several precautions should be observed. The measured liquid should not corrode the gage, the liquid must not contains solids that might enter the gage and the distance between the tank and the gage should be limited. In addition, it should be noted that locating the gage at a substantial distance below the tank would offset the effective range indication by elevating the indication span and can significantly reduce the resolution of the measurement. The following example demonstrates.

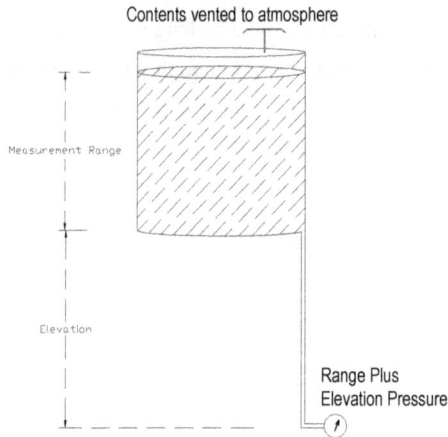

Contents vented to atmosphere

Measurement Range

Elevation

Range Plus
Elevation Pressure

Figure 21 - Tanks are commonly elevated allowing gravity to create the required pressure and subsequent outlet flow.

Example:

Assume a 16-foot tall water storage tank is elevated 40 feet above the location of a pressure gauge. The desired operating range of the level in the tank is between 2 feet and 14 feet. Determine the indicated pressure at the gauge at minimum and maximum tank levels.

Solution:

The level range is between 2 and 14 feet above the bottom of the tank. The tank is elevated 40 feet. A pressure gauge located 40 feet below the bottom of the tank will be measuring liquid pressure created from a pressure range of 42 feet to 54 feet of water. Since pressure equals height multiplied by density, the indicated pressure range is computed as follows.

$$P = Ht * D$$
$$P_{min} = 42 \text{ ft} * D_{H2O}$$
$$P_{min} = 42 \text{ ft} * 62.4 \text{ \#/ft}^3$$
$$P_{min} = 42 \text{ ft} * 62.4 \text{ \#/ft}^3$$

$$P_{min} = 2620.8 \text{ #/ft}^2$$

Converting to units of psi requires dividing by 144 in^2/ft^2,

$$P_{min} = 2620.8 \text{ #/ft}^2 \div 144 \text{ in}^2/\text{ft}^2$$

$$P_{min} = 18.2 \text{ #/in}^2$$

Determining the maximum pressure,

$$P_{max} = 54 \text{ ft} * D_{H2O}$$

$$P_{max} = 54 \text{ ft} * 62.4 \text{ #/ft}^3$$

$$P_{max} = 3369.6 \text{ #/ft}^2$$

Converting to units of psi requires dividing by 144 in^2/ft^2,

$$P_{max} = 3369.6 \text{ #/ft}^2 \div 144 \text{ in}^2/\text{ft}^2$$

$$P_{max} = 23.4 \text{ #/in}^2$$

As can be seen from the computations the pressure at the gauge located 40 feet below the tank will vary from 18.2 psi to 23.4 psi. If a common 30-psi pressure gauge is used to infer the level in the tank only a small portion of the gauge's display will be used to indicate level resulting in poor resolution (resolution is the ability to observe small changes clearly).

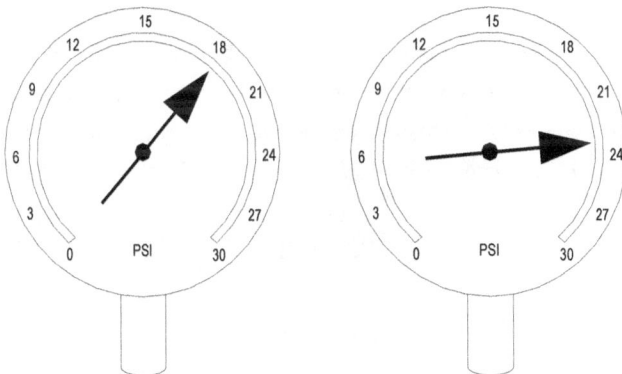

Figure 22 - A level measurement range of between 2 and 14 feet of water produces a small change in indication when the range is elevated to 40 feet.

In liquid measurement and control processes, the Pressure-Level relationship is fundamental. A pressure measurement of this type is often referred to as a 'Head' of pressure, referring to the liquid above the measurement point as creating a "head" of pressure.

Figure 23 - The water tower in downtown Chicago is an example of using a column of water to generate pressure. The water tower is a 13-story standpipe and was used to maintain pressure in the city's water system during the late 1800's and early 1900's. The resulting water pressure at ground level approached 4 atmospheres or approximately 60 psi.

Many liquid level pressures are expressed as inches of water (In H2O) or inches of Mercury (In Hg), the different material densities being responsible for creating different pressure ranges. Pressure can also be determined from Specific Gravity as in the following example.

Since,

$$P = Ht*D$$

And,

$$D = SG*D_{H2O}$$

Then,

$$P = Ht*SG*D_{H2O}$$

Figure 24 - A gauge-pressure pneumatic level transmitter.

Trapped Air Systems

Trapped air systems are closed systems that utilize the varying hydrostatic pressure associated with a changing level to yield an inferential level indication. Trapped air systems measure the static pressure at a specific liquid depth without actually containing the measured liquid. The operational concept is similar to holding a cup inverted underwater. An air pocket within the inverted cup does not allow water to penetrate the interior of the cup. As the cup is lowered the pressure increases and although the volume of air space will decrease the pressure within the airspace increases. The trapped air system of measurement contains a volume of air that compresses as the measured level increases. The resulting pressure increase is displayed on a pressure gauge and/or transmitted and used to infer liquid level.

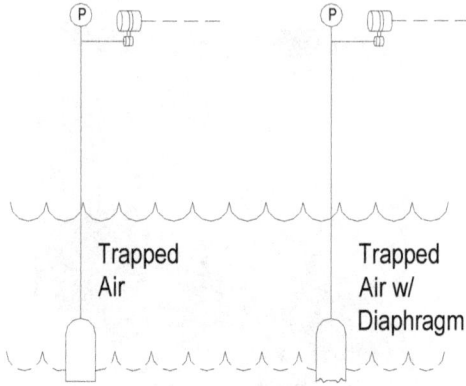

Figure 25 - Trapped air and trapped air with diaphragm systems. Diaphragm seals are commonly employed to isolate the interior of the trapped air system from a caustic (corrosive) material being measured

Purge Systems

When conventional hydrostatic level measurement systems are not feasible a purge system may need to be considered. Purge systems, also called bubbler systems, apply a gas (commonly air) pressure into and against the liquid pressure contained within a vessel.

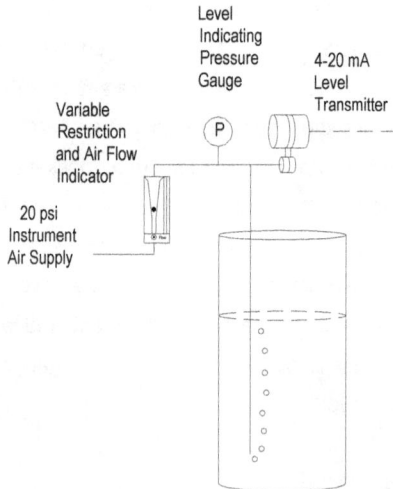

The term "purge" means to force out. Similarly, a pressure line is placed into the liquid level to be measured. Air, or other gas pressure is introduced into the line with enough pressure to slightly overcome the hydrostatic pressure created by the liquid level. Since the purge pressure is greater than the hydrostatic liquid pressure, bubbles will be seen rising to the surface of the liquid level. For this reason, purge systems are often referred to as bubbler systems.

Figure 27 - A rotameter (variable area flowmeter) and a needle valve is commonly used to measure and control the airflow rate into a purge system. Note the small airflow indicator at the 1.0 SCFH location.

Purge systems require the tanks to be vented to atmosphere to avoid a build-up of pressure above the liquid. In operation, the measured pressure is applied via an airline lowered into the tank level. As indicated in the previous and the subsequent diagram, pressure gauges and transmitters are used to measure the

321

airline pressure and to infer liquid level, volume or weight. The applied airline pressure is just enough to slightly overcome the head pressure of the column of liquid to be measured. In operation, initially assume the tank is empty. With an empty tank the measured airline pressure would approximate 0 psi since there exists no backpressure to oppose the flow of the purge air into the tank. At the opposite extreme, If the pressure line were completely blocked the pressure gauge would indicate the instrument air supply pressure of 20 psi since no air can escape through the airline. A tank holding liquid will result in an airline pressure somewhere between these two extremes.

Figure 28 – Upon inspection the bubbles can be seen escaping from the purge system's airline at a constant rate in this photo.

As the tank fills with liquid the indicated pressure at the gauge will be a function of the height of the liquid level and the liquid's weight density. Once a consistent flow of air bubbles is established, the pressure displayed at the gauge will be slightly greater than the backpressure created by the liquid level above the bottom of the purge pipe. The pressure required to maintain a flow of bubbles is slightly greater than the backpressure created by the liquid height. For proper operation the air supply (also called instrument air supply) pressure should be at least twice the maximum anticipated static pressure of the liquid to minimize variations in the purging line pressure as the hydrostatic head varies. An airflow regulator as in the following diagram (right side figure) eliminates purge line supply pressure variations by maintaining a constant flow of air as the measured liquid level changes.

Note – an airflow regulator is NOT an air pressure regulator. The difference is similar to the difference between an electrical voltage regulator and a constant current generator. Airflow regulators maintain a constant flow of air and can only do so by varying the regulator's output pressure as the opposition to airflow changes. Using an air pressure regulator in a purge application will maintain the airline pressure constant as the liquid level varies and will result in no change of indication at the pressure gauge or transmitter.

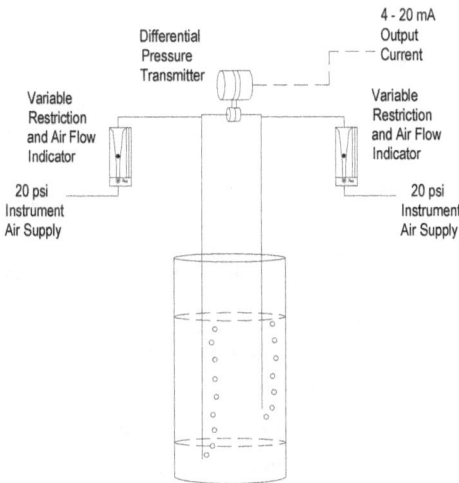

323

Figure 29 – A means of measuring the depths of separated materials - the purge system on the left measures the depth of both materials while the system on the right measures the depth of the floating material. The differential transmitter outputs an indication of only the lower material by subtracting the pressures of each purge system

As demonstrated in the following figure (left side), an alternative to an airflow regulator purge system is an airflow indicator in conjunction with a variable restriction placed into the airline. The quantity of purge air flowing into the liquid level is commonly indicated by a common rotameter flowmeter. The airflow indicator and variable restriction are used to set and retain purge airflow through the liquid, and to maintain a constant flow of air should the supply pressure vary.

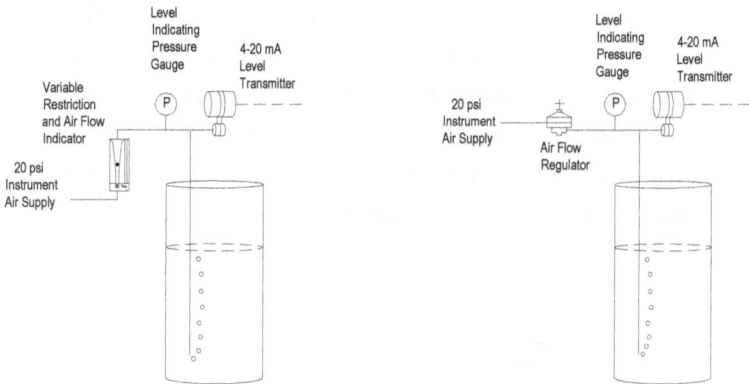

Figure 30 - Purge systems maintain a constant flow of air or other gas into liquid level while sensing the hydrostatic backpressure of the liquid level variation. The transmitter converts the sensed pressure variation into a 4-20 mA or other standard signal span. The system on the left uses a variable restriction and airflow indicator to control the bubble flow. The system on the right uses an airflow regulator to maintain a constant flow of air into the contained liquid.

With a constant flow of air, as the liquid level increases, the hydrostatic pressure at the airline bubble output location also increases. Consequently, the airline pressure required to maintain a continuous and constant flow of bubbles through the changing liquid level also changes. A pressure gauge placed anywhere downstream from the variable restriction in the purge system airline will show an increase in pressure as the level increases, and a decrease in pressure as the level decreases. Purge systems of the flow indicator and variable restriction type work best with level heights under 25 feet and liquid densities near that of water. Once set, the variable restrictions only occasionally need fine-tuning to provide an accurate level indication.

Figure 31 - A side view of a common rotameter (airflow indicator) and variable restriction as used with purge systems. The variable area between the bottom and top of the meter results in a linear airflow display.

Loss of Weight Systems - Displacers

Two common level measurement systems utilizing buoyant properties are floats, and *displacers*. Floats are less dense than the measured liquid and ride on the liquid surface. Displacers are denser than the liquid and remain submerged. Floats commonly provide an indication through a torque, gear or sheave mechanism. Typically, moving mechanisms within the tank are not desirable, as eventually the moving mechanism will require some form of maintenance. Gas tank indications in automobiles utilize a potentiometer connected to a float through a torque mechanism. The resulting display is usually dampened electronically with a capacitor to prevent the indication from "Bouncing" continuously as the level sloshes within the gas tank. Sheave systems connect the level indicator to the float through a pulley and cable.

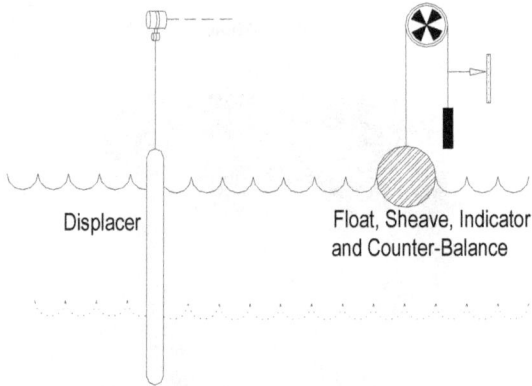

Figure 32 - The displacer (left) and float (right) both rely on principles of buoyancy to yield a level indication but operate in completely different ways.

The displacer resides submerged within a liquid. Displacers are lengthy objects of greater weight density than the level being measured. As the liquid level changes, the amount of the displacer under the surface varies. Given the weight of the displacer as constant, as the level changes the buoyant force pushing upwards on the displacer will also change. As the level rises the resulting net force (weight minus buoyant force) will appear as a loss of weight to the secondary element (the transmitter) supporting the displacer. After conditioning (conversion to 4 mA to 20 mA), the net force becomes the signal representing the liquid level. The net force will always be downward throughout the span of level variation as the displacer is substantially heavier than the liquid density. The net force resulting from the displacer is defined as follows.

$$F_{net} = F_{down} - F_{up}$$

Where,

F_{net} is the difference between the buoyant force on the displacer and the weight of the displacer,

F_{down} is created by the weight of the displacer,

F_{up} is the result of the buoyant force placed upon the displacer by the liquid.

Or,

$$F_{net} = F_{weight} - F_{Buoyant}$$

Where,

F_{net} = the resulting force and direction,

F_{weight} = the displacer weight, equal to its volume multiplied by its weight density,

$F_{Buoyant}$ = the upward force on the displacer created by the volume submerged and the density of the measure material,

$$F_{net} = F_{weight} - (\text{Displacer Volume Submerged} * \text{Liquid Density})$$

Since,

$$\text{Volume Submerged} = (\text{Area} * \text{Submerged Depth of Liquid})$$

$$F_{net} = F_{weight} - [(\text{Area} * \text{Submerged Depth of Liquid}) * \text{Liquid Density}]$$

$$\therefore \text{Submerged Depth of Liquid} = (\mathbf{F_{weight}} - \mathbf{F_{net}}) / (\text{Area} * \text{Liquid Density})$$

Or,

$$\text{Liquid Depth} = (F_{wt} - F_{net}) / (\text{Area} * \text{Liquid Density})$$

As can be seen, the displacer output (F_{net}) will be a downward force that varies with liquid level. Note specifically that the displacers' output (F_{net}) varies, as the inverse of liquid level, meaning the displacer is a reverse acting level sensor. As the liquid level increases, the buoyant force on the displacer increases and the effective weight, or net force, decreases. Again, the displacer is often considered a "Loss of weight" sensor.

The displacer is commonly applied as the primary element (sensor) to a torque balance secondary element (transmitter).

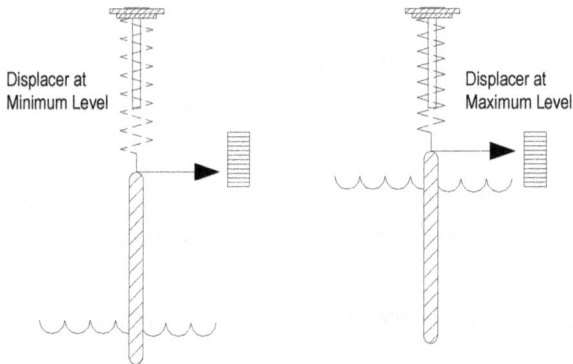

Figure 33 - As a reverse-acting sensor, the displacer loses weight (input force decreases) as the level increases. Note the force-balance nature of the level measurement operation.

Example Problem - Displacer:

Assume a 10-foot long, 4-inch diameter, aluminum (D =170 #/ft³) rod acting as a displacer is suspended by a spring into water. If the water levels varies between 1 foot and 9 feet along the displacer determine the weight of the metal rod displacer in air and the resulting force required by the spring to suspend the displacer at the water level extremes.

Solution:

First, the weight of the displacer must be determined. Since weight equals volume multiplied by density, the volume of the displacer must also be determined.

$$\text{Volume} = \text{area} * \text{length}$$
$$\text{Volume} = (\pi * r^2) * \text{length}$$
$$\text{Volume} = 12.56 \text{ in}^2 * 120 \text{ inches}$$
$$\text{Volume} = 1507.96 \text{ cu. In.}$$

Converting volume to cubic feet,

$$\text{Volume} = 1507.96 \text{ cu. In.} \div 1728 \text{ in}^3/\text{ft}^3$$
$$\text{Volume} = .873 \text{ ft}^3$$

The displacer can now be determined by multiplying the displacer volume by the weight density of aluminum.

$$\text{Wt.} = \text{Vol.} * \text{Density}$$
$$\text{Wt.} = .873 \text{ ft}^3 * 170 \text{ \#/ft}^3$$
$$\text{Displacer Wt.} = 148.35\#$$

The buoyant forces acting on the displacer can now be determined by multiplying the volume submerged by the density of water.

Finding the buoyant force at minimum level,

$$\uparrow F_B = \text{Vol.} * D$$
$$\uparrow F_B = (\text{Area*Level}) * D$$
$$\uparrow F_B = (12.56 \text{ in}^2 * 1 \text{ Ft}) * 62.4 \text{ \#/ft}^3$$

Converting units of area into square feet yields the buoyant force,

$$\uparrow F_B = (.0873 \text{ ft}^2 * 1 \text{ Ft}) * 62.4 \text{ \#/ft}^3$$
$$\uparrow F_B = (.0873 \text{ ft}^3) * 62.4 \text{ \#/ft}^3$$
$$\uparrow F_B = 5.445\#$$

The resulting apparent weight of the displacer at 1 foot of water level becomes,

$$F_{net} = F_{wt} - F_B$$
$$F_{net} = 148.35\#\downarrow - 5.445\#\uparrow$$
$$F_{net} = 142.9\# \downarrow$$

The spring would need to support 142.9# at the minimum water level. Determining the buoyant force at the maximum level yields,

$$\uparrow F_B = (.0873 \text{ ft}^2 * 9 \text{ Ft}) * 62.4 \text{ \#/ft}^3$$
$$\uparrow F_B = (.7857 \text{ ft}^3) * 62.4 \text{ \#/ft}^3$$
$$\uparrow F_B = 49.03\#$$

The resulting apparent weight of the displacer at 9 feet of water level becomes,

$$F_{net} = F_{wt} - F_B$$
$$F_{net} = 148.35\#\downarrow - 49.03\#\uparrow$$
$$F_{net} = 99.32\# \downarrow$$

The displacer's apparent weight varies between approximately 150# at minimum level and 100# at maximum level. Note the reverse acting nature of the displacer output – as the level increases the force created by the displacer decreases and vice versa. Displacers are commonly used with reverse-acting secondary elements (transmitters) to yield a signal that varies in proportion with level, unless a signal representative of empty volume is desired.

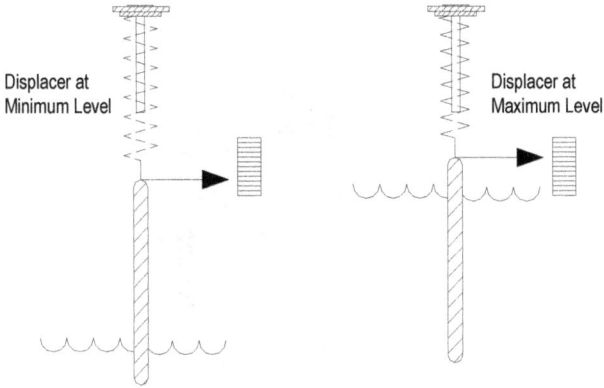

Figure 34 - Diagram of displacer operation at minimum and maximum level.

Time of Flight Systems

Ultrasonic, radar, laser and most non-contact forms of level measurement systems utilize the concept of "Time of flight" to infer a weight, volume or level measurement. The time of flight concept involves transmission of electromagnetic radio frequency (radar), sound (ultrasonic) or light (laser) energy towards the material level to be measured. As the energy is radiated a timer is started. Once the energy strikes the level to be measured, a portion of the energy will be returned, or reflected to the source. Detection sensors at the source receive the reflected energy and stop the timer. The difference in time between the transmitted and returned signals is then used to interpret the material level in the containing vessel.

Figure 35 - A radar form of level transmitter. Photo courtesy of Endress and Hauser.

Figure 36 - A Laser form of level measurement system. The display is indicating the distance to the reflecting body.

Figure 37 - An ultrasonic level transducer and display.

Radiation Sensing
-This material is under development.

Capacitance Sensing
One of the more desirable characteristics of level-gauging apparatus is that there be no moving parts in the tank or vessel. Using a capacitive probe meets the no moving parts requirement by utilizing differing dielectric characteristics of the vessel contents. Capacitive level sensing applies concepts associated with the electrical properties of a capacitor to infer level. Similar in construction to an electronic capacitor, level sensing capacitors consist of at least 2 conductors separated by a non-conductive dielectric. The capacitance probe for level measurement is normally composed of two electrodes, usually with one inside the other, and uses the stored material and air as the insulating dielectric. When a potential (usually a small AC voltage) is applied to the plates, electrons will be removed from one of the plates and accumulated on the other. The quantity of

electrons available will be contingent upon the area (A) of the plates, the distance (d) between the plates and the dielectric nature (K) of the material between the plates. Expressed as a mathematical relationship the capacitance of a two-plate capacitor in Pico-farads (10^{-12} Farads, or μμF) follows.

$$C = .225K * A \div d$$

Where,

.225 = Compatibility coefficient for units

C = Capacitance in μμF

A = Plate area, in^2

d = Distance between the plates, in.

K = Dielectric constant

From the equation, for a given size and spacing of the capacitive probe's plates note that the capacitance value depends entirely on the dielectric coefficient. The dielectric strength provides an indication of the availability of electrons to charge the capacitor. Some materials have electrons that can be made readily available while other materials share electrons very reluctantly. Assuming the vessel containing the capacitive probe will be vented to atmosphere, the capacitance exhibited by the probe will be determined by the ratio of the tank material's dielectric constant to the dielectric constant of air. The capacitance, which can be measured in a number of ways, can then be related to the height of the material in the tank. The capacitance will be minimum when the tank is filled with air and maximum when the tank material is the dielectric to be measured and the tank material fills the space between the capacitor's electrodes. The following figure illustrates.

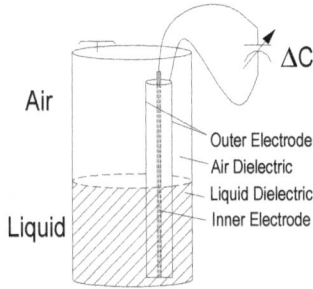

Figure 38 - The changing proportion of air and liquid dielectric varies the electrode's capacitance.

Table of Dielectric Coefficients, Relative to Air

Material	Dielectric Constant	Material	Dielectric Constant	Material	Dielectric Constant
Air	1.0	Sand	3-5	Petroleum	2-3
Water	80.0	Nylon	4-5	Asphalt	2.7
Glycerin	47.0	Cereals	3-5	Benzene	2.3
Alcohol	37.0	Rice	3-5	Chlorine	2.0
Glycol	35-40	Cellulose	3.9	Paper	2.0
Dolomite	8.0	Glass	3.7-10	Liquid CO_2	1.59
Porcelain	5-7	Sulfur	3.4	Liquid Air	1.5
Salt	6	Turpentine	3.2	Gases (typ)	1.0
Aqueous Solutions	50-80	Barium Titanate	5,000	Sugar	3.0

Signal conditioning of the capacitance probe is typically performed using an AC bridge circuit. The schematic diagram follows.

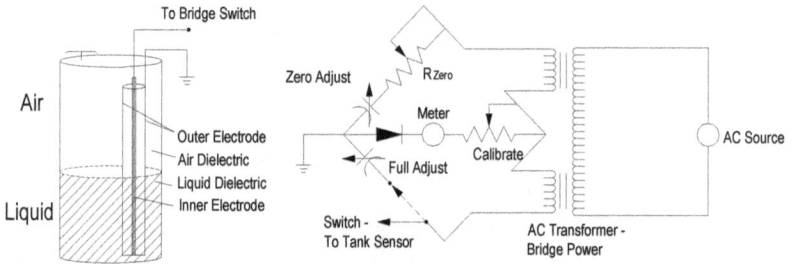

Figure 39 - A typical bridge circuit for a capacitive sensor utilizes an AC power source. The DC output signal is observed on the meter or the meter could be replaced with an amplifier to convert the output into 4-20 mA.

The bridge circuit provides an AC power source for the bridge and capacitive sensor. As the level varies in the tank and the capacitance of the sensor changes, the amount of voltage appearing across the capacitive sensor in the bridge also varies. The difference in voltage across the upper and lower halves of the bridge will be rectified and displayed on the meter. The meter could be replaced with a resistor and the resistor voltage sent to a DC amplifier for conversion into a 4 mA to 20 mA signals.

Application Notes

The following application information is taken from automation manufacturer's advertising, maintenance materials and personal experience. The manufacturer's materials are located before the homework problems at the end of this section. The materials are provided to assist the student in understanding Level and pressure sensor and transmitter specifications, to provide an overview of typical applications, and to demonstrate the technologist's responsibilities and perspective. It is suggested that the student closely examine the following information.

1 – Two-way hand-valves (also called cocks or gauge cocks) should always be installed just ahead of pressure and level measurement instruments. Manual valves facilitate instrument removal, service, testing and replacement without shutting down the pressure system or draining a tank.

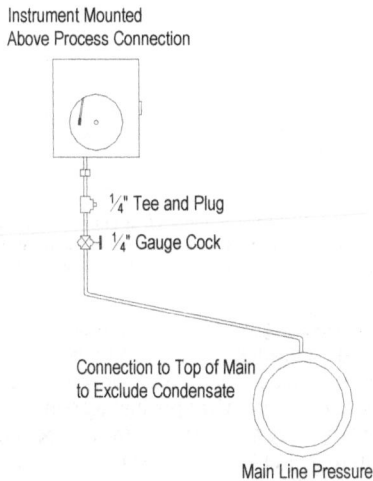

Figure 40 - Two-way hand-valves (also called cocks or gauge cocks) should always be installed just ahead of pressure and level measurement instruments.

2 – Sediment, moisture and other undesirables have a way of settling in all tanks and storage vessels. For this and other reasons hydrostatic pressure sensing, transmitting and related level detecting items are located slightly above the bottom and on the side of the containing vessel. If gauge contamination is probable provisions should be included to protect and maintain the measurement apparatus. Dirt traps, blow-down and clean-out valves should be installed and regularly maintained.

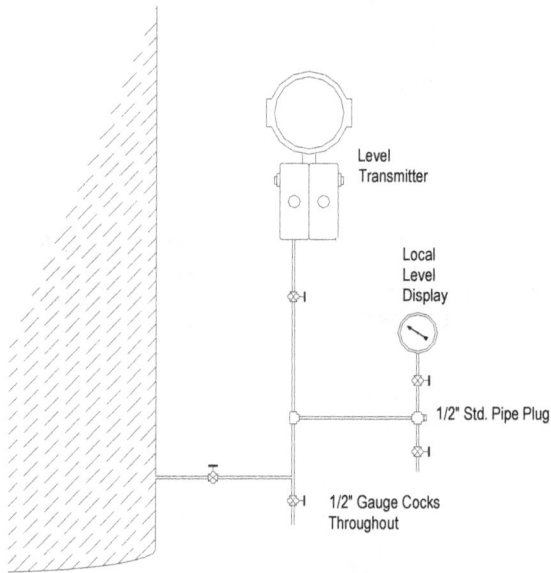

Figure 41 - Provisions for impulse line cleaning such as gauge cocks and pipe plugs should be installed and maintained periodically. Note - the elevation of the transmitter above the point of connection to the process will effect the measured pressure.

3 – All elastic member pressure gauges (Bourdon tubes, bellows, and diaphragms) measure the difference between the internal and external pressure. When using low-range (near atmospheric pressure) gauges, errors may result from variations in ambient pressure. Gauges installed in rooms ventilated by exhaust and circulating fans may also exhibit errors. Opening and closing the case doors of low-range (close to atmosphere) instruments may cause a momentary reading variation that will be especially apparent on recording instruments. Gauges should be located to be free from dust, dirt, corrosive fumes, vibrations and extremes of temperature.

4 – To minimize the possibility of equipment damage and injury, always open and close hand valves and pressure adjustment mechanisms slowly. One cannot always accurately assess the physical condition of industrial hardware by visual inspection. To reduce the possibilities of valve stem breakage, internal component damage or violent process upsets (often called "Bumps"); always proceed cautiously when making field adjustments.

5 – The cooling of system air is inevitable between the compressor and the system loads. As air temperature decreases, any residual moisture is squeezed out as a result of contraction. Moisture can cause pressure components to

corrode and even worse, can move dirt and other undesired particulates through the air system. The particulates are certain to appear in the narrow passages within air relays and regulators and may become stuck holding the pilot valve or exhaust capillary open. Should this occur, air will often be heard "Bleeding" from the exhaust ports continuously. If it is practically possible, a rapid exercising (changing) of the input or output pressure (remove the output connection and rapidly covering and uncovering the output pressure line, or adjusting the setting knob in and out several times) usually frees the particulate without removal or replacement of the component in question. To minimize the opportunity of this occurrence, always provide accommodations for moisture and dirt by installing air dryers, oilers and filters, and performing the required maintenance on the ancillary systems at regular intervals.

6 - The maximum pressure for purge systems is often limited to150 psi or about 350 feet of H_2O head. To avoid the possibility of the measured liquid entering into the purge system components, the purging gas should be turned on before the liquid is admitted and the tank should be drained before turning the purging gas off.

7 - Occasionally gases other than air are purged into liquid level as air may have a detrimental effect upon the material being measured, the measurement instruments or the supplementary hardware. Air and gas purges are most beneficial when measuring corrosive or caustic liquids that would normally damage or corrode most immersion types of level sensors.

– Following this page are a few pages of additional browsing. The first 5 pages (numbers are hand-written and circled at the top of each page) are an overview of level sensing technology. Consolidated Electric Company provides the pages and although brief, the listing is fairly comprehensive. Review the 32 systems presented in pages 1 through 5.

-Pages 6 and 7 are listings of common material densities and specific gravities. Dr. Richard Henry provides Page 6 and page 7 is provided by pump. net.

- The homework problems follow the next 7 pages of application information.

Author's note – Application material is continuously under development. Feel free to submit recommendations, corrections and clarifications.

- Include purge material from ch. 5? (pg 230)
- To be included in future:
Nuclear, Ultrasonic and Radiation Sensing, Hydrometers

Homework

Hydrostatic Pressure

1 – Water level in a tank varies between 0 and 4 feet. The tank is elevated and the pressure gauge used to represent tank water level is placed 5 feet below the tank. Determine the pressure at the gauge with 0 feet and with 4 feet of water level in the tank. How can a pressure gauge be made to designate only the level in the tank?

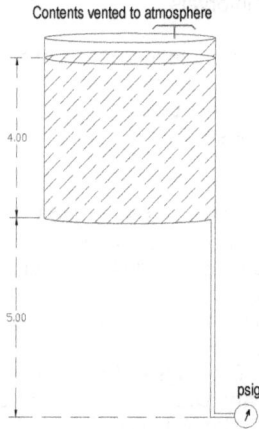

Figure 42 - Diagram of problem #1.

2 - A trapped air level measurement system is used to measure the level of a petroleum product (specific gravity =. 8). The level varies between 5 feet and 35 feet from the bottom of the tank. What pressure span would be indicated on a pressure gauge connected to the trapped airline at the minimum and the maximum petroleum levels if the bottom of the trapped air system is 3 feet above the bottom of the tank?

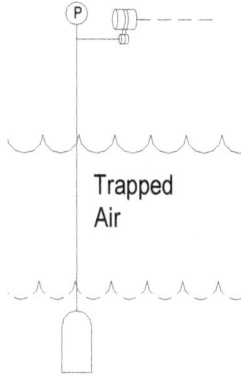

Figure 43 - Diagram for problem #2.

3 - An air purge system is used to measure water level between 20 inches and 60 inches. What should be the approximate purge system pressure?

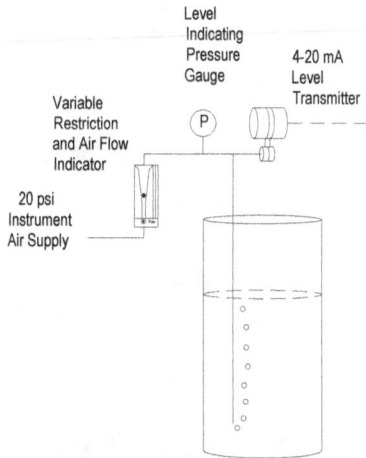

Figure 44 - Diagram for problem #3.

Buoyancy

4 – A stone weighs 82# in air and 58.5# in H_2O. Find the specific gravity and density of the stone.

5 – A barge has vertical sides, is 35 feet wide by 200 feet long and has a flat bottom. When empty, the barge sits 3 feet deep in fresh water. At maximum load the barge will sit 9 feet in the water. Find the weight of the barge and the weight of the full load.

6 – A 10,000 cubic foot hydrogen balloon is being prepared for launch on a 32 degree F day in December. Ropes anchored into the ground suspend the balloon. What is the tension in the ropes if two - 200 pound men are aboard the balloon? Assume the balloon, basket, ropes and other hardware weigh 185 pounds.

7 – A metal sphere weighs 35.2 oz. in air and 30.8 oz in water. Find the volume and specific gravity of the sphere.

8 – Determine the submerged depth of a 6-inch tall, 1-foot square cork floating in water. Assume the cork density is 5 pounds per cubic foot.

9 – Determine the area of six-inch thick ice that would be required to support a 200-pound man or woman. Assume the ice density as 57 pounds per cubic foot.

Displacers

10 – Assume the displacer shown is 5 feet in length with a circular area of 5 square inches. The SG of the displacer is 1.5. If the fulcrum is set to .4:1 and the spring force is 20# pulling up, determine the pressure output span as a liquid of SG = 1.4 submerges the displacer from bottom to top. Assume the output area is 1 square inch.

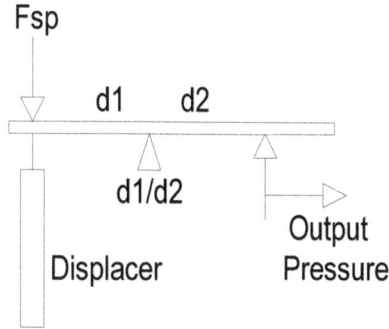

Figure 45 – Displacer and transmitter of problem #10.

11 - Assume the displacer shown is 5 feet in length with a circular are of 5 square inches. The SG of the displacer is 1.3. If the fulcrum is set to 4.84:1 and the spring force is 3.13# down, determine the pressure output span as a liquid of SG = 1.2 submerges the displacer from bottom to top. Assume the output area is 2 square inches.

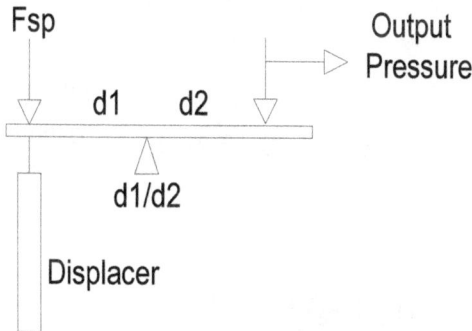

Figure 46 – Displacer and transmitter of problem #11.

12 - Assume the displacer shown is 10 feet in length with a circular area of 2 square inches. The SG of the displacer is 6.0, and the SG of the liquid is 1.3. Determine the required gain and spring force to Output a 3-5 psi span as the

liquid submerges the displacer from bottom to top. Assume the output area is 1 square inch.

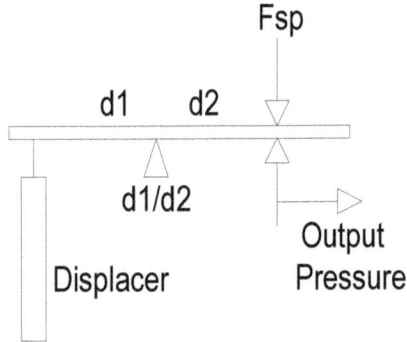

Figure 47 - Displacer and transmitter for problem #12.

13 - Assume the displacer shown is 15 feet in length with a circular area of .066 square feet. The weight of the displacer is 100 pounds. If the level of the liquid is maintained constant at 15 feet, and the SG of the liquid varies between 1.0 and 1.2, determine the gain and spring force to output a pressure span of .64-3.0 psi. Assume the output area is 5 square inches.

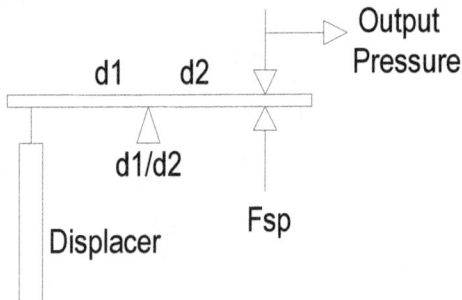

Figure 48 – Displacer and transmitter of problem #13.

Calibration and Scaling

14 – A liquid level transmitter outputs 4-20 mA over a span of 30" to 150" of liquid. The 4-20 mA signal is converted to 1-5 Volts using a 250-Ohm resistor. The 1-5 Volt signal is converted using a software program to display the liquid level as 30" to 150". Determine the scale factor and offset values to perform the required scaling function.

15 – A pressure sensor outputs –20 to +90 millivolts when measuring hydrostatic pressure over a liquid level span of 1 foot to 25 feet of water. The sensor output is applied to a signal-conditioning amplifier. The amplifier needs to be calibrated to output .1 Volts at 1 foot of water and 2.5 Volts at 25 feet of water. Determine the amplifier gain and offset to accomplish the desired output voltage span.

16 – A load cell rated at 500,000 pounds is used to measure tank level. If the material specific gravity is 5.6 and varies between a minimum volume of 100 cubic feet and a maximum volume of 1200 cubic feet, and the tank related hardware weighs 7.5 tons, determine the load cell output voltage if the rated output of the load cell is 5 mV/V. Assume a 10 Volt supply voltage is used to power the load cell.

17 – Assume the output voltage of the load cell in the previous problem varies between 4.99 mV to 43.4 mV between 100 and 1200 cubic feet of volume. Determine:
  A) The required scale factor and offset values to display the tank volume as 0% to 100%.
  B) The required scale factor and offset values to display the tank volume as 100 to 1200.

18 – An air purge system is measuring water level between 1 foot and 5 feet. A pressure transmitter senses the variation in air purge pressure and outputs a current loop signal. If the transmitter is calibrated to output 4-20 mA over 0" to

120" of water, determine the output current values at 1 foot of water and at 5 feet of water level.

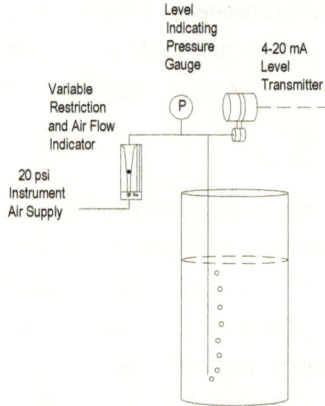

Figure 49 - Diagram of problem #18.

19 – Moisture (water) condensation accumulates at varying levels at the bottom of a gasoline storage tank. How can two purge systems and a differential pressure transmitter be used to measure only the gasoline level?

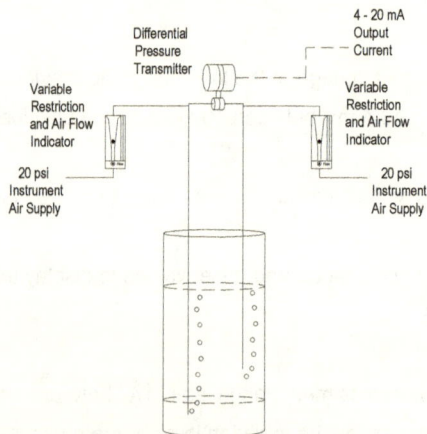

Figure 50 - Diagram of problem #19.

Chapter 7 - Fluids and Flowmeters

This chapter introduces commercially available flowmeters. The nature of liquid and gas flow is presented to support the descriptions of flowmeter operation. Flow rate and total flow are discussed; and d/p transmitters and square root extractors are applied to various flowmeters for operational and mathematical analysis. The chapter is still under development.

Objectives:
Upon completion of this chapter, you should be able to:
- Explain the characteristics of streamlined and turbulent flow
- Explain the operation of constriction flowmeters
- Explain square root extraction
- Explain the operation of positive displacement flowmeters
- Explain impinging, magnetic and Coriolis flowmeters
- Explain open-channel, vortex, impinging
- Explain ultrasonic, mass and vortex flowmeters
- Explain open-channel and dry material flowmeters
- Perform flow rate and total flow computations
- Perform differential pressure computations
- Perform differential pressure transmitter computations

Introduction

Flow measurements are among the most common and essential made by the process industries, and the same can be said for residential consumers as well. Water, electric power, natural or liquid propane (LP) gas and automotive fuel require precise metering to determine the amount delivered, charged and paid for by the consumer. As consumers, we rarely notice these instruments and never question their accuracy. In industry, flow measurement is everywhere and assumes many forms.

Of all process measurement instruments, more flowmeters or variations of flow measurement instruments exist than any other group of process measurement instruments. So many flowmeters are available that categories of flowmeters have been established. This chapter introduces the majority of available flowmeters and provides the physical principles, operational descriptions, application limitations and specifications.

Fluidics

Any material capable of flowing is considered a fluid. As a result, *Fluidics* is the study of liquids and gases in motion. Flowmeters measure the flow rate of gases and liquids in motion and "Q" is the symbol used to represent flow rate. Fluid flow rate, Q, is determined by dividing the volume of a fluid in motion by the time required for the fluid to pass through a given location. In equation form;

$$Q = \text{Volume} \div \text{Time}$$

Flow rate (Q) is expressed in dimensions of liters/hour, gallons/minute (GPM), cubic feet per second (ft^3/s) or any other units of volume over time. Flow rate is more commonly computed using an equation derived from the volume over time equation. Since volume equals area multiplied by distance, subbing into the previous equation yields the following.

$$Q = \text{Volume} \div \text{Time}$$

Since,

$$\text{Volume} = \text{Area} \times \text{Distance}$$

Subbing,

$$Q = (\text{Area} \times \text{Distance}) \div \text{Time}$$

Or,

$$Q = \text{Area} \times (\text{Distance} \div \text{Time})$$

And since,

$$\text{Velocity} = \text{Distance} \div \text{Time}$$

$$Q = \text{Area} \times \text{Velocity, units of ft}^3 / \text{sec}$$

This equation is commonly used to compute flow rate and also best represents the manner with which flow is occasionally measured. According to the equation, sensing fluid velocity through a fixed and constant area will yield a flow rate. The velocity of a fluid in motion is not always uniform across a pipe area however.

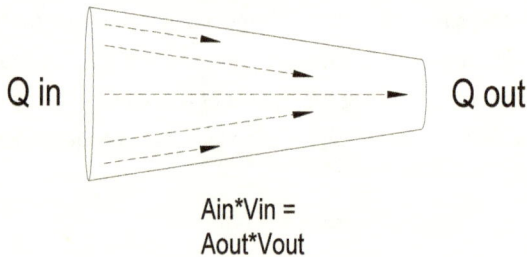

Ain*Vin =
Aout*Vout

Figure 1 - The input and output flow rates through a closed system are equal. Due to friction, the velocity of the material along the outside is less than the velocity through the center. In streamline flow situations, the flow along the inside of the pipe is stationary.

The AV equation demonstrates that although the areas may be different, <u>the flow rate remains constant</u> but the fluid velocities change. This effect is obvious when one places their thumb over the outlet of a garden hose. The volume of flow moving at any location along the hose is determined by the opening created at the thumb. The flow along the hose cannot be greater or less than at the restriction. However, the water velocity is greater within the restriction created by the thumb than upstream since the opening is smaller. As a result, fluid velocity exhibits an inverse relationship with cross-sectional area.

Since,

$$Q = \text{Area} \times \text{Velocity}$$

And,

$$Q_{IN} = Q_{OUT}$$

$$A_{IN}V_{IN} = A_{OUT}V_{OUT}$$

Rearranging,

$$A_{IN}/A_{OUT} = V_{OUT}/V_{IN}$$

In summary, as the area decreases, the fluid velocity increases.

Streamline Versus Turbulent Flow

If a fluid in motion can be closely observed, one of two extreme conditions may be noticed. Depending upon the amount of energy associated with the fluid motion, it will be seen to be flowing in a predictable *streamlined* manner, or the fluid will be flowing with much greater energy in a *turbulent* manner.

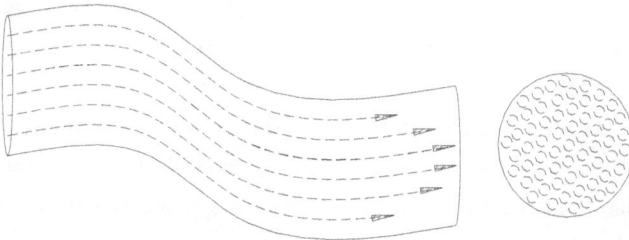

Figure 2 - Streamline, or laminar flow appears as layers or bundles of flowing material with each individual flow stream remaining in the same relative location throughout the legth of the pipe.

351

Streamline flow, also called *laminar* or *layered* flow is often described as flowing in a tube-like manner with a bullet shaped velocity profile. Imagine the fluid in motion as appearing in layers or bundles with each individual flow stream remaining in the same relative location as it travels the length of a pipe. Due to the friction of neighboring bundles, those bundles flowing near the center portion of a pipe travel at a higher velocity than those around the outside. Ideally, the bundles traveling at the outer circumference against the pipe wall do not contain sufficient energy to overcome the friction of the pipe wall and are not in motion. This effect can be observed in creeks, streams and with fluids flowing through transparent tubing when a bubble appears stuck to the inside wall and is observed to not be in motion. As can be seen in the previous drawing, the resulting cross-sectional velocity profile of streamlined flow assumes the shape of a bullet where the faster currents are towards the center and the slower currents appear at the outer regions of the pipe.

The bullet-shaped velocity profile explains the concept of "Undertow" in rivers. The flow of water below the surface of a mildly flowing river is moving at a greater velocity than the water at the bottom, sides or surface. The reasons for the different velocities are explained by friction. The water at the bottom and sides attempts to overcome the friction of earth, sediment, sand or silt. As mentioned previously, the result may well mean the water along the sides and bottom may not be able to overcome the friction and may not be in motion. Water near the surface is flowing against air, and although the motion of flow is often obvious, the surface velocity is not as great as the water velocity just below the surface. The water below the surface is flowing "against" adjoining flowing water, or adjoining bundles of flowing water. The result is a greater velocity where the friction is less below the surface.

Figure 3 - Although frozen at the surface, the Mississippi River continues to flow with substantial velocity under the ice.

Undertow is a serious and hazardous concern to those employed in river-borne occupations and leisure boaters navigating larger waterways. Undertow has a tendency to hold an object below the surface.

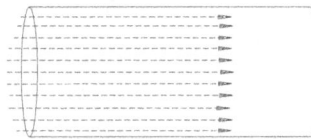

Turbulent Flow

Streamline Flow

Figure 4 – Velocity profiles of turbulent and streamline flow. The upper figure indicates a more uniform velocity of turbulent flow.

Holding all other variables constant, turbulent flow is the result of fluid moving at a much greater velocity than streamlined flow. Density, viscosity, pipe dimensions and fluid velocity all contribute to determining whether a fluid will flow under streamline or turbulent conditions. Reynolds Number is the calculation that takes into consideration the variables and is representative of the fluid's flowing nature. Reynolds Numbers under 2000 predict streamline flow and computations above 3000 are understood to be turbulent. With Reynolds Numbers between 2000 and 3000, the flow is undeterminable.

Reynolds number, R_D indicates the flow rate, for a given size of pipe and viscosity of fluid, at which the ratio of average speed to maximum speed through the pipe cross-section approaches a constant. For lower values of R_D the ratio changes so rapidly as to be of little use. The higher the Reynolds number the more stable the flow coefficients. The type of flow in which the Reynolds number is low is called *viscous, streamlined or laminar*. At the higher ratios of flow at which the flow coefficients stabilize, the flow is termed turbulent.

$$\text{Reynolds number} = \frac{K(\text{pounds of material per unit of time})}{(\text{diameter of pipe})(\text{absolute viscosity})}$$

Because turbulent flow will be obtained with certainty at high values of R_D, flow calculations based on this assumption should be accurate. If (1) the flow rate is so low, (2) the pipe diameter so large, or (3) the viscosity so high that the value of R_D is low enough to indicate that turbulent flow will not be attained, then corrections must be made because the true ratio of average speed to maximum speed will be less than that anticipated for turbulent flow.

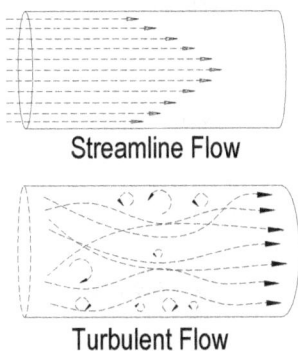

Streamline Flow

Turbulent Flow

Figure 5 - Turbulent flow exhibits a more random pattern than streamline flow.

Turbulent flow is associated with energy loss. Unions, elbows, flanges, some flowmeters and even slight obstructions provide an opportunity to create eddies, whirlpools, pressure drops or vortices which usurp energy. Turbulent flow exhibits a uniform velocity profile across the pipe cross-section. Even against the pipe wall the fluid is flowing with approximately the same velocity as that flowing in the center. Turbulent flow does not appear in bundles but assumes a more random and unpredictable flow path or pattern. As is demonstrated in the previous diagram, fluid particles appear to move throughout the pipe cross-sectional area while in motion.

As an analogy, consider a canoe flowing slowly down a wide creek or river. Where the river is wide the flow is streamline, low energy. As one continues downstream, the river narrows into rapids and the water velocity increases. The flow rate (volume ÷ time) of the river is the same upstream and downstream, yet where the river constricts the velocity increases. This is easily proven using the A×V (area × velocity) equation.

Since,

$$Q_{upstream} = Q_{downstream}$$

And,

$$Q = Velocity \times Area$$

$$Velocity_{upstream} \times Area_{upstream} = Velocity_{downstream} \times Area_{downstream}$$

$$Velocity_{upstream} \times (Area_{upstream} \div Area_{downstream}) = Velocity_{downstream}$$

The downstream velocity is increased by the ratio of the upstream to downstream areas. For those familiar with canoeing, the "rapids" always exist where the water flow narrows with eddies (circulating currents) and whirlpools usually obvious in these locations. The principle holds for fluid flowing in closed systems also, in that reducing the pipe diameter increases the Reynolds Number. Many flowmeters require or create turbulent flow conditions that result in significant energy loss, an expensive proposition when projected over the period of several years at 24 hour-a-day of operation. However, recent developments in flowmeter technology minimize obstructions in the flow path to reduce energy loss and although the newer instruments cost more, the energy savings can usually recover the additional expense within a few years.

Figure 6 - An ultrasonic flowmeter has no obstructions in the flow path.

Total Flow

Total flow, Q_t, is the accumulation of flow rate over a finite period of time. Totalizing, or *integrating*, the output of a flow rate meter over a period of time can determine total flow. Precision integrators are commonly employed to compute total flow from flow rate information. Totalizing, or integrating flowmeters are commercially available and usually contain flow rate and totalized indications, a means to reset the integrated value to zero after a period of time, a corresponding electronic output signal and some form of display. Totalizing flowmeters are commonly employed to inventory processed and raw materials used in manufacturing. Common residential water meters, natural gas meters and automotive gasoline pumps employ totalizing flowmeters.

Total flow = Flow Rate * Time

Or,

$$Q_T = Q \times^* T$$

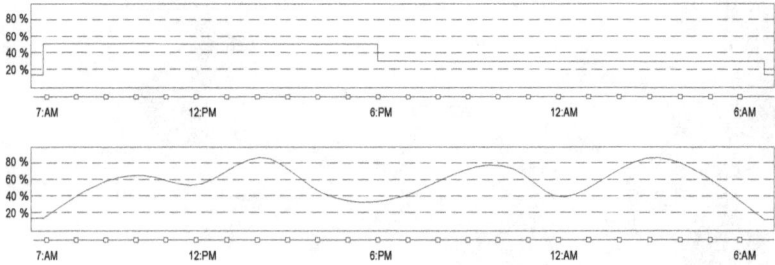

Figure 7 - Flow Rate (Q) versus Time for a 24-hour period. Total flow in the upper graph can be determined manually. Total flow in the lower graph would require a flow totalizer, or integrator.

Physical Principles

A wide variety of physical principles are employed to measure flow rate and total flow of liquids and gases in motion. A couple of the more commonly applied principles are Torricelli's and Bernoulli's, which are introduced in the following pages.

Torricelli's Equation

Gravitational acceleration causes unsupported objects to descend at velocities that can be accurately determined by a single equation. The same concept applies to liquids placed at altitude. Water towers are a convenient means of using gravity to develop the pressure required for flow and subsequent distribution.

Figure 8 - The Water Tower in downtown Chicago is an example of using gravity to establish fluid pressure and flow.

The resulting pressure at ground level can be determined using the previously introduced P = H × D equation where pressure is a function of liquid height and weight density. Liquid velocity can be approximated using Torricelli's equation. Torricelli's equation allows for a velocity determination knowing the height of the stored water. The equation is usually provided as follows.

$$V^2 = 2gH$$

Or,

$$V = \sqrt{2gH}$$

Where,

    V = velocity
    g = Gravitational acceleration
    H = the height of the stored liquid

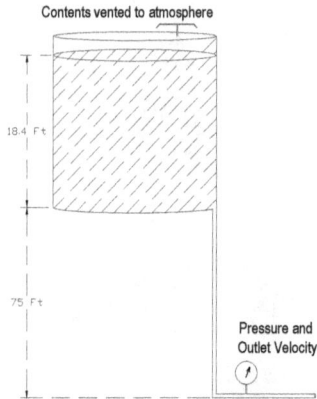

Contents vented to atmosphere

18.4 Ft

75 Ft

Pressure and
Outlet Velocity

Figure 9 - Figure for the following example problem.

Example Problem:

Water is stored in an elevated vessel. Assume the level in the vessel is 18.4 feet and the vessel is raised 75 feet above ground level. Determine the static pressure (in psi), and the maximum possible velocity (in feet/second) of the water flow at ground level. If the outlet flow is passing through a 6-inch pipe, determine the maximum possible flow rate (Q) in gallons per minute (GPM).

Solution:

The static pressure at ground level is determined using the P = H×D equation and the maximum velocity is computed using Torricelli's equation. Flow rate will be found using the flow rate, Q = A×V, equation. Dimension (units) compatibility should be addressed before proceeding and the known quantities should be collected. For reasons of consistency, it is suggested that units of feet, pounds and seconds be used for the computations. Once the computations are completed, conversion factors can then be used to determine the desired units of pressure in psi (or psig), velocity in feet/second and flow rate in GPM.

The known quantities are a total water level above ground level of 18.4 feet plus 75 feet, or 93.4 total feet of water "Head" pressure. The density of water is 62.4 $\#/ft^3$ and gravitational acceleration is understood to be 32.2 ft/sec². Flow rate will be determined in units of cubic feet per second and subsequently converted into gallons per minute using conversion factors of 7.48 gallons per cubic foot, and 60 seconds per minute.

Determining pressure,
$$\text{Pressure} = \text{Height} \times \text{Density}$$
Or,

$$P = H \times D$$
$$P = 93.4 \text{ feet} \times 62.4 \ \#/ft^2$$
$$P = 5828.16 \ \#/ft^2$$

Converting to psi,
$$P = 5828.16 \ \#/ft^2 \div 144 \ in^2/ft^2$$
$$P = 40.473 \text{ psi, or psig}$$

Determining velocity,
$$\text{Velocity}^2 = 2 * \text{Gravitational Acceleration} * \text{height of Level}$$
Or,
$$V^2 = 2gH$$
Solving for velocity,
$$V = \sqrt{2gH}$$

Subbing the given quantities,
$$V = [2 \times 32.2 ft/s^2 \times 93.4ft]^{1/2}$$
$$V = [6014.96 \ ft^2/s^2]^{1/2}$$
$$V = 77.556 \text{ ft/sec}$$

Keep in mind the velocity of the water will decrease as the level in the system drops. In order to maintain constant pressure and flow characteristics, water towers typically have a large reservoir at the top.

Author's Note – As mentioned in the example problem, the results of the computations included in this chapter are ideal and represent the maximum possible velocity and does not consider the real-world effects of standpipe friction, pressure drops at unions and elbows, downstream loading or related concerns that tend to reduce the maximum achievable velocity and flow rate of the water. It should also be mentioned that the amount of available pressure at the outlet would be reduced to facilitate high velocity flow since kinetic energy is always afforded at the expense of potential. Although the results may deviate from realistic values, the example problem does demonstrate the propensity of gravity to establish pressure and induce fluid flow.

Determining flow rate,
$$\text{Flow Rate} = \text{Pipe area} * \text{Fluid velocity}$$
Or,
$$Q = A \times V$$

Since the pipe diameter is provided, converting diameter to area requires,
$$\text{Area} = \pi \times \text{Radius}^2 = \pi \times \text{Diameter}^2/4$$
Or,
$$A = \pi r^2 = \pi D^2/4$$
$$A = \pi \,(6^2\,/4)$$
$$A = 28.274 \text{ in}^2 = .196 \text{ ft}^2$$

Determining flow rate,
$$Q = A \times V$$
$$Q = .196 \text{ ft}^2 \times 77.556 \text{ ft/sec}$$
$$Q = 15.228 \text{ ft}^3/\text{sec}$$

Converting to gallons per minute,
$$Q = 15.228 \text{ ft}^3/\text{sec} \times 60 \text{ sec/minute}$$

$$Q = 913.685 \text{ ft}^3/\text{min}$$

$$Q = 913.685 \text{ ft}^3/\text{min} \times 7.48 \text{ gallons/ft}^3$$

$$Q = 6834.365 \text{ GPM}$$

Again, the results appear bloated due to the omission of the non-ideal considerations as explained previously.

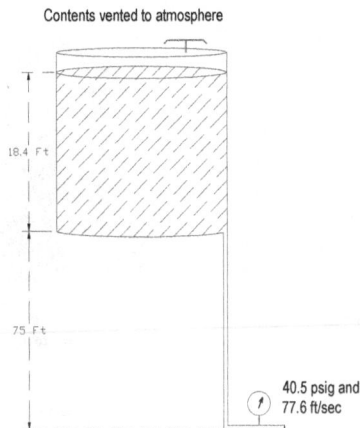

Figure 10 - The results of the Torricelli's example problem.

Bernoulli's Equation

Among the simplest and most common flowmeters (flow rate meters) are the category of constriction flowmeters. Orifice plates, flow nozzles, Venturi tubes, flow tubes and Pitot-Venturi instruments are variants of the basic constricted flow differential pressure concept. All constriction flowmeters reduce the area of the flow path to intentionally increase the velocity of the fluid. This is performed in a manner similar to partially covering the end of a garden hose to increase water velocity and distance of the water spray. The result of increasing the fluid velocity is a corresponding decrease in the static pressure associated with the fluid flow.

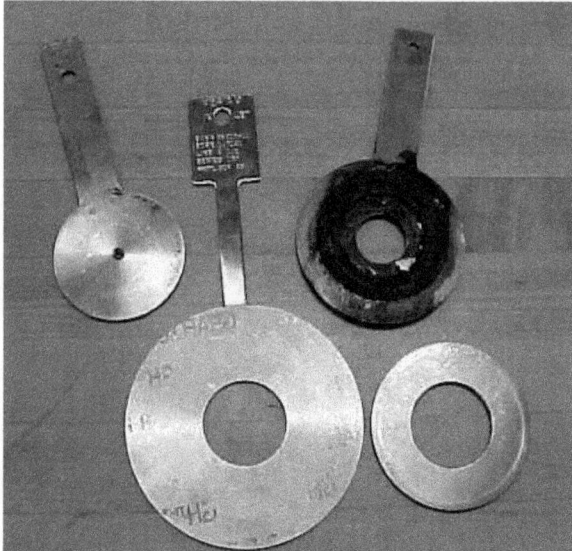

Figure 11 – Orifice plates used for liquid and gas flow rate measurement. Note the loop and orifice designations stamped on the gas plate tab.

The pressure difference between the pressure upstream, and the pressure within or just downstream from the restriction can be used to determine flow rate. The pressure difference, or differential pressure (often designated d/p or ΔP) is usually provided to a transmitter for conversion into a 4 to 20 milliamp or 3 to 15 psi signal.

When using differential pressure to determine flow rate, the unit measuring the upstream and downstream pressures, usually a transmitter, determines the differential pressure by subtracting the high pressure from the low pressure in the throat or slightly downstream. In some cases the pressure within the constriction can become low enough to create a vacuum; a condition that is frequently employed to mix materials.

Figure 12 – An orifice plate form of constriction flowmeter. The orifice plate is held between two flanges where the differential pressure taps are located. Liquid taps are commonly placed in the sides of the flanges to avoid any possibility of becoming plugged.

Occasionally a manometer is used in conjunction with a transmitter for a local indication at the flowmeter site. As with level and pressure sensing instruments, whenever a flowmeter is used with a transmitter, the flowmeter is often referred to as the primary element and the transmitter is the secondary element.

As mentioned previously, flow rate is the product of cross-sectional area multiplied by fluid velocity (A×V). Since the volumetric flow rate within the restriction is equal to the upstream or downstream flow rate away from the restriction, the volume of flow is constant throughout a closed system. In equation form this appears as,

$$Q_{upstream} = Q_{Restriction} = Q_{downstream}$$

$$A_1V_1 = A_2V_2 = A_3V_3$$

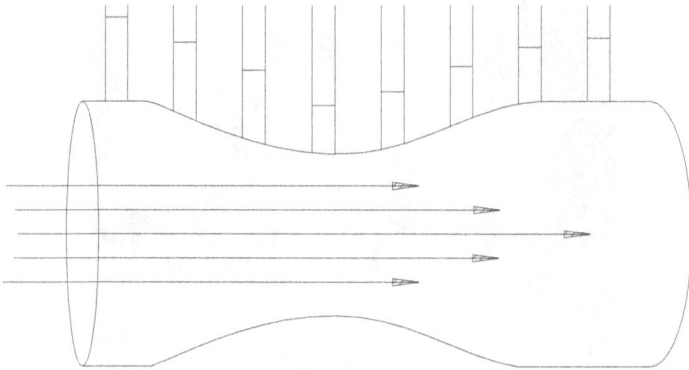

Figure 13 - Flow rate, Q, is constant throughout the length of a variable area flow path. The static pressure along the flow path is demonstrated by the manometer-like indicators above the flow section. Note that the pressure downstream does not fully recover to the value upstream and the lowest pressure is observed where the velocity is highest and the area is most narrow.

From the equation, it can be seen that as the area decreases at any point in the flow path, the corresponding fluid velocity must increase. When the fluid velocity increases, the associated fluid pressure must decrease since the velocity represents the kinetic energy and the pressure represents the potential energy form. Since the energy through the restriction must be equal to or less than the energy at the approach, if the kinetic increases it will be at the expense of the potential. Hence, the pressure will decrease within the restriction.

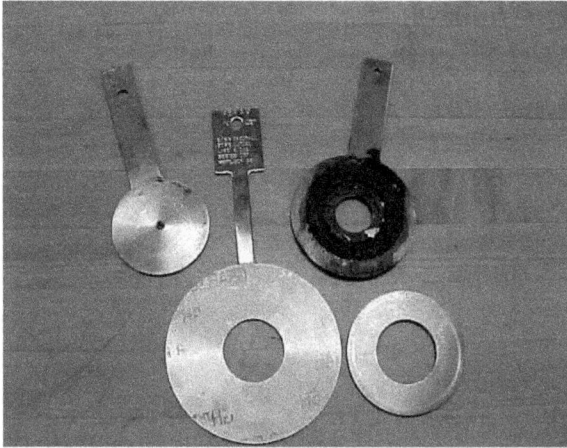

Figure 14 – Orifice plates usually contain identification tabs imprinted with information about the size of the pipe, constriction bore, material type, flow direction and location within the factory.

A low pressure is always associated with a high velocity. This effect is referred to as the *Venturi effect* and is described by Bernoulli's equation. The effect is obvious in all constriction flowmeters and can usually be observed with any object or fluid in motion. Bernoulli's equation is derived from an observation of the energy forms at the approach to a restriction and the energy forms within the restriction. The kinetic energy forms at both locations use Torricelli's equation of $V^2 = 2gh$ and the potential energy forms utilize the Pressure = Height * Density equation. By setting both equations equal to their common terms of "h" or height, the potential (PE) and kinetic (KE) energies at the approach to the restriction can be summed and set equal to the kinetic plus the potential energies within the restriction. The following equalities are created,

$$PE_{in} + KE_{in} = PE_{out} + KE_{out}$$

For the KE velocity term,

$$V^2 = 2gh$$

Solving for the "H" yields,

$$H = V^2 \div 2g$$

For the PE pressure term,

$$P = H \times D$$

Solving for the "H" yields,

$$H = P \div D$$

Bernoulli's Equation becomes,

$$(P_1 \div D) + (V_1^2 \div 2g) = (P_2 \div D) + (V_2^2 \div 2g)$$

Where the "1" subscript designates the upstream pressure and velocity characteristics, and the "2" subscript designates the pressure and velocity within the restriction. Since the pressure difference is used to represent the fluid flow rate, the equation should be arranged for differential pressure. Likewise, differential pressure represented as d/p or ΔP becomes;

$$P1 - P2 = d/2g \, (V_2^2 - V_1^2)$$

$$\therefore \Delta P = d/2g \, (V_2^2 - V_1^2)$$

Or,

$$d/p = d/2g \, (V_2^2 - V_1^2)$$

Where,

- H represents height in feet or inches,

- ΔP or d/p is the differential pressure at the flowmeter output in pounds per square foot,

- D, or "d" is the weight density in pounds per cubic foot,

- g is the gravitational acceleration constant of 32.2 feet per second squared, 2g = 64.4 ft/s^2,
- V is the fluid velocity in feet per second, computed from the given flow rate and the area of the pipe and the area of the throat (restriction)

Figure 15 – An orifice plate (bottom) and corresponding differential pressure transmitter (top).

When the above equation is used, make certain that units of feet, pounds and seconds are consistent throughout the computation. The resulting differential pressure will be computed in pounds per square foot and can then be converted into psi (by dividing by 144 in^2/ft^2), inches of H2O (by dividing the d/p by the density of water, 62.4 lbs./ft^3 and converting into inches by dividing by 1728 in³/ft³), inches of Hg (divide by the density of Mercury, 848.6 lbs./ft³) or whatever dimensions are desired.

The pipe and throat areas, used to compute the velocities in the previous equation, are rarely provided directly but can be computed from the information provided on the flowmeter. The pipe diameter and Beta ratio (throat diameter to pipe diameter ratio) is commonly available on constriction flowmeters. The indicated line diameter is the inside diameter of the connecting pipe upstream and downstream from the flowmeter. The Beta ratio, or d/D ratio, is the throat-to-pipe ratio, or the ratio of the constriction diameter to the pipe diameter. The resulting pipe and orifice (throat) areas can be computed from the diameter information.

Figure 16 - Orifice tabs containing identification information about the orifice plate. Note the control loop identifier (FRC-610E) on the tab at the left.

From Bernoulli's equation it can be seen the relationship between differential pressure and velocity is exponential, or non-linear.

$$d/p = d/2g \ (V_2{}^2 - V_1{}^2)$$

Since velocity (V) is flow rate (Q) divided by area (A), the relationship between differential pressure and flow rate, Q, is also exponential. Although the relationship can be made linear with a signal conditioning device referred to as a *square root extractor*, if left non-linear serious response problems may occur if a constriction flowmeter is used in a closed-loop control system and the flow rate extremes are large. Many constriction flowmeter manuals refer to the relationship between differential pressure and flow rate as;

$$Q = K \sqrt{\Delta P}$$

It should be noted here that this equation is used to demonstrate a relationship and unless the coefficient, K is well defined [using density, throat to pipe diameter ratio (Beta), and gravitational acceleration], the equation should NOT be used to solve for Q from ΔP unless "K" is well defined. The equation demonstrates how the differential pressure developed across the restriction will quadruple if the flow rate doubles. The non-linear relationship is indicated in the following graph.

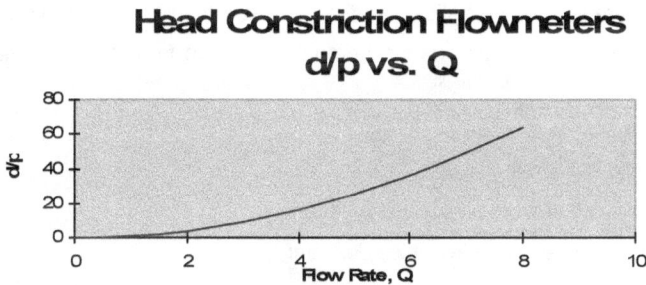

Figure 17 - Differential pressure vs. Q for head constriction flowmeters. Doubling the flow rate quadruples the differential pressure.

A closed-loop control system designed to operate with the flowmeter characteristics from the graph will respond in a sluggish manner at lower flow rates and will become very responsive and may oscillate at higher flow rates. For this reason additional components are commonly employed to establish a linear relationship between flow rate, Q, and the transmitter output signal.

Finding d/p From Q

The following problem demonstrates the use of Bernoulli's equation to determine differential pressure when flow rate (Q) and the flowmeter characteristics are known.

Example Problem:

Water is flowing through a 4-inch pipe between 50 Ft^3/Min and 100 Ft^3/Min. A head constriction flowmeter (diameter = 4", beta = .5) is used to measure the flow rate. Determine the differential pressure across the flowmeter at the minimum and maximum flow rates and demonstrate the square root relation between differential pressure and flow rate.

Solution:

First the areas and velocities need to be determined for substitution into Bernoulli's equation. Then arrange Bernoulli's to solve for differential pressure, d/p, in psid.

-Finding diameters,

Given, Beta = .5 and the pipe diameter = 4 inches,

Since, Beta = throat diameter ÷ pipe diameter

Or,

$$Beta = d/D$$

Algebraically,

$$d = Beta \times D$$

-Finding Areas,

Area = $\pi r^2 = \pi d^2/4$

Pipe Area, A_1 = 12.566 in^2 = .0873 ft^2

Throat Area, A_2 = 3.1416 in^2 = .0218 ft^2

-Finding Velocities at Q_{min},

$V = Q \div A$

V_1 = 50 ft^3/min ÷ .0873 ft^2

V_1 = 572.74 ft/min = 9.546 ft/sec

V_2 = 50 ft^3/min ÷ .0218 ft^2

V_2 = 2293.578 ft/min = 38.226 ft/sec

-Arranging Bernoulli's equation to solve differential pressure, d/p, at minimum flow rate, Q_{min}.

$d/p = P_1 - P_2 = d/2g (V_2^2 - V_1^2)$

d/p = [(62.4 #/ft^3)÷ (2*32.2 ft/s^2)] ×[(38.226 ft/s)^2 – (9.546 ft/s)^2]

d/p = (.969 #s^2/ft^4) ×(1461.227 ft^2/s^2 – 91.126 ft^2/s^2)

d/p = (.969 #s^2/ft^4) ×(1370.101 ft^2/s^2)

d/p = 1327.628 #/ft^2 = <u>9.2196 #/in^2</u> of differential at Q_{min} = 50 ft^3/min

To determine d/p at Q_{max} = 100 ft^3/min,

- Finding Velocities at Q_{max},

$V = Q/A$

V_1 = 100 ft^3/min ÷ .0873 ft^2

V_1 = 1145.48 ft/min = 19.0912 ft/sec

V_2 = 100 ft^3/min ÷ .0218 ft^2

V_2 = 4587.156 ft/min = 76.4526 ft/sec

-Arranging Bernoulli's equation to solve differential pressure, d/p, at minimum flow rate, Q_{min}.

$d/p = P_1-P_2 = d/2g\ (V_2{}^2 - V_1{}^2)$

$d/p = [(62.4\#/ft^3) \div (2*32.2\ ft/s^2)] * [(76.452\ ft/s)^2 - (19.0912\ ft/s)^2]$

$d/p = (.969\ \#s^2/ft^4) * (5845.0\ ft^2/s^2 - 364.474\ ft^2/s^2)$

$d/p = (.969\ \#s^2/ft^4) * (5480.526\ ft^2/s^2)$

$d/p = 5310.6297\ \#/ft^2 = \underline{36.879\ \#/in^2}$ of differential at $Q_{min} = 100\ ft^3/min$

Or,

Since, $Q = K\ \sqrt{d/p}$, doubling flow rate (Q) requires d/p to quadruple,

Therefore, d/p at $Q_{max} = 4 * 9.2196\ \#/in^2 = \underline{36.879\ psid}$

Unfortunately, a considerable amount of energy can be lost using head constriction flowmeters with abrupt restrictions like an orifice plate since the downstream pressure after the restriction never fully recovers to the amount of pressure before the restriction. The cost associated with such a pressure loss can be excessive, especially if one considers that many of these types of flowmeters are applied in plants operating 24 hours, every day of the year. *Flowmeters providing similar output and maintenance results without an obstruction in the flow path and corresponding energy loss can be proven to pay for themselves when compared to constriction flowmeters and are rapidly replacing them.* In today's energy preservation conscious workplace, common replacement flowmeters are the electro-magnetic (or the "mag flowmeter"), ultrasonic flowmeters and others.

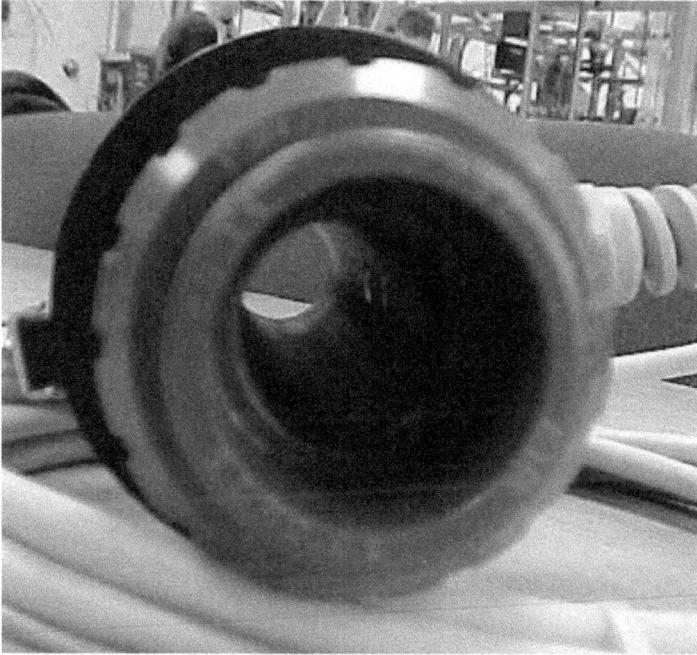

Figure 18 – Magnetic flowmeters offer no obstruction to the flow path.

Finding Q from d/p

Interpreting flow rate (Q) from the measured differential pressure is a fairly routine, one-step computation if Bernoulli's equation is rearranged and the velocities manipulated. The algebraic manipulation of the variables needed to solve for d/p follows.

Since,

$$P_1/D + V_1^2/2g = P_2/D + V_2^2/2g$$

$$(V_2^2 - V_1^2) = (2g/D)(P_1-P_2)$$

Since,

$$A_1V_1 = A_2V_2$$

Or,

$$V_1 = (A_2 \div A_1)\, V_2$$

Since area equals,

$$A = \pi r^2 = \pi d^2/4$$

Subbing for the areas yields,

$$V_1 = (\pi d_2{}^2/4) \div (\pi d_1{}^2/4) \times V_2$$

Canceling yields,

$$V_1 = (d_2{}^2/d_1{}^2) \times V_2$$

Since Beta equals the ratio of the throat (d_2) to pipe (d_1) diameters, substituting yields,

$$V_1 = Beta^2 \times V_2$$

Subbing the result into the rearranged Bernoulli's equation for V_1,

$$[V_2{}^2 - (Beta^2\, V_2)^2] = (2g/d)(P_1 - P_2)$$

Or,

$$[V_2{}^2 - Beta^4\, V_2{}^2] = (2g/d)(\Delta P)$$

Factoring V_2,

$$V_2{}^2\, (1 - Beta^4) = (2g/d)(\Delta P)$$

Solving for V_2,

$$V_2{}^2 = (2g/d)\, (\Delta P) \div (1 - Beta^4)$$

$$V_2 = \sqrt{[(2g/d)(\Delta P) \div (1 - Beta^4)]}$$

Or,

$$V_2 = [(2g/d)(\Delta P) \div (1- Beta^4)]^{1/2}$$

Since,

$$Q = AV$$

$$Q = A_2V_2 = A_2 * [(2g/d)(\Delta P) \div (1- Beta^4)]^{1/2}$$

Therefore, the equation to find flow rate (Q) from d/p is,

$$Q = A_2 * [(2g/d)(\Delta P) \div (1- Beta^4)]^{1/2}$$

Example Problem:

Differential pressure is measured at 10 psi (1440 psf) across a head constriction flowmeter with a Beta ratio of .8 and an inlet pipe diameter of 2 inches. Assuming water is flowing through the flowmeter, determine the flow rate (Q) in gallons per minute (GPM).

Solution:

To determine flow rate from differential pressure, units of feet, seconds and pounds will be used. It's best to perform the unit conversions and compute the area before substituting into the equation.

Knowing the pipe diameter (D) and Beta (d/D), the throat diameter (d) can be determined.

$$Beta = d/D$$
$$d = Beta * D$$
$$d = .8 \times 2"$$
$$d = 1.6"$$

Since the throat area (A_2) will be used to find flow rate (Q),

$$A_2 = \pi d^2/4 = 2.0106 \ in^2$$

Converting area to units of square feet,

$$A_2 = 2.0106 \text{ in}^2 \div 144 \text{ in}^2/\text{ft}^2 = .01396 \text{ ft}^2$$

Substituting values into the flow rate equation,

$$Q = A_2 \times [(2g/d)(\Delta P) \div (1 - \text{Beta}^4)]^{1/2}$$

$$Q = (.01396 \text{ ft}^2) \times [(64.4 \text{ft/s}^2/62.4 \#/\text{ft}^3)(1440 \#/\text{ft}^2) \div (1 - .8^4)]^{1/2}$$

$$Q = (.01396 \text{ ft}^2) \times [(1.032 \text{ ft}^4/\#\text{s}^2)(1440 \#/\text{ft}^2) \div (1 - .4096)]^{1/2}$$

$$Q = (.01396 \text{ ft}^2) * [(1486.08 \text{ ft}^2/\text{s}^2) \div (.5904)]^{1/2}$$

$$Q = (.01396 \text{ ft}^2) \times (2517.073 \text{ ft}^2/\text{s}^2)^{1/2}$$

$$Q = (.01396 \text{ ft}^2) \times (50.17 \text{ ft/s})$$

$$Q = .7004 \text{ ft}^3/\text{s}$$

Converting to gallons per minute,

$$Q = .7004 \text{ ft}^3/\text{s} \times (60 \text{ sec/min} * 7.48 \text{ gal's/ft}^3)$$

Or,

$$Q = .7004 \text{ ft}^3/\text{s} \times 448.8 \text{ GPM}/(\text{ft}^3/\text{sec})$$

$$Q = 314.34 \text{ GPM}$$

Flow Measurement Hardware

With the exception of the direct-volume positive displacement meters, flow measurements are all of the inferential type. This means that all properties, qualities, or conditions that are the result of fluid flow are used to indicate fluid flow rate. As mentioned in the introduction, so many types of flowmeters exist that categories of flowmeters have been established. The specific application niche of each category is governed by the operating conditions, the type of fluid, materials in suspension, electrical resistivity, boiling point, piping conditions, allowable pressure drop, the volume of flow, cost and a host of other factors. Following are the commonly available categories and flowmeters.

Head Constriction Flowmeters
Meters that deliberately insert an obstruction to create a reduction in the static pressure are commonly called head meters, or head-constriction flowmeters. Head constriction flowmeters reduce the flow path to increase the velocity of the fluid. Since kinetic and potential energy cannot both be increased without a pump or other energy transferring device, as the kinetic energy associated with fluid velocity increases the potential energy associated with the fluid pressure decreases. The differential pressure between the restriction, often called the throat, and upstream ahead of the throat is used to represent the flow rate of the fluid. This well-established principle, called the Venturi effect and described by Bernoulli's equation, is applied using several forms of flowmeters. Orifice plates, Venturi tubes, Flow tubes, Weirs, Parshall flumes, flow nozzles and Pitot tube flowmeters all squeeze the fluid flow path to create an indication of flow rate.

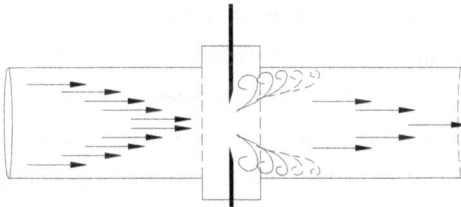

Figure 19 - A profile of a sharp edged orifice plate flowmeter. Abrupt restrictions in the flow path can severely reduce the energy recovered downstream. However, this is occasionally desirable.

Orifice Plates

This simplest form of constriction flowmeter is the orifice plate. Orifice plates are available in three forms, concentric, eccentric and segmented. Each appears as an opening in a plate that is contained between pipe flanges.

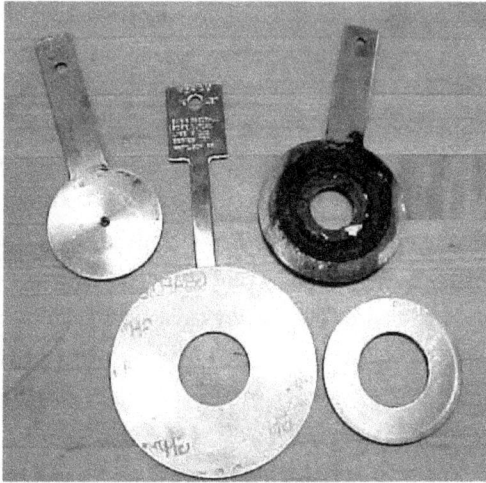

Figure 20 - The simplest type of constriction flowmeter is an orifice plate. The orifice plate in the upper right was removed after 20 years of measuring water flow.

Orifice plates are relatively inexpensive (a few hundred dollars) to purchase and if required could be made in the plant's machine shop and calibrated in the lab before application in the plant. However, their simplicity of operation and inexpensive cost comes at the expense of energy loss downstream. As can be imagined with an abrupt obstruction in the flow path, up to 50% of the upstream static pressure can be lost downstream. Since many plants operate 24 hours, seven days each week the lost energy can result in considerable expense over a period of years. In addition, turbulence, eddy currents and cavitations just downstream from the orifice plate can cause the edge around the orifice to wear resulting in a calibration shift of the differential pressure versus flow characteristics.

Orifice plates require that the flow path be maintained consistent for at least 8 pipe diameters upstream and 2 diameters downstream. Within these distances no valves, elbows, flanges or other flow components should be installed. Another requirement is that the fluid to be relatively clean and free of solids to avoid accumulating within and around the orifice. Fluids such as wood pulp, soup, sewage and slurries (solid materials using liquids for transport between locations) are not used with orifice plates. Accumulation of material near the orifice will result in a calibration shift since the physical characteristics of the restriction and flow path will be altered. Material compatibility is a major consideration when applying flowmeters. Cavitation created by the abrupt restriction in the flow path can also result in changes to the flow characteristics to change.

Three types of pressure taps, flange taps, pipe taps and Vena Contracta taps, are utilized to measure the resulting differential pressure across an orifice plate. Flange taps are pressure fittings placed through the orifice plate flanges into the flow path on each side of the orifice plate. Flange taps are convenient but may not result in as much measured differential pressure as pipe taps or Vena Contracta taps.

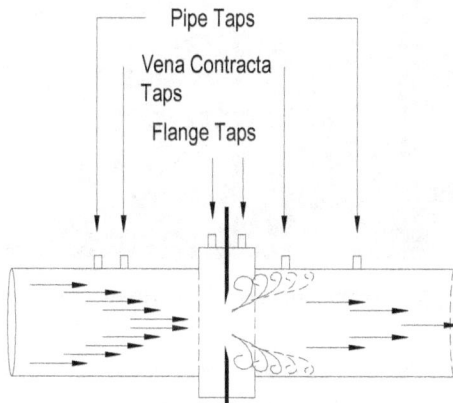

Figure 21 - Common pressure taps used with orifice plates and other head constriction flowmeters.

Pipe taps are located at equal distances, measured in pipe diameters, upstream and downstream from the orifice plate. The high-pressure tap connection is upstream and the low pressure is downstream. *Vena Contracta* pressure taps attempt to locate the largest amount of differential pressure. The Vena Contracta is the location downstream from the orifice where the fluid flow assumes the narrowest cross section. The effect can be observed when a thumb is placed over the end of a garden hose, the flow narrows just beyond the location of the thumb. Placing the low-pressure tap at the Vena Contracta measures a lower low-pressure since the velocity of the fluid is greatest at this location. Vena Contracta taps is particularly beneficial where flow rates and Reynolds Numbers are relatively low. The Vena Contracta is commonly determined by computation and is usually located one to three pipe diameters downstream from the orifice plate. High-pressure taps are typically located 1-2 pipe diameters upstream from the orifice plate.

Figure 22 - An orifice plate flowmeter with flange pressure taps.

Given the current interest in energy conservation, orifice plates are often substituted with obstruction-free, smooth bore flowmeters such as electro-magnetic or ultrasonic types. However, many applications continue to use orifice plates to minimize initial cost and limit the amount of spilled material resulting from a downstream rupture.

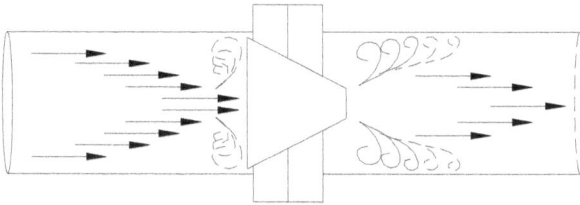

Figure 23 - Flow nozzles are similar in operation to an orifice plate.

Flow Nozzle

An improvement over the orifice plate is the flow nozzle. Although more expensive than an orifice, the constriction is more gradual resulting in less energy loss downstream. Flow nozzles with low Beta ratios (throat diameter to pipe diameter ratio) can result in significant energy loss downstream. As can be seen in the previous figure, the approach to the restriction is more gradual than the orifice plate. The tapered approach results in less upstream turbulence and less energy loss. The exiting fluid is not shaped downstream however and results in eddies and turbulence, representative of energy loss.

Flow nozzles are commonly employed in filling lines of tanks and storage vessels where a single upstream pressure gauge can be used in place of a differential pressure transmitter. When the flowing fluid exists through a flow nozzle into a ventilated storage tank the low-pressure side of the nozzle will be surrounded by atmospheric pressure. A gauge-type pressure gauge located upstream from the nozzle will vary in a non-linear square-root relation to the fluid flow rate. The following example problem demonstrates.

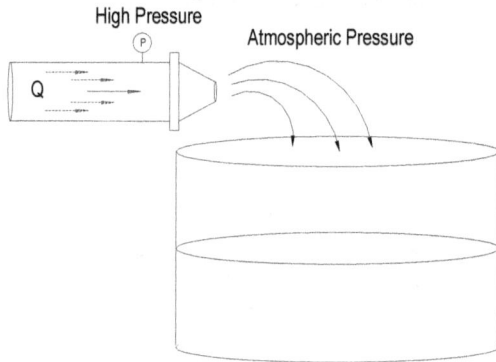

Figure 24 - A flow nozzle at the exit of a filling line utilizes a single pressure gauge to indicate flow rate.

Example Problem:

A flow nozzle, Beta = .6, 4 inch pipe, is used to measure water flow rate between 100 gpm and 400 gpm. Determine the pressure indicated on a single pressure gauge used upstream from the flow nozzle. Assume an atmospheric pressure of 14.7 psia.

Solution:

To determine the pressure indicated on the upstream gauge at the flow rate extremes, Bernoulli's is rearranged to solve for the high-pressure, P1.

Since,

$$P_1/d + V_1^2/2g = P_2/d + V_2^2/2g$$

And,

$$P_1-P_2 = d/2g \ (V_2^2 - V_1^2)$$

Solving for P1,

$$P_1 = d/2g \ (V_2^2 - V_1^2) + P_2$$

-Finding Areas,

Area $= \pi\, r^2 = \pi\, d^2/4$

Pipe Area, $A_1 = 12.566$ in$^2 = .0873$ ft^2

Throat Area, $A_2 = 4.5239$ in$^2 = .031416$ ft^2

-Finding Velocities at Q_{min},

100 gpm $= .2228$ ft^3/sec

$V = Q/A$

$V_1 = .2228$ ft^3/sec $\div .0873$ ft^2

$V_1 = 2.552$ ft/sec

$V_2 = .2228$ ft^3/sec $\div .031416$ ft^2

$V_2 = 7.092$ ft/sec

-Converting psia to psfa

14.7 psia $= 2116.8$ psfa

-Finding P1 at Q_{min},

$$P_1 = (62.4/64.4)\,(7.092^2 - 2.552^2) + 2116.8$$
$$P_1 = (.969)\,(50.296 - 6.513) + 2116.8$$
$$P_1 = (.969)\,(43.783) + 2116.8$$
$$P_1 = (42.426) + 2116.8$$
$$P_1 = 2159.226 \text{ psfa} = 14.99 \text{ psia}$$
$$P_1 \text{ at } Q_{min} = 14.99 \text{ psia} - 14.7 = .299 \text{ psig}$$

-Finding Velocities at Q_{max},

$$400 \text{ gpm} = .8913 \text{ ft}^3/\text{sec}$$
$$V = Q/A$$
$$V_1 = .8913 \text{ ft}^3/\text{sec} \div .0873 \text{ ft}^2$$
$$V_1 = 10.208 \text{ ft/sec}$$
$$V_2 = .8913 \text{ ft}^3/\text{sec} \div .031416 \text{ ft}^2$$
$$V_2 = 28.368 \text{ ft/sec}$$

-Converting psia to psfa

$$14.7 \text{ psia} = 2116.8 \text{ psfa}$$

-Finding P1 at Q_{max},

$$P_1 = (62.4/64.4)(28.368^2 - 10.208^2) + 2116.8$$
$$P_1 = (.969)(804.743 - 104.203) + 2116.8$$
$$P_1 = (.969)(700.54) + 2116.8$$
$$P_1 = (678.823) + 2116.8$$
$$P_1 = 2795.623 \text{ psfa} = 19.414 \text{ psia}$$
$$P_1 \text{ at } Q_{max} = 19.414 \text{ psia} - 14.7 = 4.714 \text{ psig}$$

Since the square root relationship between d/p and flow rate exists, an increase in flow rate of four times results in an increase in d/p of sixteen times, or,

$$Q = K \sqrt{d/p}$$

Venturi Tube

Among the more expensive and efficient means of constriction flowmeters is the Venturi tube. As can be seen in the following figure, the abrupt approach to the constriction evident in the orifice plate and flow nozzle has been replaced with a more gradual flow path. The downstream flow path just beyond the constriction has also been shaped to minimize turbulence and energy loss. Venturi's can commonly recover up to 90% of the upstream pressure downstream from the throat. As might be expected, the amount of differential pressure generated across the Venturi section may be somewhat less than with an orifice or flow nozzle. For this reason Venturi tubes are commonly associated with large diameter pipes and higher flow rates, as with flow measurement at the inlet of a municipal water treatment plant.

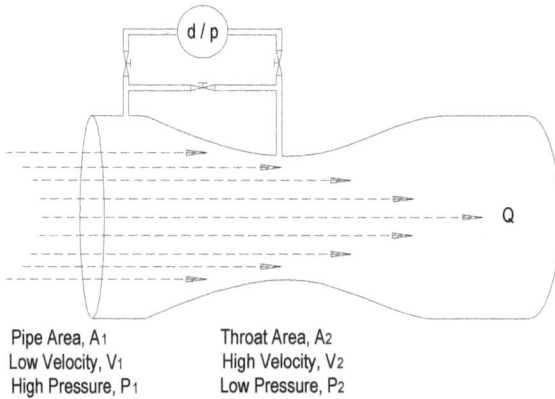

Pipe Area, A_1 Throat Area, A_2
Low Velocity, V_1 High Velocity, V_2
High Pressure, P_1 Low Pressure, P_2

Figure 25 – A Venturi tube gradually shapes the inlet and outlet fluid flow resulting in less energy loss downstream.

Many variations on the basic Venturi have appeared over the years. Some modify the means or location of the pipe taps while others place a slight ridge near the pressure taps. The modifications are generally performed to increase the amount of differential pressure (d/p) or to increase the amount of recovered downstream pressure.

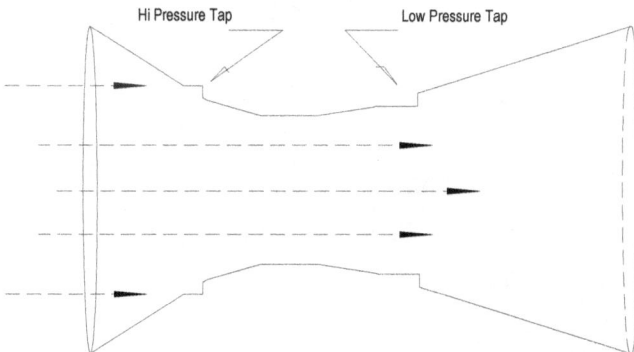

Figure 26 - The Flow Tube, a variation on the Venturi tube provides recessed pressure taps and a shortened overall length to create an increased differential pressure.

As the following example demonstrates, computing the differential pressure across a Venturi section is similar to other constriction flowmeters.

Example Problem:

A Venturi flowmeter is used to measure flow rate of water into a municipal treatment plant. Determine the differential pressure at the minimum and maximum flow rates given the following information.

Given:

Inlet pipe diameter = 4 feet

Beta = .5

Q_{min} = 5,000 gpm, H_2O

Q_{max} = 10,000 gpm, H_2O

Solution:

-Finding Areas,

Area = $\pi\, r^2 = \pi\, d^2/4$

Pipe Area, A_1 = 1809.56 in² = 12.57 ft²

Throat Area, A_2 = 452.39 in² = 3.14 ft²

-Finding Velocities at Q_{min},

5,000 gpm = 11.14 ft³/sec

V = Q/A

V_1 = 11.14 ft³/sec ÷ 12.57 ft²

V_1 = .886 ft/sec

V_2 = 11.14 ft³/sec ÷ 3.14 ft²

V_2 = 3.55 ft/sec

-Finding d/p at Q_{min},

$d/p = P_1 - P_2 = d/2g \, (V_2^2 - V_1^2)$

$d/p = [(62.4\#/ft^3) \div (2*32.2 \text{ ft/s}^2)] * [(3.55 \text{ ft/s})^2 - (.886 \text{ ft/sec})^2]$

$d/p = (.969 \ \#s^2/ft^4) * (12.6 \text{ ft}^2/s^2 - .785 \text{ ft}^2/s^2)$

$d/p = (.969 \ \#s^2/ft^4) * (11.82 \text{ ft}^2/s^2)$

$d/p = 11.45 \ \#/ft^2 = \underline{.0795} \ \#/in^2$ of differential

$d/p = 2.2$ inches Water Column (WC) at $Q_{min} = 5000$ gpm

-Finding Velocities at Q_{max},

10,000 gpm = 22.28 ft³/sec

$V = Q/A$

$V_1 = 22.28 \text{ ft}^3/\text{sec} \div 12.57 \text{ ft}^2$

$V_1 = 1.77$ ft/sec

$V_2 = 22.28 \text{ ft}^3/\text{sec} \div 3.14 \text{ ft}^2$

$V_2 = 7.1$ ft/sec

Note – Recognizing the square root relation between flow rate and differential pressure, the differential pressure should quadruple as the flow rate doubles. The following computation will demonstrate the differential pressure becomes four times the differential at minimum flow rate, or 8.8 inches of WC.

-Finding d/p at Q_{max},

$d/p = P_1 - P_2 = d/2g \, (V_2^2 - V_1^2)$

$d/p = [(62.4\#/ft^3) \div (2*32.2 \text{ ft/s}^2)] * [(7.1 \text{ ft/s})^2 - (1.77 \text{ ft/sec})^2]$

$d/p = (.969 \ \#s^2/ft^4) * (50.41 \text{ ft}^2/s^2 - 3.13 \text{ ft}^2/s^2)$

$d/p = (.969 \ \#s^2/ft^4) * (47.28 \text{ ft}^2/s^2)$

$d/p = 45.81 \ \#/ft^2 = \underline{.318} \ \#/in^2$ of differential

Or, $d/p = 8.8$ inches Water Column (WC) at $Q_{max} = 10,000$ gpm

In summary, the Venturi tube provides the smoothest flow pattern and therefore has the smallest energy loss since the flowing liquid completely fills the opening within the Venturi. The Venturi tube is also the most costly of the head-constriction type primary units. The flow nozzle approaches the Venturi tube in performance since the liquid fills about 90% of the throat area. The contraction of the flow stream as it passes through an orifice plate reduces the net effective area to about 60% of the nominal value. The orifice plate costs the least of the head-type units and is the simplest to construct but has the greatest downstream pressure loss. Most head-constriction meters are sensitive to flow conditions upstream of the flowmeter.

Pitot Tubes

Anyone that has spent time around aircraft has probably come into contact with a Pitot (pronounced "Pee-toh") tube. Traditionally used to measure fluid velocity instead of volumetric flow rate, Pitot tubes are located under the wing or near the nose of numerous military, private and commercial aircraft.

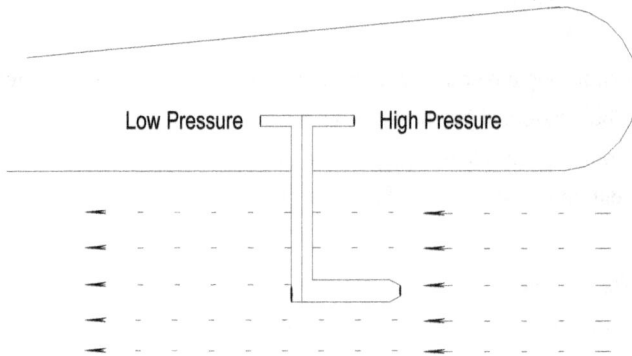

Figure 27 - Pitot tubes mounted under the wing of an aircraft develop a pressure differential as a means of determining the velocity of the aircraft.

In the case of aircraft, flowing air passing near the surface of the wing impacts upon the upstream port of the Pitot tube creating a high pressure. As the air stream passes the Pitot tube a reduced pressure is created on the backside. Both pressures are transferred to a mechanism or differential pressure sensor.

The difference between the two pressures is representative of air velocity, or in this case the aircraft velocity. Some weather stations utilize similar mechanisms to measure air velocity.

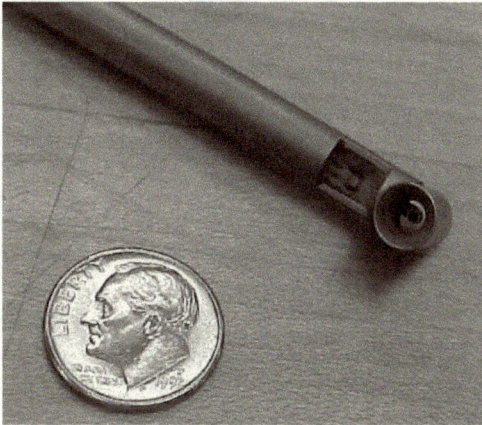

Figure 28 - Pitot tubes area available in a variety of sizes and assemblies. This Pitot tube was used to measure gas flow at a specific location in the flow path. A temperature sensing thermocouple located just above the Pitot element is used to provide a more accurate flow measurement.

Industrial Pitot tubes are available in a variety of configurations. Among the more common is the averaging Pitot tube. Rather than placing a single upstream and downstream pressure ports in the flow path, an averaging Pitot tube places multiple sets of pressure ports in the flow path. One upstream and downstream set is placed near the center of the pipe and another set can be found near the inside wall of the pipe. In this manner the upstream and downstream pressures will be the result of the high velocity near the center of the flow path and the lower velocity fluid flow along the pipe wall. The result will be an average of each of the upstream and downstream pressures. Averaging Pitot tubes are commonly recommended for applications where an orifice plate is not appropriate. As can be seen in the following figure, Pitot tubes do not obstruct the flow path as much as with constriction flowmeters.

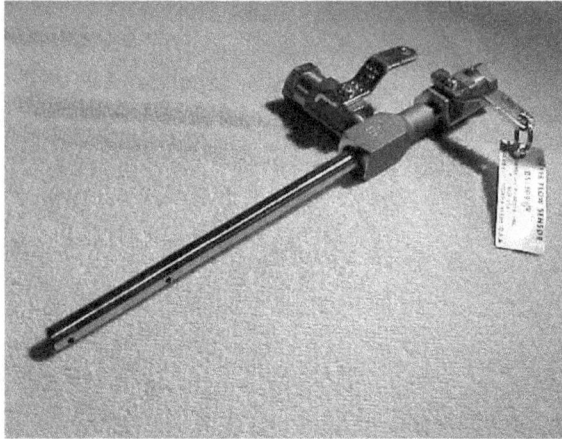

Figure 29 - The pressure sensing ports are obvious on this Dwyer averaging Pitot tube. The pressure taps at the top provide connection to the differential pressure (Phi - Plo) output.

As with constriction flowmeters, Pitot tubes develop a differential pressure from the fluid velocity. As such, the resulting differential pressure will yield a non-linear, square root relationship with the fluid velocity. In order for the averaging Pitot tube to operate accurately a uniform velocity profile must be accommodated in the vicinity of the flowmeter and liquids should be relatively clean without solids. It is suggested that 10 pipe diameters upstream and 5 downstream should be maintained free of obstructions, bends and elbows for best results with averaging Pitot tubes.

Figure 30 - Averaging Pitot tubes provide only a slight obstruction in the fluid flow path.

Square Root Extraction

As was noticed in the previous computations, the differential pressure resulting from a constriction flowmeter varies in proportion to the square of the fluid velocity. A differential pressure transmitter connected to the output of the constriction flowmeter will vary linearly with differential pressure, meaning the transmitter output will have a non-linear relation with the fluid flow rate.

Head Constriction Flowmeters
d/p vs. Q

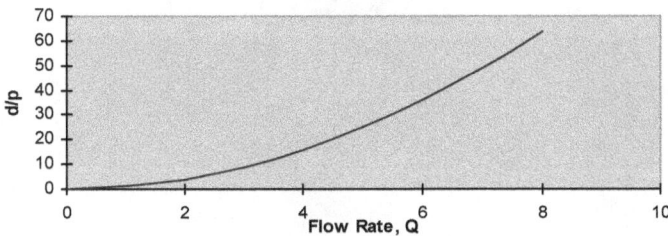

Figure 31 - The relation between flow rate (Q) and differential pressure (d/p) with constriction flowmeters is not a straight line.

Non-linear sensors result in less than desirable control characteristics. Control system response can be sluggish at low set points and overly responsive at higher set point values. If the transmitted flowmeter signal is not made linear, a flow control system will take an excessive amount of time to reach set point at low flow rates and may oscillate at higher flow rates. Converting the transmitted flowmeter signal to one that is linear with flow rate (Q) is performed by a *square root extractor*.

Figure 32 - An orifice plate flowmeter and a differential pressure transmitter (following photo). Without linearization, the 4mA to 20 mA transmitter output is exponential (non-linear) in relation to flow rate.

Figure 33 - A d/p transmitter is used to output 4-20 milliamps over a flow rate range. Square-root extractors are normally connected to the output of the transmitter but may be employed at other locations in the control loop.

The square root extractor is a component that performs a mathematical square root function on the output of the flowmeter or transmitter. When connected to the transmitter output, the 4mA to 20mA signal is applied to the square root extractor and is initially biased to eliminate the "Zero" value of the transmitter signal, usually 4mA. The square root function can then be accomplished using an application specific integrated circuit (ASIC) form of mathematical divider. The divider IC has its output tied back to one of the inputs. The resulting computation is described following.

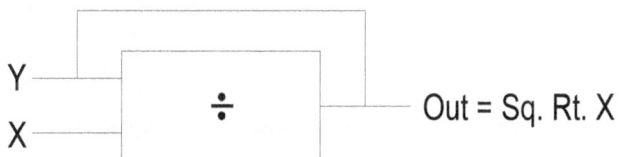

Figure 34 - A divider configured to perform a square root.

Since the divider output equals the division of the two inputs,

$$Out = X \div Y$$

With the output connected to the "Y" input,

$$Y = Out$$

Substituting Out for "Y" yields,

$$Out = X \div Out$$

Algebraically rearranging,

$$Out^2 = X$$

Or,

$$Out = \sqrt{X}$$

In this manner a mathematical divider circuit can perform the required square root function to linearize the flowmeter/transmitter output. Once linearly converted, the output from the square root circuit can be converted back into a 4mA to 20mA signal. Provisions for conditioning the output into 4-20 mA are usually contained within the computational module. The result will be a 4mA to 20mA signal that varies in a direct and proportional manner with flow rate (Q). Square root linearization is also commonly performed using software and analog input/output hardware.

Example Problem:

A constriction flowmeter develops 9.22 psid at 50 ft³/min of water flow and 36.88 psid at 100 ft³/min of water flow. The d/p is applied to a pressure transmitter with a gain of .433:1 and an output pressure offset of zero psi. Determine the output pressure from the transmitter and from a square root extractor connected to the output of the transmitter. Prove the converted signal linear with flow rate.

The following figure illustrates.

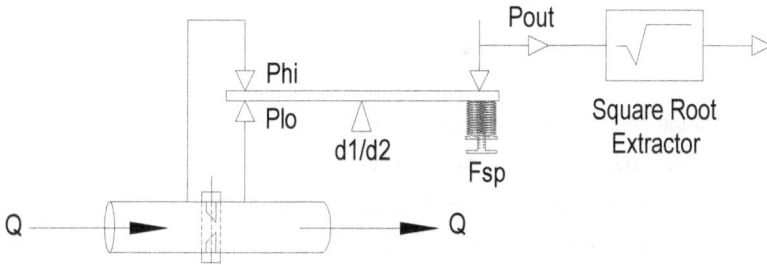

Figure 35 - A constriction flowmeter, differential pressure transmitter and square root extractor.

Solution:

The output of the flowmeter varies between 9.22 and 36.8 psid between the flow rate extremes of 50 and 100 cubic feet per minute. Note the differential pressure quadruples as the flow rate doubles. The resulting differential pressure is applied to the transmitter and is amplified by a d1/d2 gain of .433. The offset created by the transmitter spring force is set to zero to facilitate the square root computation.

The resulting transmitter output becomes,

$$\text{Transmitter Pout} = [(d1/d2) \times (Phi\text{-}Plo)] + Poffset$$
$$\text{Pout at Qmin} = [(.433) \times (9.22 \text{ psid})] + 0 \text{ psi}$$
$$\text{Pout at Qmin} = 4 \text{ psi}$$
$$\text{Pout at Qmax} = [(.433) \times (36.88 \text{ psid})] + 0 \text{ psi}$$
$$\text{Pout at Qmax} = 15.969 \text{ psi}$$

Once the transmitter output is determined, the resulting pressure range can be applied to the square root extractor. The output of the square root extractor becomes,

$$\text{Sq. Root Output at Qmin} = \sqrt{4} \text{ psi}$$
$$\text{Sq. Root Output at Qmin} = 2 \text{ psi}$$

$$\text{Sq. Root Output at Qmax} = \sqrt{15.969} \text{ psi}$$

$$\text{Sq. Root Output at Qmax} = 3.996 \text{ psi} \approx 4.0 \text{ psi}$$

Note that the square root extractor output varies in a direct proportional manner with the flow rate. As the fluid flow rate varies between 50 and 100 ft³/min the square root output varies between 2 and 4 psi. A doubling of the signal represents a doubling of flow rate. Once made linear the signal can again be conditioned to 3-15 psi, 4mA to 20mA or any other convenient signal range.

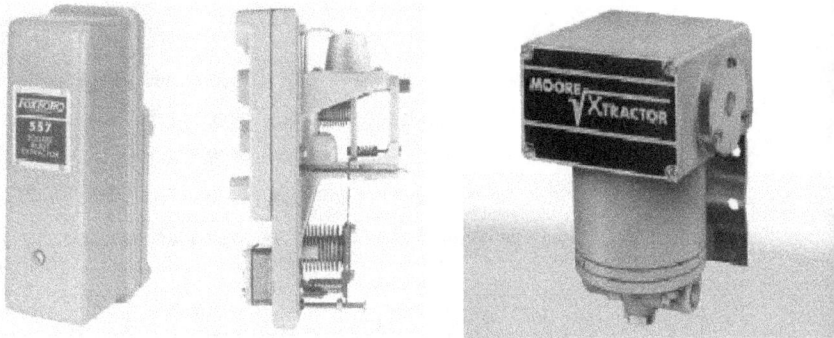

Figure 36 - Pneumatic square root extractors, courtesy of Foxboro (left) and Moore (right).

Flowmeter - Transmitter Example Problems

Flowmeters are commonly employed with a corresponding transmitter. The transmitter is required if the output signal from the flowmeter is to be compatible with commercially available controllers. A transmitter converts the differential pressure from the constriction flowmeter into a standard signal range of 4 milliamps (mA) to 20mA or in the case of pneumatic system, 3psi to 15psi. The following example problems demonstrate methods of analyzing a pneumatic flowmeter/transmitter pair and determining the required transmitter gain and offset for a given output pressure range.

Figure 38 - An orifice plate flowmeter (bottom) and transmitter (top).

399

When transmitters are employed with head constriction flowmeters such as orifice plates, Venturi tubes, Pitot tubes and flow nozzles, the resulting transmitter output signal will be a linear relation with the transmitter input pressure range. The transmitter output signal will be non-linear (square-root) when compared to the flowmeter input of flow rate, Q. As mentioned in the previous section, if the transmitter output signal is desired to be linear with the flowmeter input of flow rate (Q), a square root extractor must be employed somewhere downstream from the flowmeter output. The transmitter output is the usual location for a square-root function.

An example of a constriction flowmeter-transmitter pair is provided in the following figure.

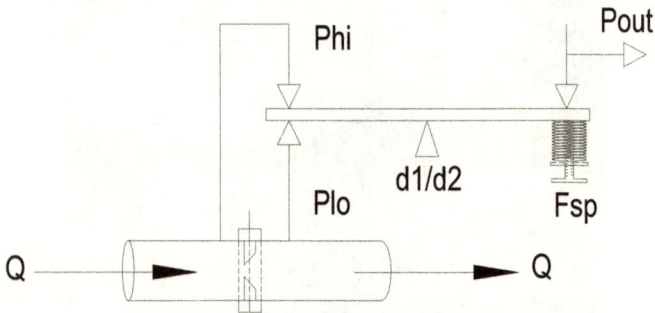

Figure 39 - A representation of an orifice plate flowmeter as the primary element and a torque-balance transmitter as the secondary element.

Example 1 – Determining Transmitter Pout Given d/p

Using the computed values from a previous example problem, water is flowing through a 4-inch pipe between 50 ft³/min and 100 ft³/min. A head constriction flowmeter (4", beta = .5) is used to measure the flow rate. Determine the differential pressure across the flowmeter at the minimum and maximum flow rates and demonstrate the square root relation between differential pressure and flow rate. Next, assume the differential pressure is applied to a torque-balance transmitter with gain (d1/d2) of .25:1and an offset spring force of 1.3 lbs. down. Assume the diaphragm areas are equal at 1 in^2.

The differential pressure extremes at minimum and maximum flow rate will be determined first followed by the output pressure range from the transmitter.

Solution:

First the areas and velocities need to be determined for substitution into Bernoulli's equation. Then arrange Bernoulli's to solve for differential pressure, d/p, in psid.

-Finding Areas,

Area $= \pi\, r^2 = \pi\, d^2/4$

   Pipe Area, A_1 = 12.566 in^2 = .0873 ft^2

   Throat Area, A_2 = 3.1416 in^2 = .0218 ft^2

-Finding Velocities at Q_{min},

V = Q/A

   V_1 = 50 ft^3/min ÷ .0873 ft^2

   V_1 = 572.74 ft/min = 9.546 ft/sec

   V_2 = 50 ft^3/min ÷ .0218 ft^2

   V_2 = 2293.578 ft/min = 38.226 ft/sec

-Arranging Bernoulli's equation to solve differential pressure, d/p, at minimum flow rate, Q_{min}.

   d/p = P_1-P_2 = d/2g ($V_2{}^2$ – $V_1{}^2$)

   d/p = [(62.4 #/ft^3)÷ (2*32.2 ft/s^2)] * [(38.226 ft/s)^2 – (9.546 ft/s)^2]

   d/p = (.969 #s^2/ft^4) ×(1461.227 ft^2/s^2 – 91.126 ft^2/s^2)

   d/p = (.969 #s^2/ft^4) ×(1370.101 ft^2/s^2)

   d/p = 1327.628 #/ft^2 = 9.2196 #/in^2 of differential at Q_{min} = 50 ft^3/min

To determine d/p at Q_{max} = 100 ft³/min,

- Finding Velocities at Q_{max},

V = Q/A

$\qquad V_1$ = 100 ft³/min ÷ .0873 ft²

$\qquad V_1$ = 1145.48 ft/min = 19.0912 ft/sec

$\qquad V_2$ = 100 ft³/min ÷ .0218 ft²

$\qquad V_2$ = 4587.156 ft/min = 76.4526 ft/sec

-Arranging Bernoulli's equation to solve differential pressure, d/p, at minimum flow rate, Q_{min}.

$\qquad$ d/p = P_1-P_2 = d/2g (V_2^2 – V_1^2)

$\qquad$ d/p = [(62.4#/ft³) ÷ (2*32.2 ft/s²)] * [(76.452 ft/s)² – (19.0912 ft/s)²]

$\qquad$ d/p = (.969 #s²/ft⁴) ×(5845.0 ft²/s² – 364.474 ft²/s²)

$\qquad$ d/p = (.969 #s²/ft⁴) ×(5480.526 ft²/s²)

$\qquad$ d/p = 5310.6297 #/ft² = 36.879 #/in² of differential pressure at Q_{min} = 100 ft³/min

Or,

Since, Q = K √d/p, doubling flow rate (Q) requires d/p to quadruple,

$\qquad$ Therefore, d/p at Q_{max} = 4 ×9.2196 #/in² = 36.879 psid

Once the flowmeter output pressure range has been computed the equation representing the transmitter can be determined.

Given the transmitter in the following diagram, the output equation becomes,

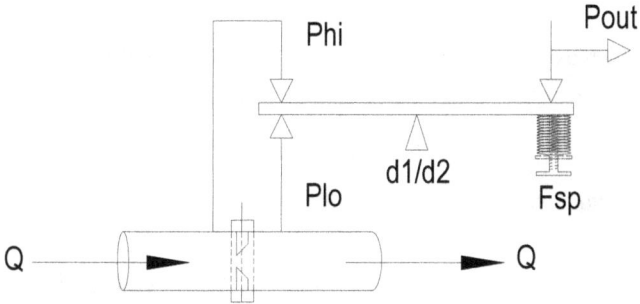

At balance, $T_{CW} = T_{CCW}$

$$F_{Hi} \times d_1 + F_{SP} \times d_2 = F_{Lo} \times d_1 + F_{Out} \times d_2$$

Since, $F = P \times A$

$$P_{Hi} \times A_{Hi} \times d_1 + F_{SP} \times d_2 = P_{Low} \times A_{Low} \times d_1 + P_{Out} \times A_{Out} \times d_2$$

$$P_{Hi} \times A_{Hi} \times d_1 + F_{SP} \times d_2 - P_{Low} \times A_{Low} \times d_1 = P_{Out} \times A_{Out} \times d_2$$

$$(P_{Hi} \times A_{Hi} \times d_1 / A_{Out} \times d_2) + (F_{SP} \times d_2 / A_{Out} \times d_2) - (P_{Low} \times A_{Low} \times d_1 / A_{Out} \times d_2) = P_{Out}$$

Or,

$$P_{Out} = d_1/d_2 \, (P_{Hi} - P_{Low}) + F_{SP}/A_{Out}$$

Since,

$$d/p = \Delta P = P_{Hi} - P_{Low}$$

The transmitter output becomes,

$$\boldsymbol{P_{Out} = d_1/d_2 \, (d/p) + F_{SP}/A_{Out}}$$

Or,

$$\boldsymbol{P_{Out} = d_1/d_2 \, (\Delta P) + F_{SP}/A_{Out}}$$

Subbing the minimum differential pressure (d/p) value into the equation yields,

$$P_{Out} = (.25/1)(9.2196 \text{psid}) - 1.3/1$$
$$P_{Out} = 2.3 \text{psi} - 1.3 \text{psi}$$
$$P_{Out} - 1.0 \text{psi}$$

The transmitter output at the minimum flow rate is 1psi.

Subbing the maximum differential pressure (d/p) values into the equation yields,
$$P_{Out} = (.25/1) (36.879psid) - 1.3/1$$
$$P_{Out} = 9.2psi - 1.3psi$$
$$P_{Out} = 7.9psi$$

The transmitter output at the maximum flow rate is 7.9psi

It is worth noting that the transmitter output is <u>linear with the differential input pressure but is non-linear when compared against the flow rate, Q</u>. This is due to the constriction flowmeters' square-root relationship between differential pressure and flow. If the transmitter output signal is required to be a linear representation of flow rate, a square root computation is required. A *square root extractor* placed on the output of the transmitter normally provides the square-root computation.

Example 2 – Determining d1/d2 and Fsp Given d/p and ΔPout

Assume the transmitter output of the previous example is to be calibrated to provide 3psi at Qmin and 15psi at Qmax. Determine the necessary gain (d1/d2) and spring force quantities.

Solution:

Recalibrating the pressure transmitter to output a 3-15psi signal requires determining and subsequently setting new values of gain (d1/d2) and spring force (Fsp). Given the differential pressure input range of 9.2196 psi to 36.879 psi and the transmitter output equation found previously the new transmitter values to provide a 3-15 psi output pressure can be found.

Since,
$$P_{Out} = d_1/d_2 (\Delta P) \pm F_{SP}/A_{Out}$$
$$\text{Where } \Delta P = d/p = \text{differential pressure}$$

Momentarily assuming the spring force to be zero, d1/d2 can be determined.

Assuming $F_{SP} = 0$,

$$P_{Out} = (d_1/d_2) * (\Delta P)$$
$$d_1/d_2 = \Delta P_{Out}/\Delta P$$
$$d_1/d_2 = (15psi-3psi)/(36.879psi-9.2196psi)$$
$$d_1/d_2 = 12psi/27.66psi$$
$$d_1/d_2 = .434 \text{ or,}$$
$$\mathit{d_1/d_2 = .434:1}$$

Finding Fsp,

Since,

$$P_{Out} = d_1/d_2 (\Delta P) \pm F_{SP}/A_{Out}$$
$$P_{Out} A_{Out} - (d_1/d_2) (\Delta P) A_{Out} = F_{SP}$$
$$F_{SP} = (15psi * 1in^2) - (.434*36.879psi*1in^2)$$
$$F_{SP} = 15\# - 16.005\#$$
$$\mathit{F_{SP} = -1\#}$$

Noting that Fsp was assumed to be upward during the output equation derivation, the resulting spring force being negative is opposite to the direction originally assumed. Therefore the biasing spring force direction is down or;

$$Fsp = 1 \text{ lb. } \downarrow$$

The proof of the resulting d_1/d_2 and F_{SP} values is determined by substituting the gain, d_1/d_2, and bias spring force, F_{SP}, values into the P_{Out} equation with corresponding input and output pressures. The substituted values should yield a torque-balanced result at the minimum and maximum input and output values.

Positive Displacement Flowmeters
It is commonly understood that the most accurate flow measuring instruments are *positive displacement flowmeters*. Traditionally, positive displacement flowmeters have been employed where accuracy of total volumetric flow (not flow rate) is required. Gasoline pumps, residential water and natural gas meters are examples of positive displacement flowmeter applications.

Figure 40 - A residential water meter is a common example of a positive displacement flowmeter. The wire connects a transmitter within the meter to a remote indicator outside the home.

Most positive displacement meters utilize a similar operating principle. A quantity of fluid is captured, an indication is incremented and the quantity of fluid is transferred to the meter output. The operation can be visualized as a water wheel.

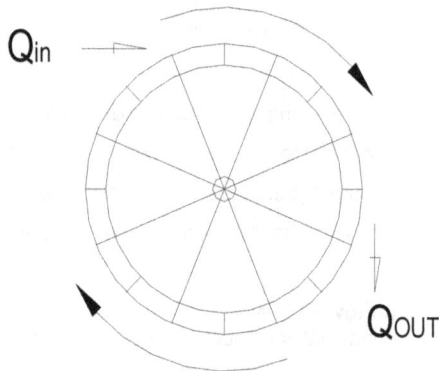

Figure 41 - Positive displacement flowmeters operate in a similar manner to a water wheel trapping quantities of water as it flows.

Residential water meters utilize a nutating disk or "Wobbler" as the primary element. The "Wobbler" traps quantities of water in transit, rotating about a vertical axis as the water enters and leaves the meter. An indication is registered using a gear arrangement and mechanical counter as the disk rotates although various sensors can also be used to detect the rotation of the primary element.

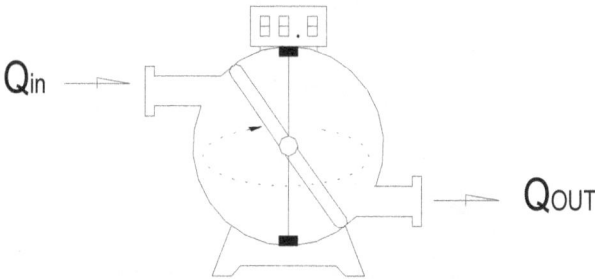

Figure 42 - As water enters a residential water flowmeter a quantity is captured. The disk then rotates about a vertical axis and passes the water to the outlet.

Figure 43 - The primary element in a residential water meter is the nutating disk. Photo courtesy of J-Tec Associates.

Given that positive displacement flowmeters are often employed in changing temperature and pressure environments, in order to assure accurate flow totals the meters must be temperature and pressure compensated.

Figure 44 - Another example of a positive displacement flowmeter is a residential gas meter. The meter cover has been removed to expose one of the internal diaphragms. Using a linkage mechanism between the front and rear (not shown) diaphragms, the diaphragms expand and contract as natural gas enters and leaves the meter. The linkage is connected to the mechanical counter at the top of the unit.

Until recently, positive displacement flowmeters were understood to be the most accurate in measuring total flow. However, other commercially available flowmeters such as ultrasonic and magnetic flowmeters currently are said to equal the positive displacement meter in accuracy.

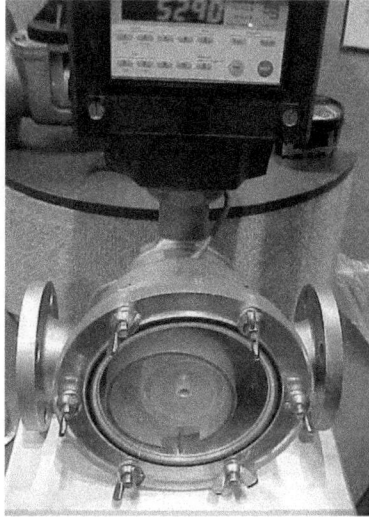

Figure 45 - An oscillating piston form of positive displacement flowmeter. Photo courtesy of J-Tec Associates.

Impinging Flowmeters

This class of flowmeter places an object in the fluid flow path in order for the fluid velocity to impact, or impinge upon a sensing primary element. Target, Turbine and Rotameters are common examples of impinging flowmeters. Impinging flowmeters utilize the "Impact head" of a flowing fluid to generate an indication of flow rate. A fluid's impact head is described by Torricelli's equation as,

$$V^2 = 2gH$$
$$H = V^2 \div 2g$$

Since,

$$V = Q/A$$
$$H = (Q/A)^2 \div 2g$$
$$H = Q^2 / A^2 \div 2g$$
$$Q^2 = H \times 2g \times A^2$$
$$Q = \sqrt{(H \times 2g \times A^2)}$$

409

Where,

     V = Fluid velocity (varies in direct proportion to flow rate, Q)

     g = Gravitational acceleration

     H = the velocity "Head"

     A = Cross-sectional pipe area

The "H" term represents the generated force or velocity "Head" that results when a fluid in motion strikes a surface area. Operation of the target flowmeter is analogous to one holding their hand out of a car window when driving at highway speeds. The impinging air velocity creates a force that is representative of the fluid, or car's velocity. Unfortunately, as can be seen by the equation the relationship between fluid velocity and the resulting force or "Head" is non-linear. A square root extractor is commonly used to generate a signal that varies linearly with fluid velocity and flow rate.

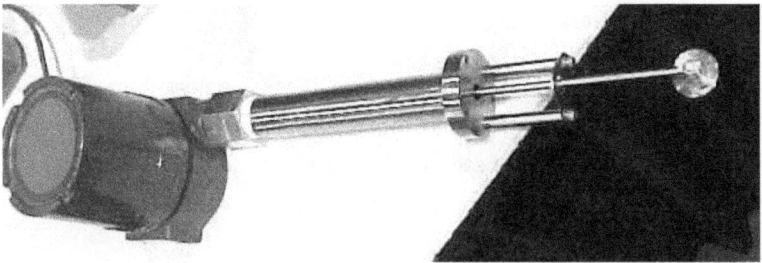

Figure 46 - A Target type of impinging flowmeter and transmitter, liquid strikes the target sensing element at the right, which is normally enclosed in a section of pipe. The transmitter at left converts the impact force of the flowing fluid into a 4-20 mA signal.

Target Meters

The Target flowmeter places a blunt object directly in the center of the flow path. The target meter employs the impact force of the moving fluid to bend an elastic member or control the position of a flapper/nozzle (orifice/baffle) combination. As the flowing fluid impacts upon the target a force is generated. The resulting force varies as the square of the fluid velocity. As the flow rate doubles the fluid velocity also doubles. According to Torricelli's equation, as the velocity doubles the resulting force upon the target quadruples. Target flowmeters are capable of operating in severe conditions that restrict the use of many differential pressure meters.

Turbine Meter

The Turbine flowmeter is composed of an impeller placed into the flow path of a fluid. The impeller, appearing similar to a propeller, rotates as the fluid to be measured impinges upon the impeller. As a Turbine's impeller rotates, the individual rotating blades are detected and are used to register an indication of the fluid flow rate.

Figure 47 - A Turbine flowmeter, transmitter and associated straightening vanes.

Impinging flowmeters often require the fluid flow to be laminar in the location of the flowmeter. To assure laminar (streamlined) flow, straightening vanes are commonly inserted ahead and downstream of the flowmeter.

Figure 48 - Straightening vanes are commonly used to assure laminar flow.

Turbine flowmeters employ mechanical and electrical read-out display systems. The rate of rotation may be measured by a potential (voltage) or by frequency, which is directly proportional to speed. Turbine flowmeters are available in a variety of sizes and flow ratings and are almost always used with straightening vanes to maintain laminar flow in the vicinity of the flowmeter.

Figure 49 – A close-up view of a Turbine flowmeter and straightening vanes.

Obviously, a major disadvantage of impinging flowmeters is the obstruction created by the primary element placed in the flow path. The sensing elements of the Target, Turbine and Rotameter not only drop a considerable amount of pressure but also limit the types of fluid materials that can be measured. Slurries and other fluids with solids in suspension prohibit impinging flowmeters from being applied for obvious reasons. The solids could accumulate and block the flow path or may damage the primary element.

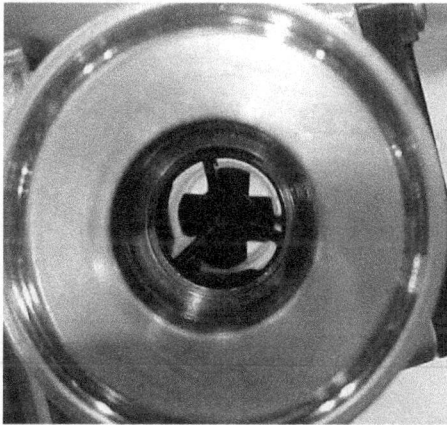

Figure 50 - A view down a section of pipe looking towards a Turbine flowmeter. The amount of flow obstruction is obvious.

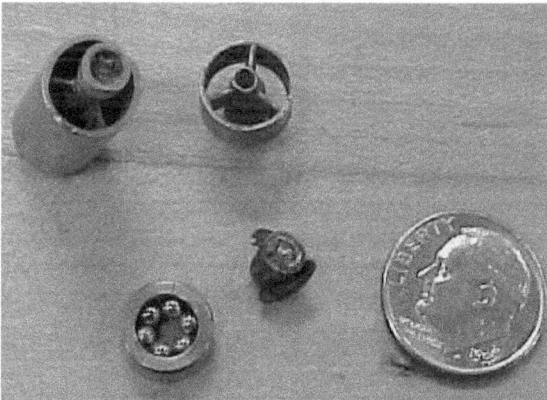

Figure 51 – A turbine flowmeter used to measure gasoline flow through an automotive fuel system. The impeller is at the center with straightening vanes at top and the bearing assembly below. A magnetic pickup was used to detect the rotation of the impeller. The straightening vanes are used to create a laminar flow condition as the fuel passed through the impeller.

Figure 52 – An impeller for use with a turbine flowmeter within a three-inch pipe. This particular impeller came into contact with the straightening vanes when the vanes moved into the impeller as a result of excessive vibration. The damage is obvious on the impeller blades.

Rotameters

Rotameters, also called variable-area flowmeters are common impinging instruments for measuring fluid flow rate. Popular with gas measurement applications, Rotameters are especially common in medical applications.

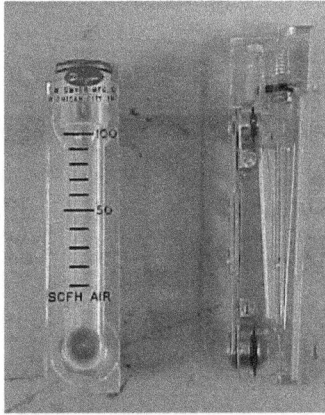

Figure 53 - Upon close inspection, the fluid flow area of the Rotameter varies between the inlet (bottom) and the outlet (top) to provide a linear measurement range.

Always mounted vertically, the Rotameter places an object in the flow path whose weight is cancelled by the impinging fluid flow upward. Originally designed to rotate when placed in the flow path, the "Float" of the Rotameter is used to indicate the fluid flow rate on an adjoining linear scale.

Figure 54 - The Rotameter, or variable area flowmeter uses a cone-shaped section to affect a linear flow rate indicating scale.

The variable area of the cone-shaped meter section is responsible for the linear indicating scale used in conjunction with the Rotameter. The area of the flowmeter varies in an exponential manner from the flow inlet at the bottom to the outlet at the top. According to Torricelli's equation, the impact head of fluid force generated upward on the "Float" varies as the square of the fluid velocity.

$$H = V^2 \div 2g$$

As the flow rate increases the float moves upward into a wider section of flow path. The wider section allows more of the fluid to flow around the float rather than impacting upon the float. Since less fluid impinges upon the float at higher flow rates, the float increments upward in a linear manner with the fluid flow rate variation. The end result is a linear indicating flowmeter.

Figure 55 - Rotameters are available in a variety of sizes and flow ratings, some are available with built-in needle valves. Note the variable areas and the bullet-shaped "Float" at the right side of the photo.

Flow meters of the rotameter class have operating ranges (also called turn-down ratio) of about 10:1, while differential pressure meters generally are limited to about 3:1 range when using round-chart instruments. Rotameters are relatively independent of upstream piping.

Magnetic Flowmeters
Whenever a conductor passes through a magnetic field a voltage is induced across the conductor. The amount of induced voltage is contingent upon the strength of the field, the length of the conductor and the velocity of the conductor motion.

$$v = KBLV$$

Where,

v = Induced (generated) voltage

K = dimensional coefficient for unit compatibility

B = flux density in Webers/meter^2

L = conductor length

V = velocity of the conductor

This basic generator principle is applied in the operation of the Electromagnetic, or Magnetic flowmeter. The Magnetic flowmeter, or "Mag-meter" has become one of the more popular flowmeters over the past 20 years primarily due to its obstruction-free bore.

Figure 56 - A view down the flow section of a Mag-meter. The electrodes are visible about halfway down the flowmeter on the right.

Electromagnetic flowmeters utilize the effect of liquids to appear conductive when flowing. By placing a coil around PVC tubing and setting electrodes into the sides of the tubing, a magnetic field will pass through the pipe. As the fluid passes through the magnetic field a potential is induced. The electrodes act as pick-ups measuring the generated potential. The resulting potential (voltage) varies in a direct and proportional manner with the fluid velocity and flow rate.

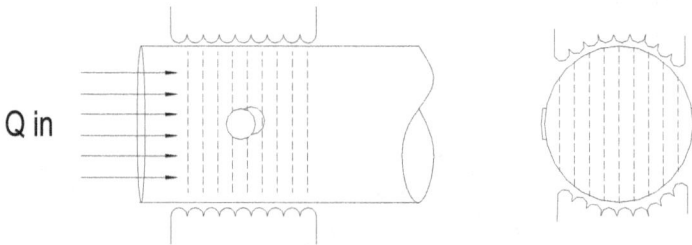

Figure 57 – Functional cross-section (left) and end (right) views of an electro-magnetic flowmeter.

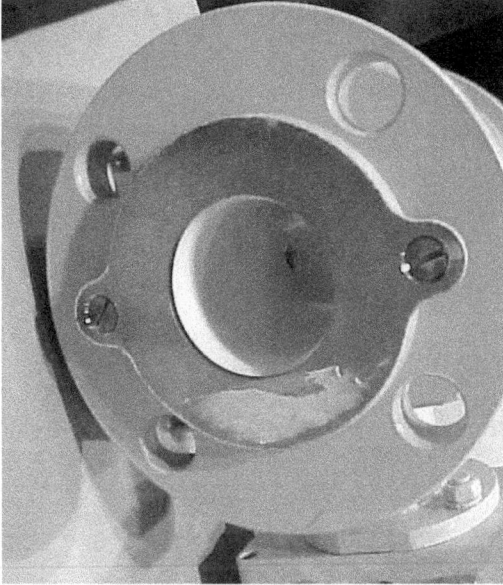

Figure 58 - An electro-magnetic flowmeter, or Mag-meter. Note the Teflon coated interior of the flowmeter. A metal pipe section would short the generated voltage. An electrode is visible within the flow section.

Coriolis Flowmeters

This material is currently under development.

Ultrasonic Flowmeters

This material is currently under development.

Mass Flowmeters

Measuring the flow of gases provides challenges that are generally not found in the measurement of liquids. The varying pressures and temperatures that gases can assume while in transit eliminate the majority of previously referenced flowmeters from use. Temperature and pressure influences result in variations of gas density. For this reason a different form of flowmeter is often required to measure gas flow.

Figure 59 - A mass flowmeter for the purpose of measuring gas flow rate. Photo courtesy of Key Instruments.

Mass flowmeters utilize operating principles that consider gas density. Among the more common methods for measuring mass flow is the use of thermal effects. In many thermal mass flowmeters an amount of the flowing gas is detoured to pass through a heated section of tubing. The flowing gas has a cooling effect. The greater the density of the gas, the greater the cooling effect. Temperature sensors located upstream and downstream from the heat source measure the temperature differential of the passing gas and from the variation in temperature the flow rate of the gas is implied.

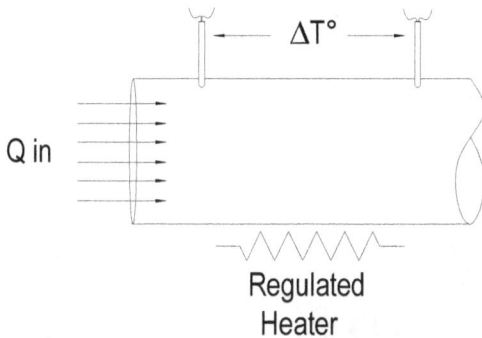

Figure 60 - A simplified diagram of a thermal mass flowmeter.

Mass flowmeters commonly equate the flowing gas to atmospheric pressure at a predefined temperature, usually 68° F. Although the actual flowing gas pressure may not be atmospheric and the temperature may not be 68° F., knowing the gas density allows an equivalency to be determined and a corresponding display and/or transmitted signal to be established.

Figure 61 - These mass flowmeters utilize a series of parallel plates in the flow path to establish a slight pressure drop. The pressure drop varies in direct proportion to the gas density.

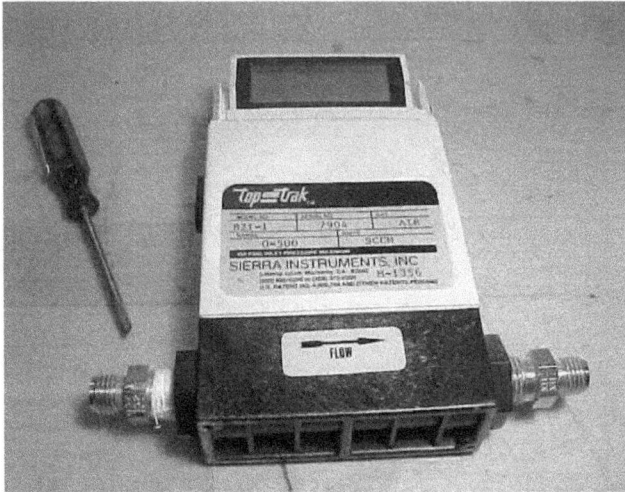

Figure 62 - A mass flowmeter for measuring 0-500 SCFM of airflow.

Vortex Flowmeters

When a fluid in motion passes an obstacle placed in the fluid flow path, vortices are shed immediately downstream from the obstacle. A vortex is whirlpool-appearing eddy current; vortices are an accumulation of eddies or whirlpools. The effect is obvious when observing a flag in a constant wind. Ripples appear to form at the flagpole and roll towards the end of the flag. The frequency of the vortices being shed is directly proportional to the velocity of the fluid impinging upon the obstacle, and inversely proportional to the fluid velocity.

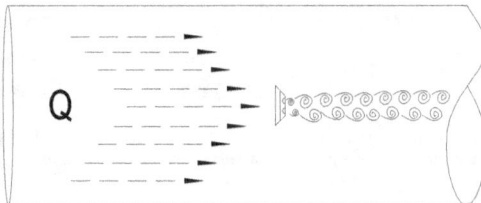

Figure 63 - An example of vortices being shed by an obstacle placed in the flow path.

$$F = K \times V \div D$$

Where,

K = dimensional coefficient for unit compatibility

F = frequency of shed vortices

V = fluid velocity

D = fluid density

Numerous variations in vortex meters have appeared in industrial flowmeters. Most of the differences are the result of variations in the shape of the obstacle and the means with which the vortices are detected. Strain gauges, pressure sensors, ultra-sound are other technologies have been applied to detect the individual vortices appearing downstream from the obstacle. The operational end result is the same with all vortex meters; the frequency of the shed vortices is proportional to the fluid velocity and the fluid flow rate.

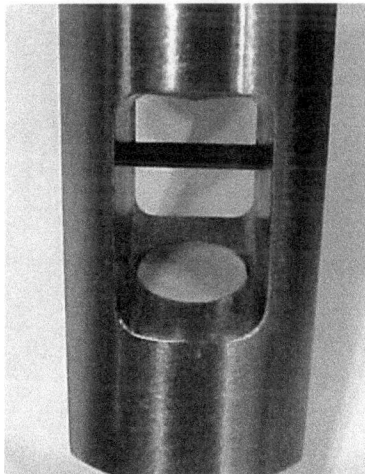

Figure 64 - An up-close view of a vortex flowmeter. The bluff-body vortex shedding obstacle is the dark horizontal element, the vortex sensor is the circular piezo-electric element below. Piezo elements generates small voltages when forces are applied. Another piezo element is located above the bluff body.

Figure 65 - The previous vortex element placed into a pipe. Photo courtesy of J-Tec.

Figure 66 - Another vortex flowmeter (bottom) and transmitter (above). Photo courtesy of Johnson-Yokogawa Inc.

Open-Channel Flow Measurement

Volumetric measurement of open-channel flow is often required for wastewater treatment, potable water filtration, storm sewage drainage and civil engineering and geological measurement projects. Open-channel volumetric flow measurements, also called measurements of discharge, are typically made where the flowing material is, or can be concentrated in a narrow stream.

Figure 67 - A Parshall Flume used to measure open-channel flow rates. Note the shape approximates a Venturi flowmeter. Photo is courtesy of the U.S Geological Survey.

Open-channel flow measurement utilizes instruments that are made to the application usually by the individuals performing the test. The volumetric measurement of the liquid, usually water or a water-based mixture, is only applicable to relatively small flow rates, but is understood to be the most accurate method of measuring such flows. The flow rate measurement involves observing the time required filling a container of known capacity, or the time required to partly fill a calibrated container to a known volume. The only equipment required, other than the calibrated container, is a stopwatch.

Figure 68 - A common technique of measuring flow and calibrating flowmeters involves a bucket and a stopwatch. Photo is courtesy of the U.S Geological Survey.

The container can be calibrated in either of two ways. In the first method, water is added to the container by known increments of volume, and the depth of water in the container is noted after the addition of each increment. In the second method, the empty container is placed on a weighing scale, and its weight is noted. Water is added to the container in increments, and after each addition the total weight of container and water is noted, along with the depth of water in the container. The measurements are made several times to assure accuracy. The equation used to determine the volume corresponding to measured weights is,

$$V = 1/d \ (W_2 - W_1)$$

Where,

   V = volume of water in container, in cubic feet or cubic meters,

   W_2 = weight of water and container, in pounds or kilograms,

   W_1 = weight of empty container, in pounds or kilograms, and

   d = density of water, 62.4 lb/ft^3 or 1,000 k g/m^3.

A *Weir* is a notch of regular form through which water flows. The term is also applied to the structure containing such a notch. Thus a weir may be a depression in the side of a tank, reservoir, or channel, or it may be an overflow dam or other similar structure. Weirs are classified in accordance with the shape of the notch; there are rectangular weirs; triangular, or V-notch weirs; trapezoidal weirs; and parabolic weirs. The edge or surface over which the water flows is called the crest of the weir. The overflowing sheet of water is termed the nappe. A broad-crested weir may be defined as one having a horizontal or nearly horizontal crest sufficiently long in the direction of flow so that the nappe will be supported.

Figure 69 - A Weir plate used for open-channel flow measurement. Photo is courtesy of the U.S Geological Survey.

Weir Plate
A portable weir plate is a useful device for measuring flow rate (discharge) when depths are too shallow and velocities too low for a reliable current-meter measurement of discharge. A 90 degrees V notch weir, as shown in the previous and following figures is particularly suitable because of its sensitivity at low flows.

The U.S. Geological Survey commonly uses three different sizes of weir plate; the recommended dimensions are provided in the following table.

The weir translates flow rate into level (depth) of the liquid to be measured. The weir acts as a dam causing liquid to backup upstream from the weir. The resulting level is representative of the liquid's volumetric flow rate. Once the liquid level is measured an equation is used to determine the volumetric flow rate. Local indication and signal transmission is possible using commonly available liquid level measurement and transmission apparatus.

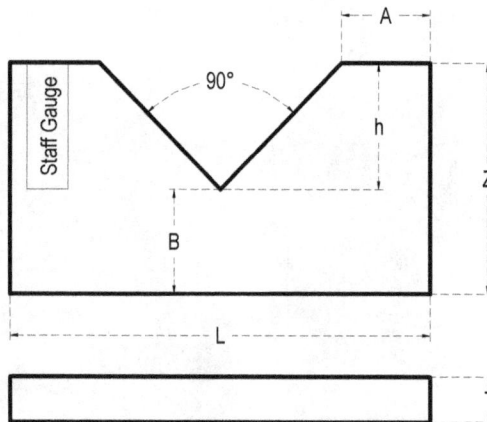

Figure 70- Portable weir plate design.

Weir	Z	h	B	L	A	T	Weight
Large	1.75	1.00	.75	4.0	1.0	16 ga.	24 lbs.
Medium	1.25	.8	.45	3.0	.7	14 ga.	17 lbs.
Small	.75	.47	.28	2.0	.53	10 ga.	8 lbs.

Figure 71 - Portable Weir dimensions for fabrication. All dimensions are in feet. Design data is provided courtesy of the U.S Geological Survey.

The weir plate is made of galvanized sheet iron, using 10 to 16-gauge metal. The 90 degrees V-notch that is cut in the plate is not beveled but is left with flat, even edges. The larger weir plates are made of thinner material than the smaller weir plates, but the medium and large plates are given additional rigidity by being framed with small angle irons fastened to the downstream face. A staff gauge attached to the upstream side of the weir plate with its zero (bottom) at the lowest elevation of the notch, is used to read the head (liquid level) on the weir. The staff gage should be installed far enough upstream from the notch to be outside the region of "drawdown" of water going through the notch. Drawdown, the downward sloping gradient of the liquid as it approaches the notch, becomes negligible at a distance from the vertex of the notch that is equal to twice the head on the notch. Consequently, if the weir plate has the dimensions recommended in the previous figure the staff gage should be installed near one end of the plate.

A large weir plate of the dimensions shown in the previous figure can measure discharges (flow rates) in the range from 0.02 to 2.0 ft³/s (0.00057 to 0.057 m³/s, or .15 to 15 GPM) with an accuracy of +/- 3 percent, if the weir is not submerged. A weir is not submerged when there is free circulation of air on all sides of the nappe. The general equation for flow over a sharp-edged 90 degree V-notch weir is,

$$Q = Ch^{5/2}$$

Where,

    Q = discharge flow rate in ft³/sec or m³/sec,

    h = static head above the bottom of the notch, in feet or meters,

    C = coefficient of discharge.

Each weir should be rated by volumetrically measuring the discharge corresponding to various values of head. In the absence of such a rating, a value of 2.47 may be used for C in the previous equation when English units are used, or 1.36 when metric units are used.

Installation concerns are addressed in the application notes at the end of the chapter. Once the weir is placed into operation after installation, the Weir obstructs the flow path until a pool is formed upstream from the notch. The accumulated upstream liquid will pass through the notch as the upstream level rises. As the flow approaching the weir increases, the flow through the notch varies and the level upstream from the weir will also vary. The resulting upstream level, or "head" is then used to determine the discharge flow rate through the notch using the provided equation.

However, weir measurements are less accurate than the other types of flowmeter measurements described in this chapter. For this reason multiple tests during field calibration is highly suggested. Weighing the liquid passed by a weir is the most accurate check on weir performance.

Parshall Flume
An alternative to the weir is the Parshall flume. A flume is a tapered chute that is frequently used to channel fluids. The Parshall flume measures open-channel flow rate utilizing a sophisticated, custom-made apparatus that approximates a Venturi tube cross-section in appearance. Ralph Parshall at Colorado State University developed the Parshall flume early in the 20th century and although complicated to install and build, Parshall flumes are scientifically accepted and commonly used for open-channel flow measurement.

A Parshall flume is composed of three sections, a tapered inlet, a throat, and a diverging section at the outlet. The measured head (level) in the upstream section ahead of the throat is used to determine the volumetric flow rate, or discharge.

Figure 72 - A Parshall flume. Photo is courtesy of the U.S Geological Survey.

A portable Parshall flume can be assembled to determine flow rate when depths are shallow and velocities low. The portable flume in the following figure is used by the U.S. Geological Survey and is a modified form of the standard Parshall flume having a 3-in (0.076 m) throat. The modification consists, primarily, of the removal of the downstream diverging section of the standard flume. The purpose of the modification is to reduce the weight of the flume and to make it easier to install.

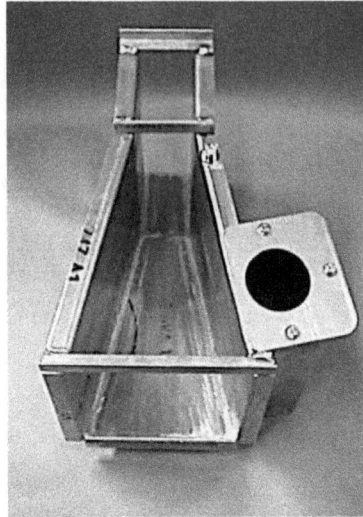

Figure 73 - A portable Parshall flume. Note the diverging outlet section has been removed and a stilling well has been attached to the inlet section. Photo is courtesy of the U.S Geological Survey.

Since the portable Parshall flume has no downstream diverging section, it cannot be used for measuring flows when the submergence ratio exceeds 0.6. Submergence is the ratio of the downstream head on the throat to the upstream head on the throat. Although a submergence ratio of 0.6 can be tolerated without affecting the rating of the portable flume, in practice the flume is usually installed so that the flow passing the throat has virtually free fall. That is usually accomplished by building up the streambed a couple of inches under the level converging floor of the flume. Portable Parshall flume specifics are included in the application notes at the end of the chapter.

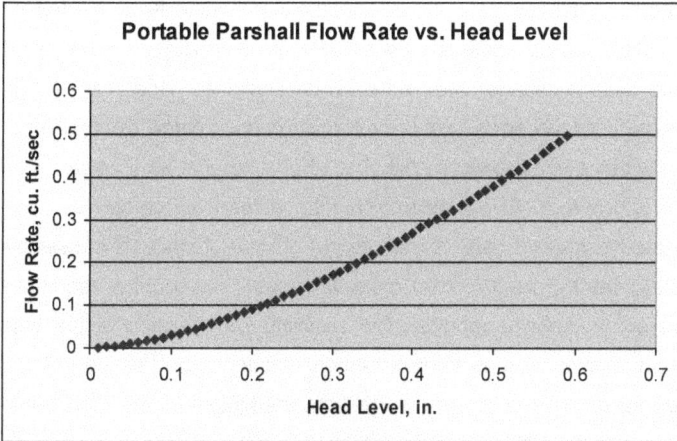

Figure 74 - Head versus flow rate for the portable Parshall flume. Note the similarity to other constriction flowmeter input/ouput characteristics.

Dry Flow Measurement

Flow measurement of dry material employs many of the basic factors involved in other measurements. Common primary element measurement techniques employ strain, forces of acceleration, torque and force-balance sensing technologies.

Figure 75 - Belt scales and storage bins provide opportunities for measuring the flow of dry materials.

433

Flow rate and total flow can be determined either electrically or mechanically. In many applications dry material is directed to a conveyor belt or other material transport system where the material is contained and various sensing technologies can be employed. A belt scale is a common example of such a system. A belt scale measures the flow of dry material as it travels along the length of a conveyor. At a location within the distance of the conveyor, a spring and torque mechanism balances the weight of the material as it passes over the mechanism. As the weight of the material varies the amount of required spring force varies in direct proportion. The resulting compression of the balancing spring can be measured using displacement (linear position) sensors. The output of the displacement sensor can then be converted into 4-20 mA representative of the material weight within the region of the mechanism on the conveyor belt.

Additional dry flow material is currently under development.

Application Notes

Please Note – The following material is in an ongoing state of construction and revision. The material is acquired from personal experience, manufacturers sales brochures and maintenance manuals. Feel free to provide suggestions for improvement, applications and related items as necessary.

Differential Pressure
Constriction flowmeters should be sized to produce a differential head within the 100-inches of water (3.6 psid) differential pressure range. Variations caused by temperature, specific gravity, or filling liquid in the impulse lines tend to be minimized and undesirable effects on the operation of associated equipment are minimized within this range. In general, the best results are obtained in the differential pressure range of 20 to 100 in. of water.

Orifice, Pitot, Venturi
The selection of the primary device – orifice plate, nozzle, or Venturi should be guided by the following:
-The orifice plate has been the most commonly used primary device and is commonly the first choice unless other factors apply. Orifice plate characteristics are well known and can be readily fabricated and field calibrated with a bucket and timing device if necessary.
-The flow nozzle is recommended in place of the orifice plate when there are moderate amounts of solids in suspension.
-Since the Pitot tube measures velocity at a single location within the flow path, the Pitot tube is recommended when the fluids are clean, the line is large, and the speed is relatively high.
    a. The Venturi tube is recommended when:
        1) The permanent pressure drop is to be reduced to a minimum.
        2) There may be large amounts of suspended solids.
        3) The maximum accuracy of measurement is desired when the flow involves highly viscous liquids.

Flow Nozzles
Flow nozzles will handle about 60% more fluid than will an orifice plate for any given pressure drop. This is because the full area of the flow nozzle will be filled with liquid, whereas the orifice plate will have a constricted stream passing through the orifice. Aside from the increased capacity, the flow nozzle can handle liquids with greater amounts of suspended materials. Also, a shorter amount of straight pipe is required before the nozzle than before the orifice plate and downstream piping requirements are minimum.

Orifice Plates
    b. The upstream edge of the orifice plate must be sharp and square with the sharp edge always installed upstream and the plate must be plane. Orifice plates are not suited to two-directional flow. If the tee plate has

a small hole to permit the escape of gases, the hole must be at the top of the pipe.

    c. The orifice opening must be accurately centered in the pipe. Improper location can cause errors that may be in either direction; the sense of the error depends upon how the plate is centered with respect to the taps.

    d. If the pipe taps are installed with the pipe screwed so far in that it interferers with the smooth flow of fluid, or if burrs are left in, the coefficients will be erroneous.

    e. Whenever steam, a heated gas or vapor that might condense in the piping is to be metered for flow rate by using sealed chambers, the piping must be insulated to ensure that the steam reaches the reservoir. One-inch pipe is suggested for connecting the supply main with the metering section. This will provide for a free interchange of vapor and condensate.

Pressure Taps

Pressure taps should never be made into the bottom of the line. The taps, regardless of the type used, must always be positioned to prevent dirt and sediment from clogging the lines and to keep the instrument sealed with fluid. Provision must also be made for permitting gases to return to the line. Steam and liquid measurements require taps in the side of the pipe. Gas taps should be made to the top of the pipe.

    f. *Vena Contracta taps.* In order to function properly, these taps are located one pipe diameter upstream and about one-half pipe diameter downstream. The exact position depends upon the value of the orifice ratio, which may be between 0.1 and 0.8. Because of mechanical problems, these taps are not used with pipe smaller than 4 in. in diameter.

    g. *Flange taps.* The correct position of these taps is 1 in. from both the upstream and the downstream faces of the plate. For pipes up to and including 3-½ in., the permissible orifice ratio is 0.20 to 0.70. The larger sizes of pipe may employ ratios of 0.10 to 0.75.

    h. *Pipe taps.* In order to obtain the maximum pressure drops, pipe taps are made 2 ½ pipe diameters upstream and 8 diameters downstream. The best results will be obtained when the orifice ratios are limited to 0.20 to 0.50 for pipe sizes of 2 in. and under and to ratios or 0.20 to 0.60 for pipes between 2 and 3 in.

Condensate Traps

Condensate traps must both be at the same height. Inequalities in level will cause significant errors. When the two traps are not at the same elevation, a false head will be developed. A somewhat different type of error is created when the bent nipple fills with condensate. This may either add or subtract from the pressure differential, depending upon the direction of flow. To prevent such action, the nipple should be adequately insulated so that condensation will not occur in it. The performance is also aided by using 1-in. pipe, with ¾ in. as a minimum. Any such pipe should be well insulated. Should it be desirable to mount the meter and the condensate traps above the steam line, the lines should be 1 in. in diameter and be provided with adequate insulation.

Weir and Parshall Flume

To install the weir, the weir plate is pushed into the streambed. Remove stones or rocks that prevent even penetration of the plate. Use a level to insure that the top of the plate is horizontal and that the face of the plate is vertical. Through eyebolts that are attached to the plate, rods are driven into the streambed to maintain the weir in a vertical position. Soil or streambed material is packed around the weir plate to prevent leakage under and around it. Canvas or a similar material should be placed immediately downstream from the weir to prevent undercutting of the bed by the discharged fluid.

When the weir is installed it will cause a pool to form on the upstream side of the plate. No readings of head on the notch should be recorded until the pool has risen to a stable elevation. The head should then be read at half-minute intervals for about 3 min, and the mean value of those readings should be the head used in the provided equation to compute discharge.

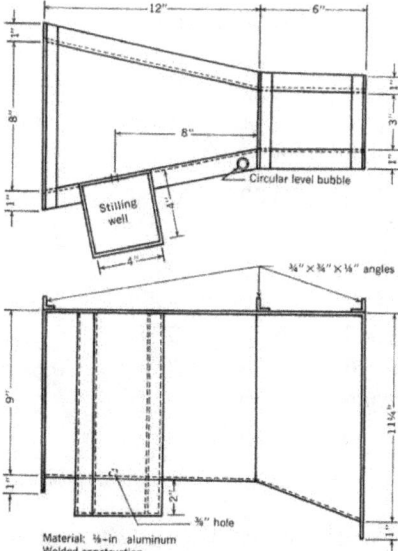

Material: ¼-in aluminum
Welded construction
Note: This stilling well can accommodate a 3-in float and be used with a recorder if continuous measurement is desired for a period.

The previous figure shows the design plan and elevation of the portable Parshall flume. The gage height, or upstream head on the throat is read in the small stilling well that is hydraulically connected to the flow by a $3/8$ in hole. Once the stilling well level has been determined the flow rate can be found using the following table.

Gauge height (ft)	Discharge (ft^3/s)	Gauge height (ft)	Discharge (ft^3/s)	Gauge height (ft)	Discharge (ft^3/s)
0.01	0.0008	0.21	0.097	0.41	0.280
0.02	0.0024	0.22	0.104	0.42	0.290
0.03	0.0045	0.23	0.111	0.43	0.301
0.04	0.0070	0.24	0.119	0.44	0.312
0.05	0.010	0.25	0.127	0.45	0.323
0.06	0.013	0.26	0.135	0.46	0.334
0.07	0.017	0.27	0.144	0.47	0.345
0.08	0.021	0.28	0.153	0.48	0.357
0.09	0.025	0.29	0.162	0.49	0.368
0.10	0.030	0.30	0.170	0.50	0.380
0.11	0.035	0.31	0.179	0.51	0.392
0.12	0.040	0.32	0.188	0.52	0.404
0.13	0.045	0.33	0.198	0.53	0.417
0.14	0.051	0.34	0.208	0.54	0.430
0.15	0.057	0.35	0.218	0.55	0.443
0.16	0.063	0.36	0.228	0.56	0.456
0.17	0.069	0.37	0.238	0.57	0.470
0.18	0.076	0.38	0.248	0.58	0.483
0.19	0.083	0.39	0.259	0.59	0.497
0.20	0.090	0.40	0.269		

Figure 77 - Rating table for 3-in modified Parshall flume

When the flume is installed in the channel, the floor of the converging inlet section is set in a level position by using a bubble level attached to one of the braces. Soil or streambed material is then packed around the flume to prevent leakage under and around it. After the flume is installed, water will pool upstream from the structure. No gage-height readings should be recorded until the pool has risen to a stable level. After stabilization of the pool level, gage-height readings should be taken at half-minute intervals for several minutes. The mean value of the readings is used in the previous table to obtain the flow rate of discharge. A carefully made measurement should have an accuracy of +/- 2 to 3 percent.

Remote reading open channel flow measurement applications require that the (liquid level) sensor be mounted over the centerline of the weir or flume and at a height exceeding the sum of the highest flow level and the sensor's blocking distance.

Measurement Notes

1 - Symptom. The chart of the recording meter indicates many random or noisy measurements when the flow is constant.

Avoid using constriction flowmeters to measure flow of any liquid near the liquid's boiling point. A liquid's boiling temperature reduces with ambient pressure and although the liquid may be removed in temperature from its normal boiling point, the turbulence created as the liquid passes through the orifice may cause it to boil, and the work that is done on the also helps to induce boiling. The tendency toward boiling is increased by using a sharp-edged orifice. Venturi meters are less subject to these disturbances than are meters using orifice plates. For example, a liquid which boils at 120° under the usual conditions of pressure but which is being measured at 70°F may well cause trouble by boiling.

2 - Symptom. The measurement error increases as the flow increases.

Inaccuracies may be caused by the misalignment of pipe. A raised section that was improperly positioned when welded to the flanges and rough particles or edges on the inside of a weld may also cause inaccuracies.

REMEDY. Use extreme care when welding, fabricating, and assembling pipe that is to be used near the sensing elements of the flowmeter.

2 - Symptom. Meter reads high and shows periodic erratic behavior.

Orifice plates, Venturi sections, and other constriction flowmeters should always be located upstream from any valve or source of a major pressure drop. Valves may cause gas or vapor to separate from the liquid. Because the mixture is no longer homogeneous, readings are in error.

REMEDY. Locate valves and related items downstream.

3   - Symptom. For a known constant value of flow, the meter readings become smaller and smaller.

Improper selection of the primary measuring element can cause improper metering performance. Orifice plates and similar constriction flowmeters must be selected on the basis of the material to be measured and the amount of material in solution which may drop out, thereby changing the approach to the constriction.

REMEDY. Remove material from in front of the plate or replace plate with a flow nozzle or Venturi if this problem is serious.

4   *Improper arrangement of piping to a differential pressure unit can create reading errors or give an indication of negative flow of steam when the flow is known to be positive.* The indication of negative steam flow can be caused by an inadequate exchange of steam and condensate in the piping to the differential pressure transmitter. Correct operation of these instruments requires that there be a constant flow of steam into both sections and of condensate out of both sections. The presence of a burr or dirt in the pipe can impede the flow of condensate back to the pipe. When this happens, the pipe fills with water and has the effect of creating a negative head, which might indicate a negative flow.

REMEDY. (1) Examine carefully differential pressure piping and remove obstruction. (2) Use nothing but gate valves in piping. (3) Insulate pipe to condensate traps; use ¾ - to 1-in. pipe to permit free flow of condensate.

5   - Symptom. If a metal object such as a rod or screwdriver is held against the condensate chambers while the ear is held to the other end, and a slurp or gurgle is heard, the indication is that the chamber is full of water.

Steam flowmeter give negative readings for positive steam flow. The preceding operating difficulty has the same symptoms as this one, but the cause and cure are quite different. During a start-up operation, boilers sometimes discharge water into the steam lines. Water also may be present in the lines because of condensation. As this water comes down the pipe, it enters and clogs the differential meter piping.

REMEDY. Install needle valves on the top of each condensate chamber and bleed until live steam comes through whenever this trouble occurs.

6   - Symptom. *Differences in the specific volumes of steam at various measuring points will give values that do not add up to the volume generated at the boiler.* The sum of the recordings of the various flowmeters does not add up to the flow indicated by the master meter. Errors may be as much as 50%.

REMEDY. If measurements are to be accurate, they should be made only at the point of use. Temperature measurements should also be made at the instrument so that the proper factors may be applied. The assumed temperature may be grossly incorrect. The condensate that may have been accumulated at various points along the line should always be considered in determining the performance of the meters.

7 - Symptom. When gases of low molecular weight are being measured, the results are erroneous.

Erroneous density values of a gas can cause serious errors.

REMEDY. The presence of water vapor is generally neglected. A measure of the importance of this term is the fact that dry hydrogen gas has a density of 0.0695, while saturated hydrogen has a density of 0.112.

8 - Symptom. Meter readings of gas flow are thought to be inaccurate.

Inadequate analysis data results in incorrect flow measurements.

REMEDY. The density of the gas being measured must include all of the components, not just one or two of them. To the degree that the presence of some of the gases is neglected, the final answers will be in error. To assume a gas composition that is not true can only lead to incorrect values.

9 - Symptom. The flow rate governs the error in certain gas measurements.

The errors of the gas flowmeter are a function of flow rate.

REMEDY. *The valves governing flow rate must always be downstream from the orifice or sensing device.* This tends to keep the pressure at the orifice more nearly constant, and when there may be a vigorous reaction between the chemicals involved, the downstream location of the valve tends to isolate the measuring section from the ensuing pressure drops.

10 *Improved flow valve performance.*

REMEDY. *Always use a valve Precisor ™or similar valve positioner* with a flow valve to eliminate dead time and friction. The benefits exceed the disadvantages.

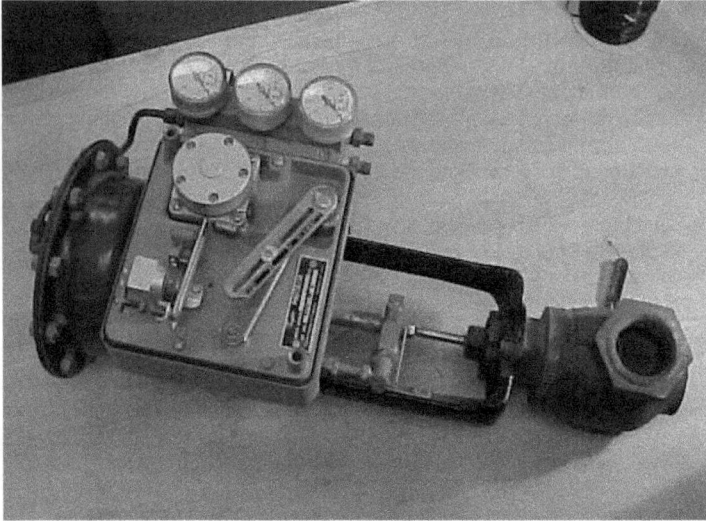

Figure 78 - A *Precisor* is a self-contained control system mounted on an actuator to provide precise valve position.

11 - Symptom. No controller setting provides stable operation.

Selection of the correct valve size.

REMEDY. It is quite probable that the wrong size of valve has been selected. Proper flow control system operation requires that the valve just be wide open for maximum flow. *Never install a larger valve to anticipate future needs.* The valve must never be on the "over" side. Experience often indicates that only a bevel-disk valve should be used. Constant percentage valves, or similar contour valves, are not recommended for flow control by some manufacturers, but are specified by others. There is a marked difference of opinion.

EXCEPTIONS. When flow control is achieved by the use of a pump bypass, the pump characteristics enter the picture and frequently require a valve other than the bevel-disk type.

12 *Flowmeter ranges.* The usual range for differential pressure flowmeters is 30% to 100% of rated flow. Low sensitivity of instruments is conducive to errors. If the differential pressure is too small to ensure good performance, alter the primary element or change the instrument in order to bring the readings upscale. Rotameters and Turbine meters can measure down to 10% of their maximum flow and retain good accuracy.

13 *Some installation inaccuracies are caused by the use of pipe sizes that are assumed.* Measure metering-pipe internal diameters before assembly; they may be quite different from the nominal or assumed values.

14 *Measuring pulsating flow.* If possible, pulsating flow should not be measured because of the errors that are likely to arise. However, if there is no way to avoid this type of measurement, a bellows-type of differential unit with hydraulic damping furnishes uniform damping. Mercury manometers are not generally recommended, because they are too slow in their response. Feedback types of differential pressure or force-balance units are not recommended, because feedback pressures can easily be out of phase with the measured variable.

Chapter Problems

Flow Rate vs. Differential Pressure

1 - Water at 1 cubic foot per second is flowing through a 4" diameter pipe and orifice flowmeter (Beta = .5). Determine the differential pressure developed across the flowmeter.

2 - Double the flow rate in the previous problem and determine the differential pressure. Approximate the relationship between differential pressure and flow rate with head constriction flowmeters.

3 – A Venturi flowmeter (5" pipe, Beta = .6) develops a differential pressure output. If the water flow rate varies from 2 to 4 Ft^3/sec., determine the differential output pressure range in psid.

4 – A flow nozzle (B =. 4, 6") is installed as a water inlet to a ventilated tank. A gauge-type pressure indicator is tapped into the pipe upstream from the flow nozzle. If the flow range is 390 to 870 gpm (1ft^3 = 7.48 gallons), and the low-pressure side of the flow nozzle is atmosphere, determine the minimum and maximum pressures upstream from the flow nozzle flowmeter.

Differential Pressure vs. Throat Diameter (Beta)

5 - Find the ΔP at Qmin and Qmax, given:
d/D = .5, 8" pipe
Qmin = 50 ft^3/min
Qmax = 100 ft^3/min
Material flowing is H2O

6 - Given the following, determine min and max d/p:
d/D = .4, 8" pipe
Qmin = 50 ft^3/min

Qmax = 100 ft³/min

Material flowing is H2O

Finding Flow Rate Given d/p

7 – A 1" Venturi flowmeter (Beta = .25) outputs 15 inches of Hg (about 7.5 psid) of differential pressure. Determine the flow rate in gallons per minute.

8 - Given a d/p of 10 psid across an orifice plate flowmeter with a Beta of .6 and a 2" pipe, determine the flow rate in gallons per minute (GPM). Assume water as the flowing material.

9 - Given a d/p of 40 psid across an orifice plate flowmeter with a Beta of .6 and a 2" pipe, determine the flow rate in gallons per minute (GPM). Assume water as the flowing material.

10 - Do the preceding problems support the relationship between differential pressure and flow rate of $Q = K\sqrt{\Delta p}$?

Constriction Flowmeters and d/p Transmitters

11 –Given the following diagram and specifications:

Beta = .5, 8" pipe

Qmin = 1.1 ft³/sec

Qmid = 1.65 ft³/sec

Qmax = 2.2 ft³/sec

Xmtr ΔPout = 4-5psi

Find:

-d/p min, mid and max from flowmeter

-d1/d2 and Fsp of Xmtr (transmitter)

-Plot d/p vs. Q of flowmeter

-Plot d/p vs ΔPout of Xmtr

Pout

Phi

Plo

d1/d2

Fsp

Q ⟶ ⟶ Q

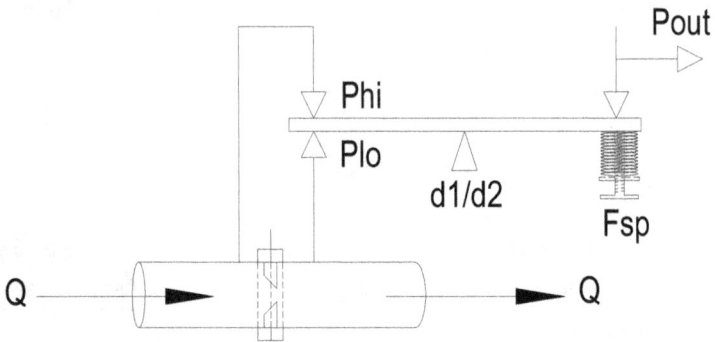

Figure 79 – Flowmeter and transmitter for problems #11 and #12.

12 – Given an 8" orifice plate flowmeter with a .8 throat-to-pipe diameter and water, determine the differential pressure range over a 1 to 3 cubic feet per second flow rate range. Apply the pressure range to a transmitter, and determine the necessary gain (d1/d2) and spring force to output a pressure range of 4-5 psi. Assume all transmitter areas are 1 square inch.

13 – Given the included diagram following, assume an 8", Beta = .8 flowmeter and water. The flow varies from 900.0 to 5400.0 gallons per minute. The transmitter has input areas of 2 in^2 and an output area of 1in^2. The transmitter is followed by a Moore GC-661 air relay to convert the transmitter output into a standard 3-15 psi signal range. Assume the transmitter output is 4 to 6 psi and determine the minimum and maximum flowmeter output pressure, the transmitter gain and spring force, and the required gain and bias spring force of the air relay. Assume the air relay has an output area of one in^2.

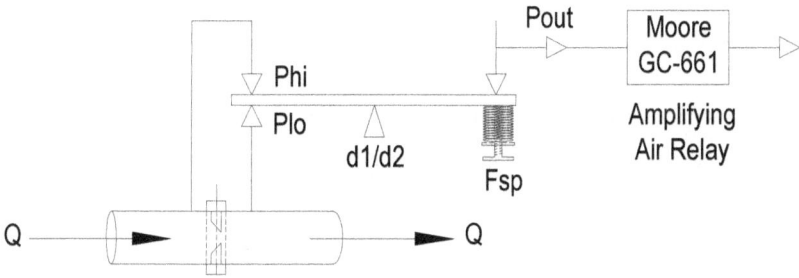

Figure 80 – Flowmeter, transmitter and air relay for problem #13.

14 – A transmitter is connected to the output of a six inch diameter, Beta = .8 orifice plate flowmeter. If the water flow range is 1 to 3 Ft^3/sec., and the gain of the transmitter is .815 with a spring force of 4.4# (up), determine the output of the flowmeter and the transmitter. Assume all transmitter areas equal.

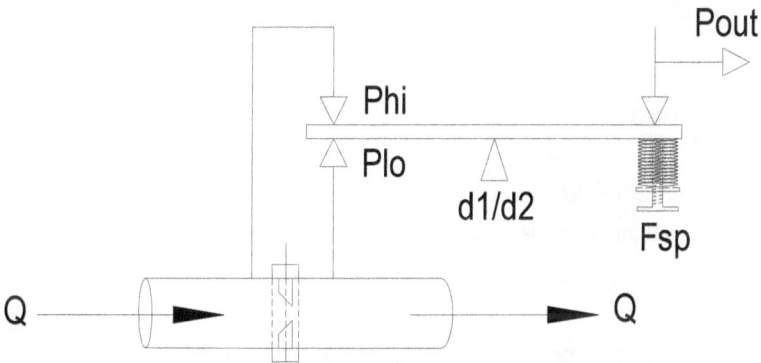

Figure 81 – Flowmeter and transmitter for problem #14.

Constriction Devices, d/p Transmitters and Square Root Extractors

15 - Given the differential pressure extremes from the previous problem, perform the mathematical square root of each output pressure. Determine the relationship between square root output and flow rate, Q.

16 -Given the included diagram, assume water is flowing through a 10" pipe with a Beta = .5. The flow range is from 1.7 to 3.9 cubic feet per second, d1/d2 = 3:1 and the spring force is zero, assume all areas are 1 in.^{2} Find: ΔPout of the flowmeter in psid, ΔPout of the transmitter in psi, ΔPout of the square root extractor,

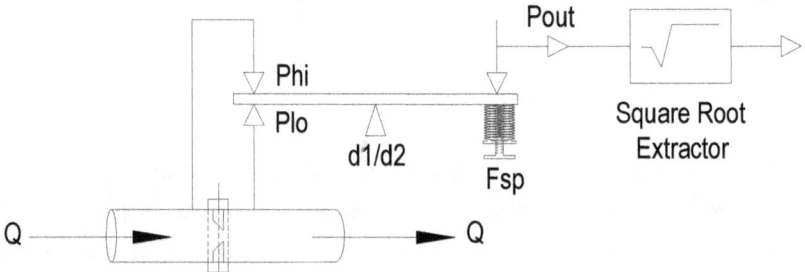

Figure 82 – Flowmeter, transmitter and square root extractor of problem #16.

17 – A Venturi flowmeter (5" pipe, Beta = .6) with water flowing between 900 and 1800 gallons per minute provides a differential pressure range to a torque-balance transmitter. The transmitter provides a signal to a square root extractor. The input differential pressure areas on the transmitter are 1 in.^{2} and the output area is 2 in.^{2}. The transmitter gain is .8:1 and the spring force is 2 lbs.$\downarrow$

Determine the flowmeter output range, the transmitter output range and the square root extractor output range. Does a linear relationship exist with Q? See next problem.

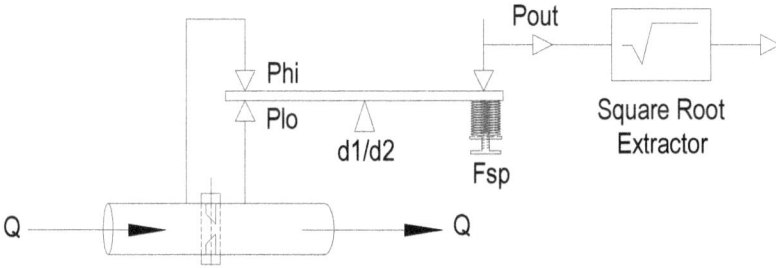

Figure 83 – Flowmeter, transmitter and square root extractor of problem #17.

18. From the square root extractor specifications included in the manuscript, an input offset (bias) is available for adding or subtracting a constant value to/from the input signal to the square root extractor. If the input offset were adjusted to null (remove) the effect of the bias spring force on the torque balance transmitter, what would be the output from the square root extractor? What is the relationship between the extractor output and flowmeter input (Q)?

This page intentionally retained blank.

Chapter 8 - Temperature

Note – The material in this chapter is currently under development.

Temperature scales and material properties are introduced. Mechanical and electrical temperature measurement apparatus and sensors are investigated. Topics include,

Objectives:
Upon completion of this chapter, you should be able to:
- Explain the operational characteristics of common temperature sensors
- Explain the operational characteristics of common non-contact temperature sensors
- Interpret the operational specifications of common temperature sensors
- Apply routine and unique temperature sensor signal conditioning concerns
- Suggest calibration techniques for common temperature sensors

Introduction

The study of temperature measurement begins with an investigation into electron-level molecular activities. As any student of engineering is aware, all materials are composed of molecules, atoms and orbiting electrons. Molecules and their associated atoms occupy a constant amount of space as long as the temperature is constant. Changing the ambient temperature has a directly proportional effect upon the space occupied by the molecule and the energy with which the electrons orbit the nucleus. As the ambient temperature increases, the electrons orbit the nucleus with greater energy causing the atom to expand and consuming more space. An increase in applied temperature causes all materials to attempt expansion. Liquids and solids expand consuming more volume and confined materials such as gases exhibit increased pressure. The opposite holds for a temperature decrease.

Over the years temperature has been measured using a number of scales. Most are currently, or have been previously associated with the effect of a changing temperature upon a material, such as the effect of a changing temperature on alcohol or water. Currently the Fahrenheit and Centigrade, or Celsius scales are the most common. Both are attributed to scientists of the same name.

Gabriel Daniel Fahrenheit and Andes Celsius both lived in Holland during the first half of the eighteenth century however, there's no evidence to indicate they were drinking buddies or even ever spoke. Fahrenheit introduced the alcohol thermometer in 1709, and the mercury thermometer and the temperature scale bearing his name around 1714. At the time, the Roemer scale was used for temperature measurement however Fahrenheit found the Roemer scale awkward since it utilized fractional temperatures assigning the freezing point of water at 7.5 degrees and boiling at 60 degrees. Fahrenheit divided each of the Roemer scale increments by 4 to eliminate the fractions. The modification resulted in the 200-plus divisions associated with the current Fahrenheit temperature scale.

Fahrenheit is credited with determining that water could remain a liquid at temperatures below the freezing point if salt were added, and above the boiling temperature if the ambient pressure were increased. Using the Mercury thermometer, Fahrenheit established three reference points on his 200-plus-increment temperature scale. The coldest temperature Fahrenheit could obtain in his lab was assigned zero degrees and was composed of a combination of ice, water, and salt. Fahrenheit realized that salt reduced the freezing temperature of the solution below the freezing temperature of water. When he omitted salt from the combination, the temperature increased to the second reference temperature that Fahrenheit assigned at thirty degrees – based upon the amount of alcohol expansion. His final reference temperature was assigned using the temperature of the human body. The final reference temperature, to be located in the mouth or under the armpit of a living man in good health, was assigned at ninety degrees – again, based upon three times the volumetric expansion of the previous reference temperature. After Fahrenheit's death, a scientific commission would modify his temperatures to 32° for water freezing and normal body temperature as 96 degrees. Accepted body temperature was again modified to 98.6 during a subsequent commission when the boiling point of water was more accurately established at 212 degrees.

The Centigrade, or Celsius temperature range divides the span of temperatures between the freezing point of water and the boiling point into 100 increments, thus the term Centi-grade or 100 units. Unbelievably, the Celsius temperature scale originally placed the freezing point of water at 100 degrees and the boiling point of water at zero. Andes Celsius, a Swedish astronomer, lived during the same period as Fahrenheit in the early 1700's and is attributed with associating 100 divisions to the temperature span between the freezing and the boiling temperatures of water, even though the 100 division temperature scale apparently existed before he assigned zero to boiling and 100 to freezing. The temperature scale was eventually inverted to make better sense. Interestingly, Andes Celsius introduced his scale is 1742 but wasn't recognized until 1948 when the Ninth General Conference of Weights and Measures declared degrees "Centigrade" should be referred to as degrees "Celsius."

Temperature Scale Conversion

Conversion between Fahrenheit and Celsius temperature scales is a relatively simple process if one retains the reference temperatures associated with the freezing and boiling points of water and if one recalls the fundamental concepts of linear relationships. Since each scale varies in a proportional and linear manner, a graph can be plotted and the corresponding slope and intercept values determined. Assembling an equation with the determined slope and intercept values forms the resulting conversion equation. The following example illustrates.

F° to C°

Graphing both temperature ranges results in the characteristics of the following figure. Once recognized, determine the slope and Y-axis intercept values.

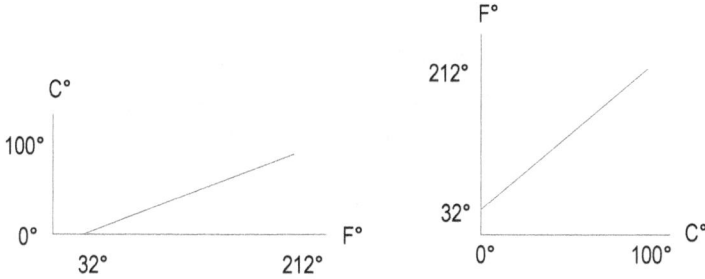

Figure 1 – Temperature scale conversion plots.

For conversion of Fahrenheit to Centigrade, assume Fahrenheit is the input and Celsius is the desired unknown or output quantity. The graph on the left in the previous figure is used for this conversion. From the graph, the slope is found by dividing the Y-axis span by the X-axis span – note that the span is computed as the mathematical difference between end points.

$$\text{Slope} = \Delta Y / \Delta X$$
$$\text{Slope} = 100° / 180°$$
$$\text{Slope} = 5/9 = .555556$$

The Y-intercept value can then be determined by rearranging the Y=mX+b equation and substituting known values from the graph.
Since,

$$Y = mX + b$$
$$b = Y - mX$$
$$b = 100 - (5/9 * 212)$$
$$b = 100 - (117.777778)$$
$$b = -17.77778$$

The resulting Fahrenheit to Celsius equation is assembled as,

455

$$C° = .56F° - 17.78$$

If the resulting equation doesn't look familiar it's because the slope is expressed as a decimal and the intercept term utilizes the Y-axis intercept value instead of the X-axis intercept value. Algebraically rearranging the expression assembles a more familiar conversion equation.

Since,

$$C° = .56F° - 17.78$$

Subbing 5/9 for .56,

$$C° = 5/9F° - 17.78$$

Dividing each term by 5/9 yields,

$$C°/(5/9) = 5/9F°/(5/9) - 17.78/(5/9)$$
$$C°/(5/9) = F° - 32$$

Multiplying each side by 5/9 yields the more familiar,

$$C° = 5/9(F° - 32)$$

Example Problem

Determine the Celsius equivalent of normal body temperature of 98.6° F.

Solution

Using the previously derived relationship between Fahrenheit and Centigrade degrees, the resulting temperature can be determined.

Since,

$$C° = 5/9(F° - 32)$$
$$C° = 5/9(98.6° - 32)$$
$$C° = 5/9(66.6°)$$

$$C° = 5/9(F° - 32)$$
$$C° = 37$$

Hospitals adopted the Centigrade temperature span decades ago. The normal body temperature in Celsius is 37°.

C° to F°

Referring to the graph at the right of the previous figure allows determination of the slope and intercept.

$$\text{Slope} = \Delta Y/\Delta X$$
$$\text{Slope} = 180°/100°$$
$$\text{Slope} = 9/5 = 1.8$$

The Y-intercept value can then be determined by rearranging the Y=mX+b equation and substituting known values from the graph.
Since,

$$Y = mX+b$$
$$b = Y-mX$$
$$b = 212 - (9/5 * 100)$$
$$b = 212 - (900/5)$$
$$b = 212 - (180)$$
$$b = 32$$

The resulting conversion equation becomes,

$$Y = mX+b$$
$$F° = 1.8C° +32$$

Or,

$$F° = (9/5C°) +32$$

Example Problem

During a conversation with a foreign student it is mentioned that the winter temperature often exceeds 20 below zero Celsius. Determine the Fahrenheit equivalent.

Solution

Twenty below zero is damn cold be it in Celsius or Fahrenheit! However, in order to reference the temperature to the more familiar units of Fahrenheit, the previous conversion equation is applied.

$$F° = (9/5C°) +32$$
$$F° = (9/5*(-20C°)) +32$$
$$F° = (-180/5) +32$$
$$F° = (-36) +32$$
$$F° = -4$$

It's to be expected that one will forget the individual conversion equations. However, understanding the concept behind the derivation of the individual conversions is far more beneficial than remembering the equations themselves.

Absolute Temperature Scales

When observing the effects of temperature variations upon solids, liquids and gases, an *absolute temperature* scale is often used. An absolute temperature scale is one that begins at the temperature where electron motion begins and like absolute pressure scales - contains no negative numbers. In other words, if the temperature of a substance is lowered to several hundreds of degrees below zero, C° or F°, the electrons orbiting the nucleus of the atoms within the molecules will cease motion. This temperature will also result in the solids, liquids and gases consuming their smallest volume. The temperature that causes the cessation of electron activity is *absolute zero temperature*. Every temperature above this value can be interpreted as increased heat energy, in an electron motion and volume expansion sense. Absolute zero temperature equals minus

459.6° Fahrenheit and minus 273° Celsius. Absolute zero temperature is considered the lowest possible temperature although the exact value has never been achieved.

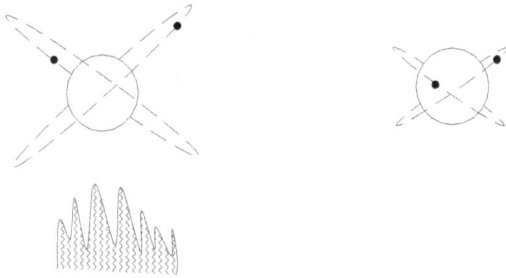

Figure 2 - The application of any temperature above absolute zero (left) causes electrons to travel in larger orbits and higher velocity.

As the temperature increases above absolute zero, solids and liquids expand. Gases, which also tend to expand, are usually contained in confined areas and exhibit the increased electron energy as an increase in pressure. All of these events occur in proportion to the materials' absolute temperature.

Conversion to an absolute temperature scale involves adding an offset to the Centigrade and Fahrenheit temperature scales. When converting common temperature scales of Fahrenheit and Celsius into absolute temperature, Fahrenheit is converted in Rankin degrees and Celsius is converted into Kelvin degrees. Since absolute zero temperature is defined as -459.6 degrees Fahrenheit, conversion of Fahrenheit into an absolute temperature scale involves adding 459.6, often approximated as 460, to the given Fahrenheit temperature. The resulting temperature scale of Rankin degrees is determined as follows.

0° Rankin = -459.6° Fahrenheit

And,

R° = F° + 459.6

459

Or,

$$R° \approx F° + 460$$

In a similar manner Centigrade degrees can be converted into an absolute scale of Kelvin degrees by adding 273. Since absolute zero degrees Kelvin equals minus 273° C, adding 273 performs the mathematical conversion of Celsius degrees into Kelvin degrees.

$$0° \text{ Kelvin} = -273° \text{ Centigrade}$$

And,

$$K° = C° + 273$$

The resulting conversions appear as follows.

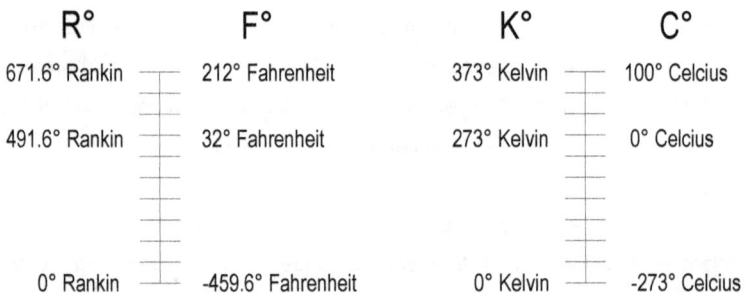

R°	F°	K°	C°
671.6° Rankin	212° Fahrenheit	373° Kelvin	100° Celcius
491.6° Rankin	32° Fahrenheit	273° Kelvin	0° Celcius
0° Rankin	-459.6° Fahrenheit	0° Kelvin	-273° Celcius

Figure 3 - A comparison of commonly used temperature scales at absolute zero temperature, water freezing and water boiling points.

Many other temperature scales exist under a variety of names. Various melting and boiling points of materials are used as the references for these not so common temperature scales. Some are used for proprietary manufacturing purposes to keep a process secret, while others have been used for centuries without adapting to the "newer" Fahrenheit or Celsius temperature scales. Whatever the temperature reference or scale used, the physical process of temperature variation is probably the most commonly applied, measured and controlled process variable in manufacturing. This is ironic however when compared against the relatively small variety of transducers, especially electronic sensors.

Figure 4 - Two common temperature sensors where both utilize the expansion/contraction of solids to yield an indication of the measured temperature.

Lineal Expansion

Among the more commonly applied concepts is that of single dimension expansion called the lineal, or linear expansion coefficient. The lineal expansion coefficient, also called the Alpha coefficient (α) of linear expansion observes a single dimension change in shape, such as object length, when the ambient temperature is varied.

Alpha Coefficient of lineal expansion,

α = Fractional change in length ÷ Change in Temperature

α = (Change in length/Original length) ÷ Change in Temperature

$$\alpha = (\Delta L / L_0) \div \Delta T$$

Or,

461

$$\alpha = \Delta L / L_o / \Delta T$$

The lineal expansion coefficient provides a convenient means for determining an object's change in length when subjected to a changing temperature environment. Although the object's entire volume will change with a corresponding temperature change, the Alpha coefficient observes only a single dimensional variation. A good example of this characteristic is temperature-sensing switches utilizing lineal expansion coefficients to open or close electric circuits as an applied temperature varies. The amount of lineal expansion due to temperature increase, or contraction due to temperature decrease varies depending upon the type of material used. The following table provides the Alpha coefficients (in inches per inch of sample per degree) of some common materials.

Material	Lineal Coef. – In/In per C^o	Lineal Coef. – In/In per F^o
Aluminum	.000022	.000012
Brass	.000019	.000011
Copper	.000017	.0000094
Glass, ordinary	.0000095	.0000053
Glass, Pyrex	.0000036	.0000020
Invar (nickel-steel alloy)	.0000009	.0000005
Iron	.000012	.0000067
Steel	.000011	.0000061

Figure 5 – Average lineal expansion coefficients for selected materials.

The temperature characteristics associated with changes in dimension are well documented for common materials. The previously provided Alpha coefficients are averages taken from a specific range of temperatures and the resulting measured changes in lengths. Unfortunately, The specific expansion coefficients used to determine changes in physical dimensions corresponding to changes in applied temperature are non-linear. In other words, the change in a given material's dimension is not uniform throughout a given temperature range but is greater at some temperatures than at others.

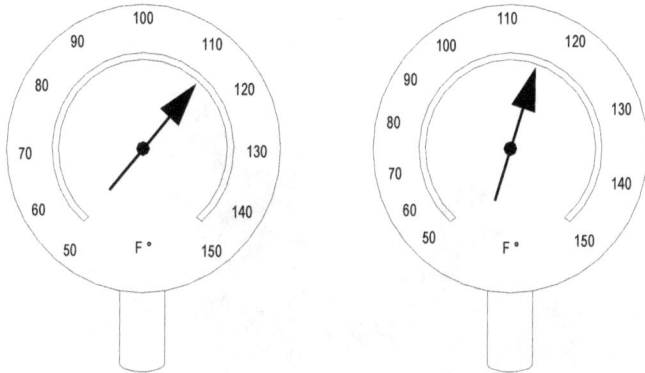

Figure 6 - A linear gauge (left) and non-linear gauge (right). Note the difference in space between temperatures differences of equal increments on the non-linear gauge at the right.

If a thermometer were made of one of the previously provided materials and used to measure temperature over a wide range of applied temperature, the indicating scale on the thermometer may not be non-linear. The distance between equal divisions of temperature would not be equal in length. However, thermometers utilizing linear expansion exist and are commonly used in cooking and home heating/cooling applications. In these applications the amount of non-linearity is usually minimal due to the limited temperature range.

Figure 7 - The common Honeywell residential thermostat utilizes a bimetallic sensing element. The element is located in the lower portion of the photo and appears as a coil.

Among the more common home instruments utilizing the lineal expansion concept is the Honeywell T87 thermostat. Two materials with different lineal expansion coefficients are bonded to form a *bimetallic element*. As the applied temperature increases the material with the greater Alpha coefficient expands more than the adjoining material. The result is a bending of the bimetallic element if the overall length is relatively short. Longer bimetallic elements are usually coiled or formed in a helix shape to yield greater distances of displacement when the temperature is varied. The following figure illustrates.

$$\alpha_1 > \alpha_2 \qquad \alpha_1 > \alpha_2$$

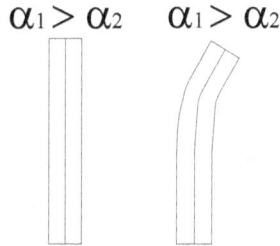

Figure 8 - An increase in temperature applied to a bimetallic element causes the element to bend (right). The effect is amplified with longer, coiled elements causing rotation at the unconstrained end.

Example Problem

Determine the change in length of a 15-inch copper rod when heated from 50°F to 650° F.

Solution

The Alpha coefficient of Copper is 9.4×10^{-6} inches for each inch of sample for each degree of temperature applied. To determine the overall change in length, the Alpha coefficient is multiplied by the distance of the Copper rod and the difference in temperature.

Since,

$$\alpha = \Delta L / L_o / \Delta T$$

Rearranging to solve for ΔL,

$$\Delta L = \alpha * L_o * \Delta T$$
$$\Delta L = (9.4 \times 10^{-6} \text{ in/in/F}°) * (15 \text{ in}) * (600° \text{ F})$$
$$\Delta L = (9.4 \times 10^{-6} \text{ in/in/F}°)$$
$$\Delta L = .0846 \text{ in}$$

Although the resulting change in length does not appear to be significant only .03 to .06 inches is required to close a switch contact, move an indicator or position an orifice/baffle for control purposes.

Cubical Expansion

As mentioned previously, solids and liquids undergo a three-dimensional, volumetric change with a corresponding change in temperature. Three-dimensional expansion occurs as a result of the atoms within the material's molecules consuming greater space when the ambient temperature increases. Contraction occurs with a decrease in the applied temperature. With solids, a single dimension change is usually required to perform a control function. Liquids however require observation of a three-dimensional, volumetric change. For this purpose cubical expansion coefficients are utilized. Cubical expansion coefficients are also called Beta coefficients (β) and allow determination of the volumetric amount of expansion of liquids and solids within defined temperature limits. Some common Beta coefficients for solids and liquids are included in the following table; the coefficients are expressed in $in^3/in^3/degree$.

Beta Coefficient, β, of cubical expansion,

$$\beta = \text{Fractional change in volume} \div \text{Change in Temperature}$$

$$\beta = (\text{Change in volume/Original volume}) \div \text{Change in Temperature}$$

$$\beta = (\Delta V / V_o) \div \Delta T$$

Or,

$$\beta = \Delta V / V_o / \Delta T$$

The following table provides a few of the more commonly applied cubical expansion coefficients for liquids.

Material	Temp Span, C°	Beta Coefficient per C°
Ethyl Alcohol, 99%	27-46° C	.0012
Benzene	11-81° C	.001237
Carbon Tetrachloride	0-76° C	.001236
Mercury, Hg	0-100° C	.000181
Petroleum (.846 #/Ft³)	24-120° C	.000955
Water, H_2O	0-33° C	.000207
Iron	0-100° C	.0000355
Platinum	0-100° C	.0000265
Silver	0-100° C	.0000583
Zinc	0-100° C	.0000893
Sulfur	13-50° C	.000223
Quartz	50-60° C	.0000353
Paraffin	20° C	.000588

Figure 9 - Average volumetric expansion coefficients for selected materials.

Example Problem

Determine the volumetric expansion of a cubic foot of water for a 20° C increase in temperature.

Solution

To determine the amount of volumetric expansion, the Beta coefficient is multiplied by the volume and the temperature variation.

Since,

$$B = \Delta V / V_o / \Delta T$$

$$\Delta V = B * V_o * \Delta T$$

$$\Delta V = (.207 \times 10^{-3}) * (1 \; Ft^3) * (20° \; C)$$

Since the Beta coefficient is expressed in inches, the volume is converted from cubic feet to cubic inches,

$$\Delta V = (.207 \times 10^{-3} \text{ in}^3/\text{in}^3/\text{C}) * (1728 \text{ in}^3) * (20° \text{ C})$$

$$\Delta V = 7.1539 \text{ in}^3$$

It should be noted that the vast majority of liquid-filled temperature sensing applications occur when the liquid is contained within a sealed volume. The resulting expansion/contraction of the liquid will appear as an increase/decrease in pressure since the closed system will allow only a very limited amount of volumetric expansion or contraction. The amount of pressure change can be predicted and is explained following.

Pressure and Temperature
Anyone that has inadvertently, or intentionally, disposed of an aerosol can into a fire has observed the effect of varying temperature upon pressure. The relationship is direct and proportional, as temperature increases – pressure increases. Examples of this association are obvious in all applications of science from weather to manufacturing. The relationship is described in the following equation.

$$P1/T1 = P2/T2$$

Where,

$P1$ = the pressure before a temperature change

$T1$ = the initial temperature

$P2$ = the pressure after a temperature change

$T2$ = the final temperature

As an aid to understanding the application of the equation, it is often written using the following subscripts.

$$P_{old}/T_{old} = P_{new}/T_{new}$$

Where,

     P_{old} = the pressure before a temperature change

     T_{old} = the initial temperature

     P_{new} = the pressure after a temperature change

     T_{new} = the final temperature

The equation and associated pressure-temperature relation is called *Charles's law* and as can be observed, a change in applied temperature results in a directly proportional change in pressure. However, all quantities must be in absolute terms before using the equation. In addition, use of the equation assumes that the volume of any vessels remains constant as the applied temperature varies.

Example Problem

An automobile tire was set to 32 psi prior to a trip from northern Wisconsin to southern Florida in January. The tire temperature is equal to the outdoor temperature of 15° F when the trip is begun and equals 105° F upon arrival in sunny southern Florida. Determine the tire pressure at the 105° F temperature.

Solution

Given the initial and final temperatures and the initial pressure, a final pressure can be determined. All values must first be converted to absolute quantities.

Since,

psia = psig + 14.7, the initial tire pressure equals,

32 psi = 32 psig

psia = 32 psig + 14.7 = 46.7 psia

P_{old} = 46.7 psia

Since Rankin equals Fahrenheit plus 460, the initial and final temperatures equal,

$15° \text{F} + 460 = 475° \text{R}$

$105° \text{F} + 460 = 565° \text{R}$

Subbing the quantities into Charles's equation yields,

$$P_{old}/T_{old} = P_{new}/T_{new}$$
$$(46.7\text{psia}) / (475° \text{R}) = P_{new} / (565° \text{R})$$
$$P_{new} = (565° \text{R}) (46.7\text{psia}) / (475° \text{R})$$
$$P_{new} = 55.548 \text{ psia}$$

Converting the tire pressure into gage yields,

$$P_{new} = 55.548 \text{ psia} -14.7$$
$$P_{new} = 40.848 \text{ psig}$$

Charles's law demonstrates the direct proportional relation between temperature and pressure.

Example Problem

The following temperature transmitter contains a filled capillary system as a temperature sensor. When a temperature variation is applied to the sensing bulb an output pressure range will be developed. Given the following information, determine the output pressure range when 100°F and 500°F is applied to the sensing element.

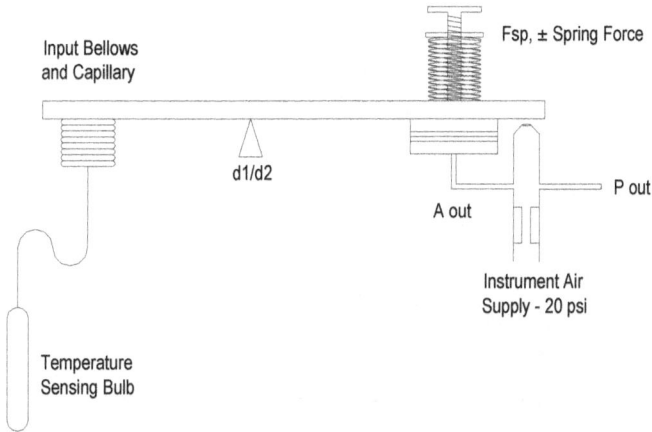

Given:

-The sensing bulb, capillary and input bellows exhibits 10 psig at 0° F.

-The input bellows area is 1 in²

-The output diaphragm area (Aout) is 1 in²

-The fulcrum (d1/d2) is located at .2:1

-The spring force is 3 pounds down

Solution

The problem requires two computations to be performed. First, the pressure in the sensing bulb, capillary and input bellows must be determined using Charles's law. Second, the output equation for the secondary element (transmitter) must be determined. Once these are known the input pressure of the sensing capillary system can be applied to the transmitter equation to determine the output pressure.

Capillary Pressure

Given the pressure of the capillary is 10 psig at 0° F., the pressure at 100°F and 500°F can be determined. The temperatures and pressure must be converted into absolute quantities before applying to the equation.

Since Rankin equals Fahrenheit plus 460, the initial and final temperatures equal,

0° F + 460 = 460° R

100° F + 460 = 560° R

500° F + 460 = 960° R

10 psi + 14.7 = 24.7 psia

Subbing the quantities into Charles's equation yields the pressure at 100° F = 560° R,

$$P_{old}/T_{old} = P_{new}/T_{new}$$

$$(24.7 psia) / (460° R) = P_{new} / (560° R)$$

$$P_{new} = (560° R) (24.7 psia) / (460° R)$$

$$P_{new} = 30.07 \text{ psia} = 15.37 \text{ psi @ } 100 °F.$$

If the transmitter were being operated in a vacuum the absolute pressure would have been applied to the transmitter output equation. However, assuming the transmitter will be operated in standard atmospheric pressure of 14.7 psi the resulting input pressure is converted to gauge since only pressure above atmosphere will result in a force applied to the transmitter beam.

Subbing the quantities into Charles's equation again to determine the pressure at the maximum temperature yields,

$$P_{old}/T_{old} = P_{new}/T_{new}$$

$$(24.7 psia) / (460° R) = P_{new} / (960° R)$$

$$P_{new} = (960° R) (24.7 psia) / (460° R)$$

$$P_{new} = 51.548 \text{ psia} = 36.848 \text{ psi @ } 500 °F.$$

The input pressure developed within the temperature-sensing bulb, capillary and input bellows varies between 15.37psi and 36.848psi between 100°F and 500°F.

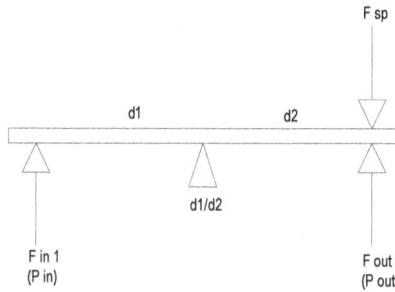

The output pressure equation for the secondary element is determined next. The previous diagram represents a simplified equivalent of the transmitter and is used to demonstrate the forces working upon the transmitter mechanism. The resulting torque-balance equation is written as follows.

$$@ \text{ Balance, Tccw} = \text{Tcw}$$

Since, $T = F * D$

$$(Fin*d1) + (Fsp*d2) = (Fout*d2)$$

Since, $P = F * A$

$$(Pin*Ain*d1) + (Fsp*d2) = (Pout*Aout*d2)$$

Solving for the output pressure, Pout, yields,

$$(Pin*Ain*d1)/(Aout*d2) + (Fsp*d2)/(Aout*d2) = Pout$$

Or,

$$Pout = Pin*(Ain/Aout)*(d1/d2) + Fsp/Aout$$

Subbing known values at 100°F yields,

$$Pout = (15.37psi)*(1in^2/1in^2)*(.2/1) + (3\#/1in^2)$$
$$Pout = 3.074\ psi + 3\ psi$$
$$Pout\ @\ 100\ °F = 6.074\ psi$$

Subbing known values at 500°F yields,

$$Pout = (36.848psi)*(1in^2/1in^2)*(.2/1) + (3\#/1in^2)$$
$$Pout = 7.37\ psi + 3\ psi$$
$$Pout\ @\ 500\ °F = 10.37\ psi$$

The output pressure from the transmitter will vary between 6 psi and approximately 10.4 psi over an applied input temperature range of 100°F and 500°F.

Example Problem

Using the previous primary and secondary elements (sensing bulb and capillary, and transmitter), input temperatures and pressures, determine the required gain (d1/d2) and offset (Fsp) to establish an output pressure span of 3 psi to 15 psi over the input temperature span of 100°F to 500°F.

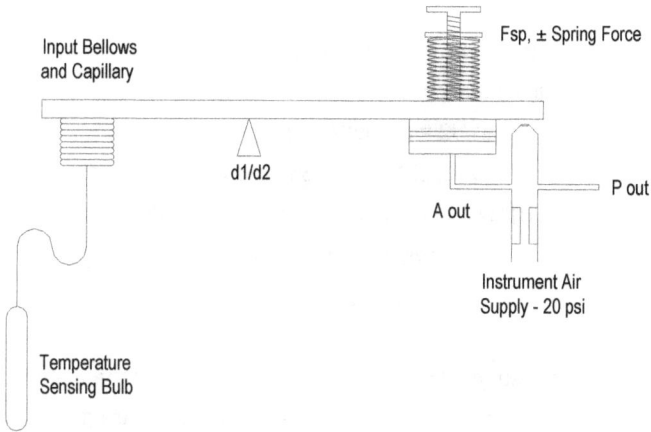

Figure 10 - The transmitter has only two adjustments, gain (also called "Span," d1/d2) and offset (also called "Zero," Fsp). The input/output characteristics can be calibrated by the adjustments.

Solution

Everything remains the same in this problem except for the fulcrum position and the spring force. The solution requires a new d1/d2 and spring force need to be determined. For problems such as this where the gain and offset are to be computed the gain (d1/d2) should be established prior to the offset (Fsp) since d1/d2 is required top compute Fsp. The first step will be to arrange the output pressure (Pout) equation to solve for d1/d2. Next, the output equation will be rearranged again to solve for Fsp.

Given,

$$Pout = Pin*(Ain/Aout)*(d1/d2) + Fsp/Aout$$

Since Fsp is set to zero to compute d1/d2,

$$Pout = Pin*(Ain/Aout)*(d1/d2)$$

Solving for the gain, d1/d2, yields,

$$d1/d2 = (Pout*Aout)/(Pin*Ain)$$

Since the d1/d2 gain is similar to a slope computation, only the input and output spans are used in the equation, or,

$$d1/d2 = (\Delta Pout*Aout) \div (\Delta Pin*Ain)$$
$$d1/d2 = (15psi-3psi)*(1in^2) \div (36.848psi-15.37psi)*(1in^2)$$
$$d1/d2 = (12psi)*(1in^2) \div (21.478psi)*(1in^2)$$
$$d1/d2 = (12\#) \div (21.478\#)$$
$$d1/d2 = .5587:1$$

Knowing the required d1/d2 gain, the spring force can be determined by rearranging the Pout equation to solve for Fsp and substituting corresponding input and output pressures.

Since,

$$Pout = Pin*(Ain/Aout)*(d1/d2) + Fsp/Aout$$
$$Pout - Pin*(Ain/Aout)*(d1/d2) = Fsp/Aout$$
$$(Pout*Aout) - (Pin*Aout)*(Ain/Aout)*(d1/d2) = Fsp$$

Or,

$$Fsp = (Pout*Aout) - (Pin*Aout)*(Ain/Aout)*(d1/d2)$$
$$Fsp = (Pout*Aout) - (Pin*Ain)*(d1/d2)$$

Subbing for a maximum output pressure of 15 psi at a maximum input pressure of 36.848psi,

$$Fsp = (Pout*Aout) - (Pin*Ain)*(d1/d2)$$
$$Fsp = (15 psi*1in^2) - (36.848psi*1in^2)*(.5587/1)$$
$$Fsp = (15\#) - (36.848\#)*(.5587/1)$$
$$Fsp = (15\#) - (36.848\#)*(.5587/1)$$
$$Fsp = (15\#) - (20.587\#)$$
$$Fsp = -5.587\#$$

The negative sign associated with the spring force indicates the direction of the required force will be opposite to that assumed in the original diagram. The required force will be pulling up instead of pushing down but will be equal to the computed value of 5.587#.

In summary, to establish an output pressure of 3 psi to 15 psi over an input temperature range of 100°F to 500°F requires a fulcrum position at .5587:1 and an upward spring force of 5.587#. (Note - the values of spring force and fulcrum position will rarely share the same numerical values as in this example.)

Proof

To determine if the computed values are correct, substitute the pressures, gain and spring force into the original diagram and verify the system is at a condition of torque-balance. The following demonstrates.

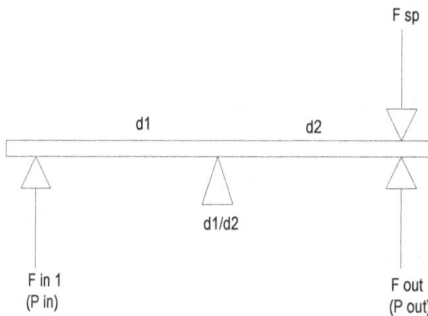

Figure 11 - The original system diagram. Note the spring force direction in the following diagrams.

Using the following diagram, the input and output pressures have been converted to forces by multiplying the pressure by the areas of the components at their respective locations (F=P*A). Once the forces are known the clockwise and counter-clockwise torque quantities must be determined (T=F*D). The individual moment arm lengths have been indicated on the beam in the following diagram to assist in computing the torque values.

$$Tcw = 15.37\#^*. \, 5587in$$
$$Tcw = 8.587 \, \#in$$
$$Tccw = (3\#^*1in) + (5.587\#^*1in) = 8.587\#in$$

Torque balance is verified at the minimum value of output and input pressure (100°F).

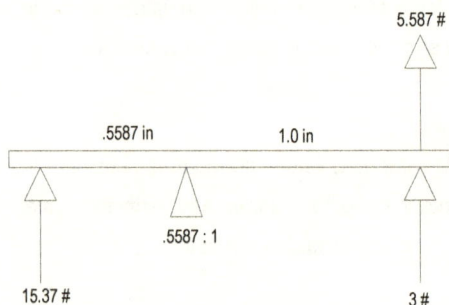

Figure 12 - The system at minimum input and output pressures.

The procedure is repeated at the maximum input and output pressures. If the unit is found to be at balance after this computation the computed results are correct. The following diagram represents the operation of the transmitter at the maximum input and output pressures.

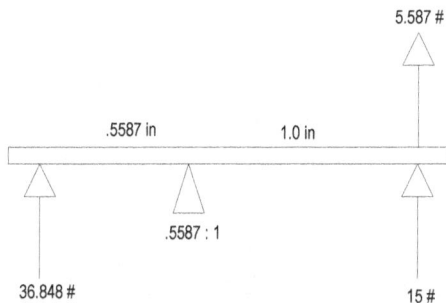

Figure 13 - The system at maximum input and output pressures.

$$Tcw = 36.848\#^*. 5587in$$

$$Tcw = 20.587\#in$$

$$Tccw = (15\#^*1in) + (5.587\#^*1in) = 20.587\#in$$

Torque balance is also verified at the maximum value of output and input pressure (500°F). The computed values are correct.

Mechanical Thermometers

The operation of mechanical temperature instruments is founded upon the molecular principles of expansion/contraction of solids, liquids and gases. American instrument manufacturers (Taylor, Fischer and Porter, Foxboro, Honeywell and others) categorize mechanical temperature measurement apparatus into the four groups or classes contained in the following table.

Division	Type	Temperature Range
Class I	Liquid Filled (but not Hg)	-300° to +600° F
Class II	Vapor-Pressure	-300° to +600° F
Class III	Gas filled	-400° to +1000° F
Class IV	Mercury (Hg) Filled	-100° to +1200° F

The sensing media of the four classes is enclosed within a bulb, capillary and Bourdon tube or similar component for conversion of pressure into force. The resulting temperature sensitive force is applied to a secondary element such as a moment arm for torque development and subsequently used to position an orifice-baffle, temperature indicator or other signal-transmitting device.

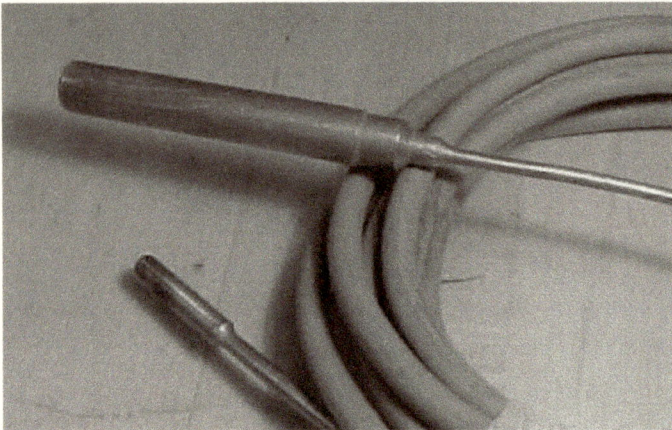

Figure 14 - A pair of temperature sensing filled capillary bulbs. The sensing bulb near the bottom of the photo is a Class I (liquid-filled) and the upper bulb is class II (gas filled). The sensing bulb in the bottom of the photo is attached to the transmitter in the previous photo.

Class I and IV instruments (liquid and Mercury filled capillary) utilize the expansion/contraction of liquids contained within a closed capillary system with temperature variation. Although normally considered a solid, mercury exhibits expansion/contraction characteristics similar to liquids. Mechanical measurement and transmitting instruments composed of a filled sensing bulb, signal transmitting capillary and force generating component such as a bellows, Bourdon tube or diaphragm usually contain some form of local temperature indication such as in the following photo. Note that the unit in the following photo indicates the correct bulb temperature even though the transmitter was powered-down when the photo was taken. Filled capillary systems do not require an external power source for operation. The transmitter power source, 20-psi in this case, is required only to generate a 3-15 psi output signal in proportion to the applied temperature at the sensing bulb.

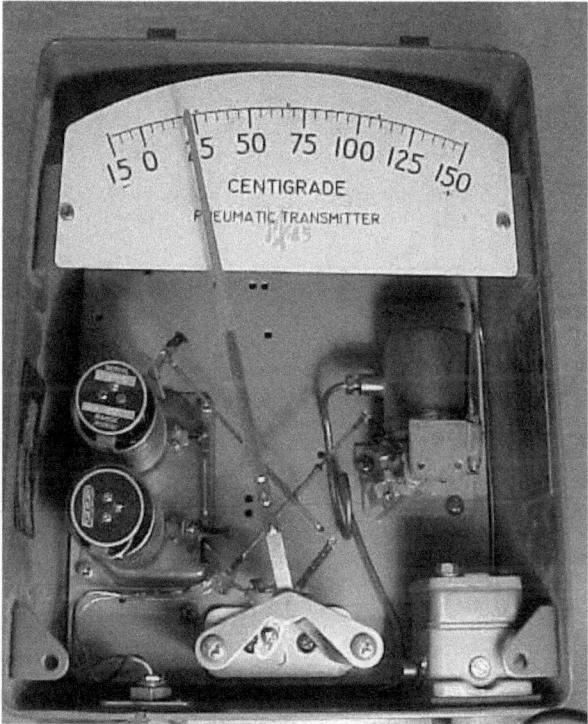

Figure 15 - A Foxboro pneumatic temperature transmitter. This unit outputs a 3-15 psi signal over an input temperature range of -15 to +150 degrees Celsius and applies torque-balance concepts to do so. The sensing and compensating capillaries enter the unit in the bottom left of the photo.

To assure the output signal is an accurate representative of the sensing bulb temperature, an output signal feedback device usually opposes the input force-generating component. Operationally, the measurement/transmitting system is similar to the following diagram.

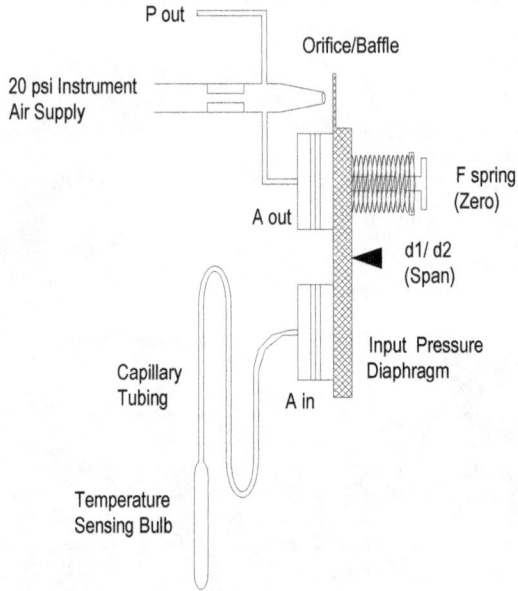

Figure 16 - Common components of a mechanical temperature transmitter.

In operation, as the temperature applied to the liquid filled sensing bulb increases, the electrons within the liquid's atoms move in larger orbits. The increased temperature creates a volume expansion of the liquid within the sensing bulb, capillary and force component such as a diaphragm or bellows.

Since the liquid is contained within a closed system, the resulting expansion causes the force generating bellows (labeled as the input pressure diaphragm in the previous diagram) to increase absorbing the expanded liquid volume. The extended bellows volume appears as a force when the bellows surface area is applied to the torque beam. In the previous diagram the resulting input force torques the beam counter-clockwise moving the baffle closer to the orifice and causing the output pressure to increase.

Figure 17 - A Class I liquid-filled temperature controller. The sensing bulb and capillary system in this unit is hydrocarbon filled. It was removed from service after the oil-like filling material was observed to be diffusing through the internal components.

The output diaphragm area (Aout) converts the output pressure increase into a force and applies the increased force against the beam in a clockwise manner opposing the input force. The output diaphragm in applying the output pressure created force against the input force is providing negative feedback to balance the transmitter and to assure that the output pressure is an accurate analog of the input temperature – for each individual input temperature there should exist a unique value of output pressure.

After the transmitter achieves torque-balance, the initial increase in temperature will result in a proportional increase in output pressure. The unit's input/output relationship can be offset with a spring force. Depending upon the direction of the applied spring force, the spring aids or opposes the output in balancing the input. When the spring force is adjusted the output pressure span will be the same but the specific output pressure associated with a given input temperature is elevated or suppressed by the new spring force. As with all other mechanical and electronic instruments, the gain (input/output slope) determines the span of the output signal over a given input range. When a standard signal span such as 3-15 psi or 4-20 mA is used as the output signal, varying the span determines the amount of input change required for the transmitter to develop the full output range.

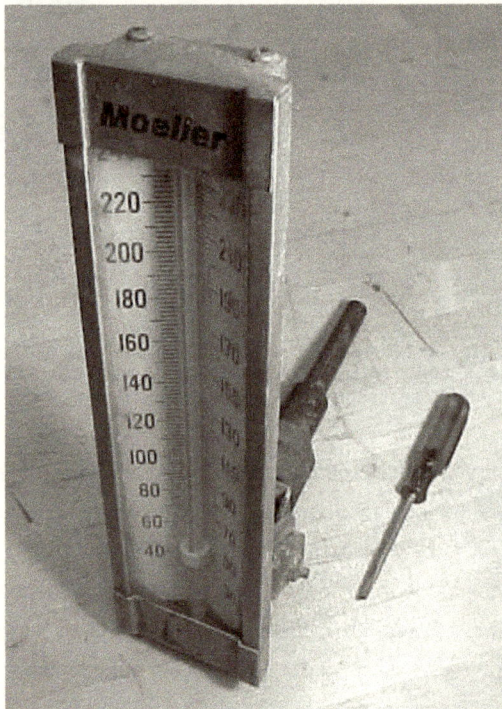

Figure 18 - A Class I alcohol-filled thermometer.

Class I liquid-filled systems commonly employ alcohol and hydrocarbons as the filling media used to measure temperature. An advantage to liquid-filled systems is the greater speed of response when compared to vapor-pressure and gas-filled temperature measurement systems. Liquid-filled systems can have capillary lengths approaching 20 feet with a minimal loss of response time. However, remote measuring temperature systems with long capillaries should be compensated against sensitivity to ambient temperatures applied to the capillary or instrument housing away from the sensing bulb and process.

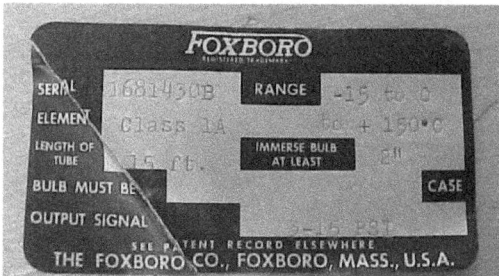

Figure 19 - An instrument tag removed from the Foxboro transmitter in a previous photo. Note the class I input element.

In fact, all quality temperature measuring systems should contain some form of temperature compensation to minimize the influence of ambient temperature variations. Given that most mechanical temperature sensing instruments employ some form of a filled capillary at the input, and that a temperature extreme applied at any point along the capillary will influence the pressure within the capillary. It seems reasonable to expect that an industrial grade temperature instrument would contain temperature-compensating components.

Temperature compensation can be employed in several ways. The most effective form of temperature compensation is to run an identical capillary within the sheath and along side of the active capillary connected to the temperature-sensing bulb. The compensating capillary does not contain a sensing bulb but does connect to a force generating bellows, diaphragm or Bourdon tube at the instrument. The force-generating component will be identical to the component connected to the active capillary and sensing bulb except that the compensating hardware opposes the actual sensing hardware.

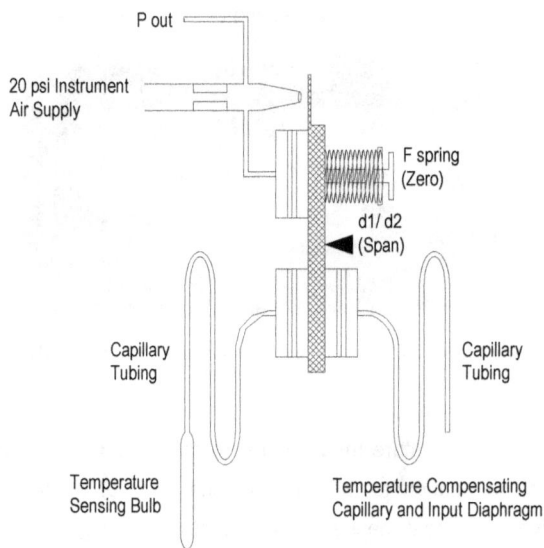

Figure 20 - Ambient temperature compensation commonly employs an identical measurement system minus the sensing bulb. The two capillaries are contained in the same sheath.

In this manner, the compensating capillary and bellows cancel temperature effects that would normally alter the temperature display and transmitter output. Bimetallic elements can also be employed to sense the ambient temperature and offset the indicated temperature display but typically do not offer the degree of compensation of the dual capillary system previously described.

Figure 21 - Temperature compensation in the Foxboro temperature transmitter is accomplished by using two capillaries and force components. In this photo the active and compensating capillaries enter the unit in the bottom center of the photo, and connect to two helix-type Bourdon tubes. The Bourdon tubes work in opposition to eliminate ambient temperature influences on the measured temperature.

Class III gas-filled mechanical thermometers operate in a similar manner to liquid-filled systems. A temperature increase results in molecular expansion of the gas. The orbiting electrons strike the sensing bulb with increased energy resulting in an increased pressure within the sensing bulb, capillary and force generating components. The pressure change is again converted into a force by applying the gas pressure to a Bourdon, diaphragm or bellows surface area. The pressure change corresponding to a temperature variation is easily computed with Charles' Law, where;

$$P_1 / T_1 = P_2 / T_2$$

Where,

      P_1 = the original pressure before the temperature change

      T_1 = the original temperature before the change

      P_2 = the new pressure after the temperature change

      T_2 = the new temperature after the change

Note: All pressure and temperature values are in absolute quantities

(Include Photo- Gas type Sensing bulb transmitter.)

Although no gas obeys Charles's law perfectly, the deviation is so small that gas thermometers are one of the most accurate means to measure temperature within the range of approximately −400° to +1000° F. Helium and Nitrogen are commonly employed in industrial measurement systems due to their chemically inert properties, good coefficients of expansion and low specific heats.

Gas-filled systems utilize larger sensing bulbs than liquid-filled systems with bulbs commonly measuring 5-20 inches in length and 1-inch in diameter. Although the larger sensing bulb mass may reduce the measurement system's time of response, it helps in providing immunity against ambient temperature induced errors. Like liquid-filled systems, gas–filled systems should also be compensated against variations in ambient temperature. Larger sensing bulbs reduce the influence of ambient temperature variations along the capillary. Without compensation, the gas-filled capillary and other internal components are sensitive to variations in the temperature outside of the process being measured.

Class II Vapor-pressure thermometers are a filled with a combination of both liquid and gas. Vapor systems contain a mixture of part gas and part liquid. The liquid resides in the sensing bulb with gas filling the capillary and force component. At any temperature within the operating range the percentage of liquid and gas is different. The sensing bulb acts as a liquid reservoir for the 'boiled' gas resulting from a temperature increase. At low temperatures the amount of gas vapor is low, as the temperature increases more molecules are released as vapor from the liquid and leaving less liquid within the bulb, capillary and force element.

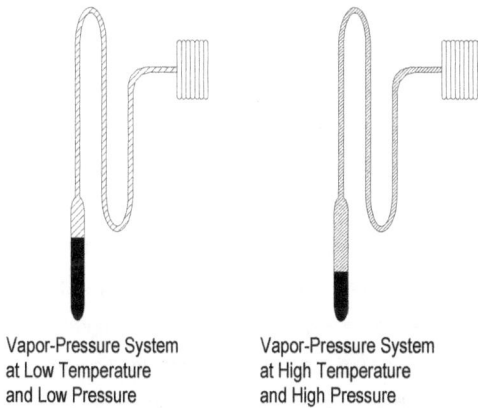

Vapor-Pressure System
at Low Temperature
and Low Pressure

Vapor-Pressure System
at High Temperature
and High Pressure

Figure 22 - The percentage of liquid and vapor varies for each temperature applied to the sensing bulb.

Because of the combined liquid and gas composition, vapor pressure thermometers are the only non-linear temperature measurement apparatus. A local temperature indication will usually exhibit the non-linearity on the scale. Vapor pressure thermometers are very repeatable because of the inherent self-regulating qualities of the liquid-vapor combination within the closed system; each specific temperature applied has a unique ratio of liquid and vapor, and a unique pressure.

Electrical Transducers

Four electronic sensing elements exist currently as discrete, in-contact temperature sensing transducers. Thermocouples, Resistive Temperature Detectors (RTD's), solid-state sensors (current generator IC's), and thermistors compose the variety of electronic temperature sensing elements. Among the four types each has found a place in industry yet only two, and primarily one, is commonly accepted among the manufacturing industries.

Thermistors

Until recently thermistors were rarely found in manufacturing processes. As the name implies, a *thermistor* is a thermally sensitive resistor. Commonly composed of metal alloys and semiconductor materials, thermistors have a reputation of not being easily matched and exhibiting highly non-linear negative resistance-temperature characteristics. Within recent years thermistors have appeared composed of nickel, a pure metal, to allow both positive temperature coefficients (PTC - as temperature goes up, resistance goes up) and negative temperature coefficients (NTC - as temperature goes up, resistance goes down). Thermistors are most often found mounted on circuit boards for limited range temperature measurement and ambient temperature compensation of circuits whose operating characteristics may vary with changes in the ambient temperature.

Thermistors are very sensitive however, changing tens of thousands of ohms within a few dozen degrees Fahrenheit. The corresponding non-linearity, poor replace-ability, required signal conditioning and narrow temperature range make thermistors somewhat undesirable for most industrial applications.

Thermistors are usually designated in accordance with their resistance at 25°C. The most common of these ratings is 2252 ohms; among the others are 5,000 and 10,000 ohms. If not specified, most instruments will accept the 2252 type of thermistor.

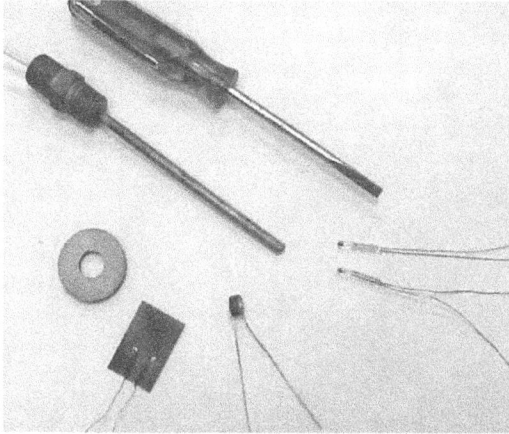

Figure 23 – Thermistors are available in a variety of styles. Shown here are a washer, a probe, a pocket screwdriver for perspective, two beads, a washer and an adhesive patch.

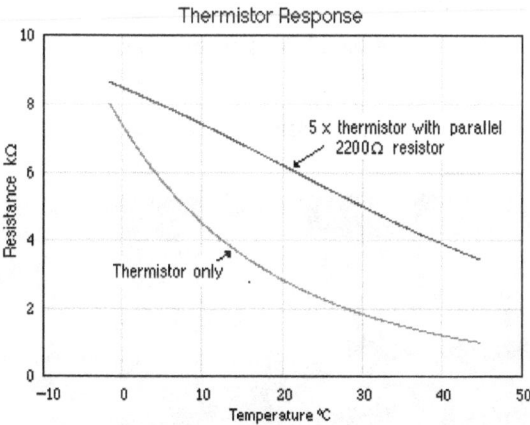

Figure 24 - Resistance versus temperature characteristic for a typical thermistor, note the non-linear negative temperature-resistance coefficient and the "Linearized" thermistor resulting from paralleling a 2.2K resistance with a thermistor. Courtesy – Capgo Thermistor.

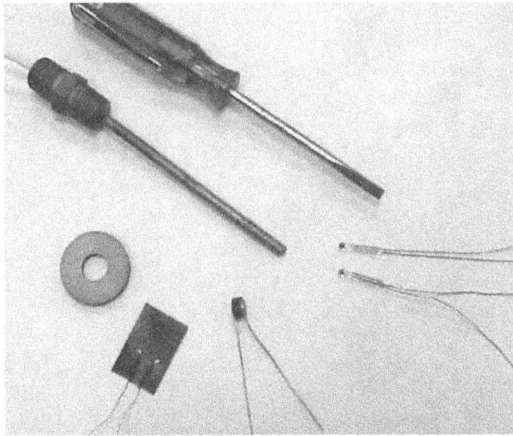

Figure 25 – Thermistors are available in a variety of styles. Shown here are a washer, a probe, a pocket screwdriver for perspective, two beads, a washer and an adhesive patch.

Solid-State Transducers

The solid-state, integrated circuit (IC) temperature sensor generates 1 μAmp per degree Kelvin (1μA/K°) over a rather limited temperature span of -50° to 150° C. It is sold in 8-pin DIP, two lead transistor (TO-92) packages and 2 wire stainless steel probe assemblies.

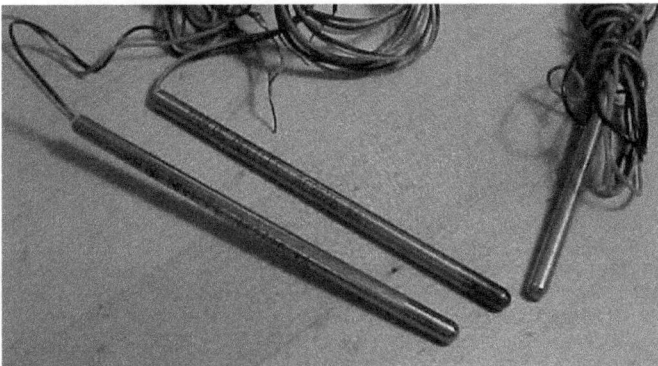

Figure 26 - The Analog Devices AC 2626 is a solid-state temperature sensor enclosed with a stainless steel sheath.

A linear sensor, the solid-state temperature transducer requires only a single 10K resistor and a voltage supply to condition its current output into a high-level voltage signal. The sensor can be purchased in very accurate versions but the limited measurement range and frail industrial perception keeps the solid-state sensor relegated to laboratory applications.

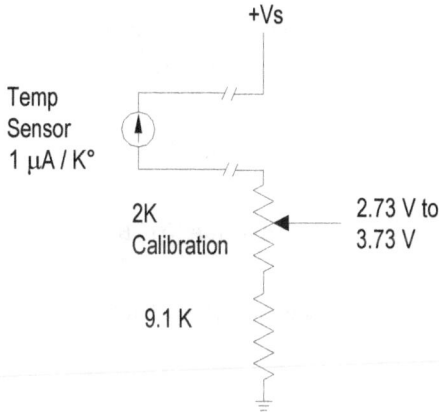

Figure 27 - The solid-state temperature sensor signal conditioning circuit utilizing a 10K resistance and solid-state current generator temperature sensor. Applying zero to 100 degrees Celsius results in a current range of 273 to 373 micro-amps and an output voltage of 2.73V to 3.73V representative of the Kelvin equivalent of the applied Celsius temperature. The variable resistor allows for single point system calibration.

The solid-state temperature sensor requires the previous circuit configuration to convert the temperature-controlled 1 μAmp per degree Kelvin (1μA/K°) current into a temperature sensitive voltage. The sensor operates as a 1μAmp per degree Kelvin current generator. To convert the current signal into a useable voltage requires passing the temperature-controlled current through a resistance. A 10K resistor is normally used to convert the 1μAmp per degree Kelvin current into 10 millivolts per degree Kelvin. For calibration purposes the 10K dropping resistor has been replaced with a 9.1K fixed resistance and a 2K potentiometer in the previous diagram.

In operation, applying 0° Celsius (273° Kelvin) to the sensor will result in 273 μA of current flowing through the sensor and resistors between the supply (+Vs) and ground. If the combined resistor and pot are set to about 10K, the resulting output voltage from the wiper of the pot will be about 2.73 Volts. The following demonstrates.

According to Ohm's law, the output voltage taken from the wiper of the pot is,

$$Vout = I * R$$

Since,

$$I = 1 \mu A/K°$$

Subbing,

$$Vout = (1 \mu A/K°) * R$$

If R = 10K,

$$Vout = (1 \mu A/K°) * 10K$$

$$Vout = 10 \, mV/K°$$

Applying a temperature of 0°C (273°K) results in an output voltage of,

$$Vout = 10 \, mV/K° * (273°K)$$

$$Vout = (.01 \, V/K°) * (273°K)$$

$$Vout = 2.73 \, V \text{ at } 0°C \, (273°K)$$

If the output voltage does not quite equal 2.73 Volts, the wiper of the potentiometer can be adjusted for an output voltage equal to 2.73 Volts. In this manner the sensor requires only a single point of calibration, an offset adjustment. The sensor gain is determined during the integrated circuit fabrication process at 1μA/K° and cannot be adjusted externally.

If the temperature applied to the sensor is raised to equal the boiling temperature of water (100°C = 373°K), the resulting current will equal 373 μA and the output voltage should approximate 3.73 Volts.

Given its ease of use, good accuracy and limited temperature range (approximately −50° to +150° C), the solid-state temperature sensor has found a niche in laboratory temperature sensing applications.

Resistance Temperature Detectors

The RTD, Resistive Thermal Device, or Resistive Temperature Detector is a precise and mostly linear resistive sensor that frequently appears in industrial applications and is commonly used for experimental testing purposes. Copper, nickel, tungsten and platinum varieties exist with platinum being the most popular, probably due to its ability to resist oxidation.

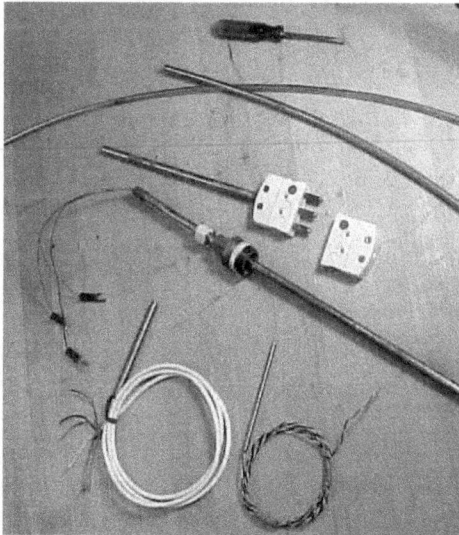

Figure 28 – An accumulation of two-wire, three-wire and four-wire RTD's. The additional lead wires are provided to compensate for thermally induced resistance variations of the leadwire connections between the RTD and the signal conditioner or transmitter.

RTD's exhibit mostly linear resistance-temperature characteristics however; each has a slight bend when their characteristics are plotted over the entire range of possible measured temperatures. Nickel exhibits the greatest non-linearity and is understood to be the only non-linear RTD. RTD's exhibit offset resistance values of 100, 500 and 1000 ohms at zero degrees Celsius, and possess a variety of gain coefficients depending upon the material composition.

RTD gain coefficients associate a change in resistance to a change in temperature and are referred to as Alpha (α) coefficients. Multiple Alpha coefficients exist for each material type of RTD. With platinum RTD's, the newer "American" Alpha equals .003902 Ohms/Ohm/C° and the older "European" Alpha equals .00385 Ohms/Ohm/C°. Ironically, the European characteristic (.00385 Ohms/Ohm/C°) with a 100-Ohm offset appears to be the most prevalent in American industry currently.

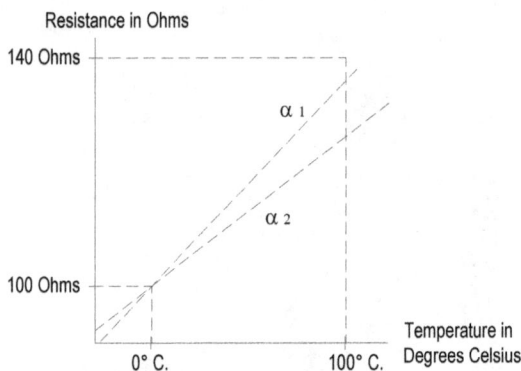

Figure 29 - The two common forms of platinum RTD exhibit slightly different resistance-temperature characteristics but an identical offset of 100 Ohms at zero degrees Celsius.

From the previous graph it can be observed that the RTD exhibits a linear resistance-temperature characteristic between 0° and 100° Celsius. Throughout the platinum RTD's measurable temperature range of approximately –250°C to +850°C a slightly reduced slope in the resistance-temperature characteristic can be observed beginning just above 100° Celsius. The actual gain of the RTD varies from .385 Ohms per degree Celsius to about .300 Ohms per degree at +850°C however the amount of non-linearity is gradual and is usually ignored.

Among the more esoteric RTD qualities, the units of RTD gain relate a fractional change in resistance to a change in temperature. For this reason RTD gain coefficients carry units of Ohms/Ohm/C°. RTD gain is normally expressed as *the amount of resistance change for each Ohm of resistance exhibited at zero degrees Celsius, for each degree of applied Celsius temperature.* Using the European Alpha coefficient as an example, .00385 Ohms/Ohm/C° = .00385 Ohms of variation for each Ohm exhibited at 0°C. Since the RTD exhibits 100 Ohms of offset at 0°C, 100 Ohms is multiplied by the gain coefficient.

$$(Gain\ Coefficient) * (Resistance\ at\ 0°) = Resistance\ change\ per\ C°$$
$$.00385\ Ohms/Ohm/C° * (100\ Ohms) = .385\ Ohms/C°$$

Once the gain is multiplied by the offset resistance value, the RTD's temperature sensitivity appears as .385 Ohms for each degree of Celsius temperature change. The resistance value at any temperature can then be determined using Y=mX+b concepts.

Example Problem
Determine the resistance values offered by a European alpha RTD at 0, 50 and 100 degrees Celsius.

Solution
To determine the RTD resistance at 0, 50 and 100 degrees Celsius the RTD characteristic is first expressed as a Y=mX+b equation and the given temperatures are then substituted into the equation.

Given,

$$Y=mX+b$$

Subbing the specifics from the input/output graph,

$$\text{Resistance} = (\text{Gain})*(\text{Temperature}) + \text{Offset}$$
$$\text{Resistance} = (.385 \text{ Ohms/C}°)*(\text{Temperature}) + 100 \text{ Ohms}$$

Applying the given temperatures,

$$\text{Resistance at } 0° = (.385 \text{ Ohms/C}°)*(0°\text{C}) + 100 \text{ Ohms}$$
$$\text{Resistance at } 0° = (0 \text{ Ohms}) + 100 \text{ Ohms}$$
$$\text{Resistance at } 0° = 100 \text{ Ohms}$$

The offset resistance is verified in the result.

Applying 50° yields,

$$\text{Resistance at } 50° = (.385 \text{ Ohms/C}°)*(50°\text{C}) + 100 \text{ Ohms}$$
$$\text{Resistance at } 50° = (19.25 \text{ Ohms}) + 100 \text{ Ohms}$$
$$\text{Resistance at } 50° = 119.25 \text{ Ohms}$$

Substituting 100° yields,

$$\text{Resistance at } 100° = (.385 \text{ Ohms/C}°)*(100°\text{C}) + 100 \text{ Ohms}$$
$$\text{Resistance at } 100° = (38.5 \text{ Ohms}) + 100 \text{ Ohms}$$
$$\text{Resistance at } 100° = 138.5 \text{ Ohms}$$

The American alpha RTD (.003902 Ohms/Ohm/C°) would exhibit slightly more resistance at the given temperatures. Using the American alpha RTD the resistance values would compute to 100 Ohms at 0° C., 119.51 Ohms at 50° C. and 139.02 Ohms at 100° C. RTD's are very precise, stable (meaning the output exhibits minimal "drift" - a varying output when the temperature is constant) and repeatable. The measured values would be very close to the computed values if an RTD were assembled into a calibration test fixture, the given temperatures applied and the resistance measured.

Figure 30 – A RTD and intelligent 4-20mA transmitter in an explosion-proof housing. Photo courtesy of Foxboro Controls.

Being a resistive sensor, the RTD normally requires a resistive bridge as the first stage of signal conditioning however, numerous varieties of integrated 2-wire RTD transmitters are commercially available for industrial purposes. Selling for about $100 the transmitters contain span and zero adjustments and accommodate two or three wire RTD's. Platinum RTD's in stainless steel sheaths (probes) sell for about $100 however, less expensive varieties have recently appeared on the market selling for under $50. The less expensive RTD's are housed in glass beads, thin-film elements, ceramic probes, cement-on patches and weld-able assemblies. One serious consideration must be retained and addressed when applying RTD's. As with strain gauges, the RTD must have a current passed through it to make it's varying resistance output signal useful. Current passing through resistance creates heat, and too much current creates self-heating effects that will alter the measured results and the stability of the RTD. It is imperative that the current and self-heating be minimized for the RTD temperature measurement to be accurate.

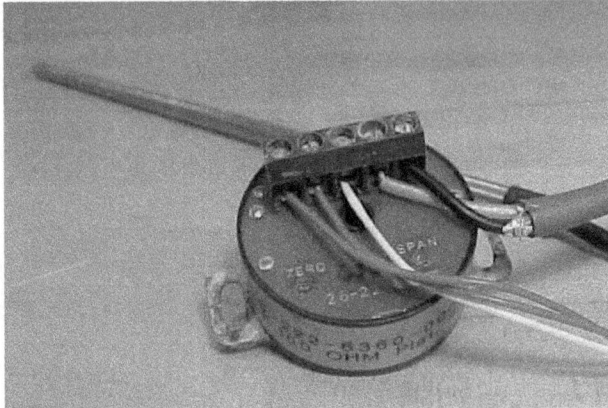

Figure 31 - A stainless steel sheathed RTD and 4-20 mA loop-powered transmitter. Note the integral span and zero adjustments for calibration.

Thermocouples

By far the most common industrial temperature sensor is the thermocouple. Rugged, predictable, repeatable, inexpensive and available in a large variety of temperature ranges and media compatibility, the thermocouple is universally accepted among industrial users. Simply stated, thermocouples generate a DC millivoltage signal when heated. The measured millivoltage can be taken to a set of thermocouple temperature-millivolt tables and the sensed temperature can be determined. Look-up data tables are commonly used in computer software with thermocouple analog inputs.

In reality, the operation of a thermocouple is a bit more complex. Thermocouples are composed of two wires yet do not require any form of power supply. Each wire is composed of a different metal. Soldering, welding, twisting or clamping joins the two wires. The connection forms the *Hot Junction*, the end that is normally placed into the process to be measured. Ideally, the same two wires are then run from the hot junction at the process to the instrument requiring the millivoltage. This is not always the case however and thermocouple extension wire is commonly used to connect the thermocouple hot junction, located in a sheathed probe, to the measurement instrument. The connection of the thermocouple extension wires to the measurement apparatus forms another junction called the *Cold Junction* or *Reference Junction*. The millivoltage appearing at the measurement instrument input terminals is contingent upon the difference in temperature between the hot and cold junctions.

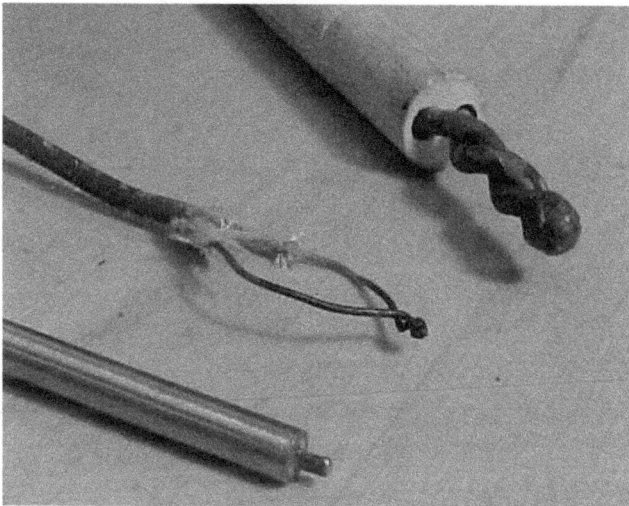

Figure 32 - Thermocouples are composed of two different types of metal wires joined at the hot junction. The hot junction is not usually exposed however; response time can improve significantly when the hot junction is not placed into a probe and thermowell.

As an example of thermocouple operation, assume the temperature of the hot junction (HJ) is the freezing temperature of water, or 0° C. Momentarily assume the temperature of the cold junction (CJ) is also 0° C. The resulting millivoltage available at the input terminals of the measurement instrument or transmitter will be zero since the temperature difference is also zero. However as the temperature of the hot junction increases a millivoltage builds at the instrument input, assuming the cold junction remains at a colder temperature than the hot junction, such as 0° C.

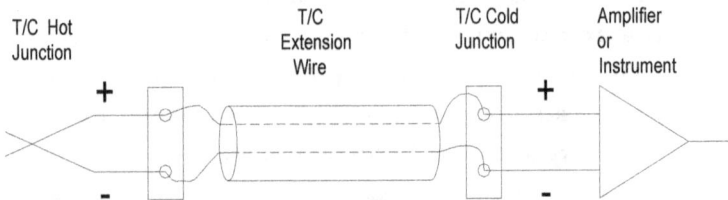

Figure 33 - Determining the operation of a thermocouple requires consideration of all system components.

The amount of generated millivoltage is determined by the temperature difference between the two junctions and the type of thermocouple used. Thermocouple millivoltage-temperature tables exist for each of the dozen types of available thermocouples. In the previous example, assume the hot junction temperature was allowed to rise to the boiling temperature of water at 100° C, while the cold junction remains at 0° C. Also assume that the type of thermocouple used in the example is iron-constantan. An iron-constantan thermocouple has one of its two leads composed of iron while the other is composed of constantan, a copper-nickel alloy.

The resulting millivoltage, often referred to as an electro-motive force (EMF) or thermoelectric voltage, can be determined by finding the hot junction temperature on the following table. The millivoltages are distributed within the table according to the temperature at the hot junction. A cold junction, or reference junction temperature is always specified when using thermocouple millivolt-temperature tables, and is usually specified at 0° C (32° F). The millivoltage values correspond to the provided hot junction temperatures in 10° increments along the far left column, and in single digit values across the top. To read the millivoltage at 100° C, locate "100" along the left column and move to the right stopping under the "0" column. The resulting thermoelectric voltage is 5.269 mV, as long as the reference junction temperature is at 0° C.

Thermoelectric Voltage in mV										
°C	0	1	2	3	4	5	6	7	8	9	10
0	0.000	0.050	0.101	0.151	0.202	0.253	0.303	0.354	0.405	0.456	0.507
10	0.507	0.558	0.609	0.660	0.711	0.762	0.814	0.865	0.916	0.968	1.019
20	1.019	1.071	1.122	1.174	1.226	1.277	1.329	1.381	1.433	1.485	1.537
30	1.537	1.589	1.641	1.693	1.745	1.797	1.849	1.902	1.954	2.006	2.059
40	2.059	2.111	2.164	2.216	2.269	2.322	2.374	2.427	2.480	2.532	2.585
50	2.585	2.638	2.691	2.744	2.797	2.850	2.903	2.956	3.009	3.062	3.116
60	3.116	3.169	3.222	3.275	3.329	3.382	3.436	3.489	3.543	3.596	3.650
70	3.650	3.703	3.757	3.810	3.864	3.918	3.971	4.025	4.079	4.133	4.187
80	4.187	4.240	4.294	4.348	4.402	4.456	4.510	4.564	4.618	4.672	4.726
90	4.726	4.781	4.835	4.889	4.943	4.997	5.052	5.106	5.160	5.215	5.269
100	5.269	5.323	5.378	5.432	5.487	5.541	5.595	5.650	5.705	5.759	5.814
110	5.814	5.868	5.923	5.977	6.032	6.087	6.141	6.196	6.251	6.306	6.360

Figure 34 - Type J (Iron-Constantan) Thermocouples -- thermoelectric voltage as a function of temperature (°C); reference junction at 0 °C. Data and table courtesy of the National Institute for Standards and Technology (NIST).

Example Problem

Assume an iron-constantan thermocouple measures 3.30 mV with a reference junction temperature of 0° Celsius. Determine the hot junction temperature.

Solution

Given the cold junction temperature is equal to the specified reference junction temperature of the iron-constantan millivolt-temperature tables, the hot junction temperature can be found by locating the measured millivoltage and determining the associated temperature. Unfortunately, 3.30 mV is not on the table but appears between the values of 3.275 and 3.329 millivolts. 3.275 and 3.329 millivolts appear in the row corresponding to "60," and below the column values of "3" and "4." Likewise and using interpolation, the hot junction temperature appears about halfway between 63 and 64 degrees Celsius.

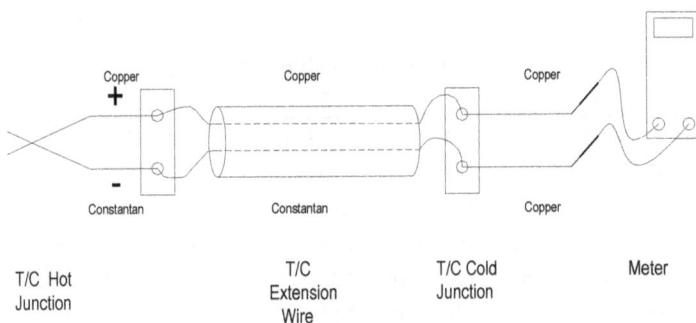

Figure 35 - The components of a typical thermocouple temperature measurement system.

Reference Junction Conversion

The reference junction in thermocouple applications is located where the thermocouple wires, or the thermocouple extension leadwires become non-homogenous, or where they connect to a material different than the thermocouple. Reference junctions are commonly located at the input terminals of the instrument, transmitter, recorder, controller, data acquisition system or other measurement device. If the reference junction temperature of the application is different from the specified value associated with the millivolt-temperature tables, and it usually is, *reference junction conversion* must be performed.

Reference junction conversion requires locating the actual reference junction in the application, measuring the reference junction temperature and finding the millivoltage corresponding to the measured reference junction temperature. The resulting millivoltage, representative of the difference between the actual reference junction temperature and the table's reference junction temperature, will be the amount the measured EMF will be offset from the millivolt-temperature tables. This value should be added to the measured value read on the meter in the previous diagram. The resulting millivoltage sum can then be used to determine the hot junction temperature from the millivolt-temperature tables.

Example Problem

A digital multimeter indicates a measured thermocouple EMF of 5.08 millivolts and a measured reference junction temperature of 72° F. Determine the temperature of the thermocouple hot junction.

Solution

To determine the hot junction temperature it should be first noticed that the reference junction temperature of the application does not meet the specified reference junction temperature of the millivolt-temperature tables. In addition, the actual reference junction temperature needs to be converted to degrees Celsius.

Since,

$$C° = 5/9 \, (F° - 32)$$
$$C° = 5/9 \, (72° \, F - 32)$$
$$C° = 5/9 \, (40°)$$
$$C° = 22.22° \, C ≈ 22° \, C.$$

Knowing the reference junction temperature, the corresponding millivoltage can be found from the millivolt-temperature tables. The millivoltage associated with 22° C is 1.122 mV. The millivoltage value of 1.122 must be added to the measured result. Summing the measured millivoltage from the meter and the millivoltage associated with the actual reference junction temperature yields,

$$5.08 \, mV + 1.122 \, mV = 6.202 \, mV$$

Why add? Because the actual reference junction temperature is warmer than the reference junction temperature of the table. If the actual reference junction temperature could be lowered to 0° C, the measured voltage at the meter would increase since the temperature difference between the thermocouple hot and reference junctions is increasing – the greater the temperature difference between the two junctions, the greater the EMF generated. Since the measured value is being adjusted to the table, the millivoltage resulting from the reference junction differences is always added – even when the actual reference junction temperature is less than the table's (below 0° C).

Once the millivoltage values are added, the actual hot junction temperature can be taken from the millivolt-temperature table. From the table, the hot junction temperature corresponding to 6.202 mV is approximately 117° Celsius.

Example Problem

An engineer is designing an amplifier to be used with an iron-constantan thermocouple over a temperature range of 35° C to 85° C. The reference junction is to be maintained at room temperature, 75° F (22° C). Determine the amplifier input voltage range corresponding to the given temperatures.

Solution

In this case the table is being adjusted to the application instead of the previous example where the measured voltage was adjusted to the table. In this example a millivoltage is to be determined from given temperatures. In the previous example a temperature was to be determined from a given millivoltage. The process of determining the millivoltage is similar to the method for finding temperature except for one key procedure – the millivoltage associated with the actual reference junction temperature must be subtracted.

The millivoltages associated with the expected hot junction temperatures are taken from the table as follows.

Millivoltage at 35° C = 1.797 mV, (RJ at 0° C)
Millivoltage at 85° C = 4.456 mV, (RJ at 0° C)

The millivoltage associated with the expected reference junction (RJ) temperature (22° C) is taken from the table as 1.122 mV. Once the millivoltage values have been determined the reference junction millivoltage is subtracted from the hot junction millivoltages.

Millivoltage at 35° C = 1.797 mV – 1.122 mV, (RJ at 22° C)
Millivoltage at 85° C = 4.456 mV – 1.122 mV, (RJ at 22° C)

The reasoning behind subtraction is due to the actual reference junction being at a higher temperature than the reference junction temperature of the table. Since the temperature difference between the actual hot junction and actual reference junction will be less than the temperature difference of the table's junctions, the resulting voltage applied to the amplifier will be less than that predicted by the table. Similarly, the millivoltage range taken from the table at 0° C must be subtracted from the millivoltages of the expected hot junction temperatures. The converted values, and expected amplifier input voltage range corresponding to hot junction extremes of 35° C and 85° C with a reference junction temperature of 22 becomes,

Actual Millivoltage at 35° C = .675 mV, (RJ at 22° C)
Actual Millivoltage at 85° C = 3.334 mV, (RJ at 22° C)

One need not remember the specifics of whether to add or subtract millivoltage values under predetermined conditions. Instead, one should retain the basic operating concept of the thermocouple – the greater the temperature difference between the hot and reference junctions, the greater the generated millivoltage.

Ice Bath Reference
Obviously, the reference junction conversion procedures described in the previous example problems are not simple or easy to recall. For this reason electronic ice bath references have been developed. An ice bath reference

measures the ambient temperature of the reference junction and automatically offsets the generated hot junction voltage by an amount equal to the millivoltage of the actual reference junction temperature.

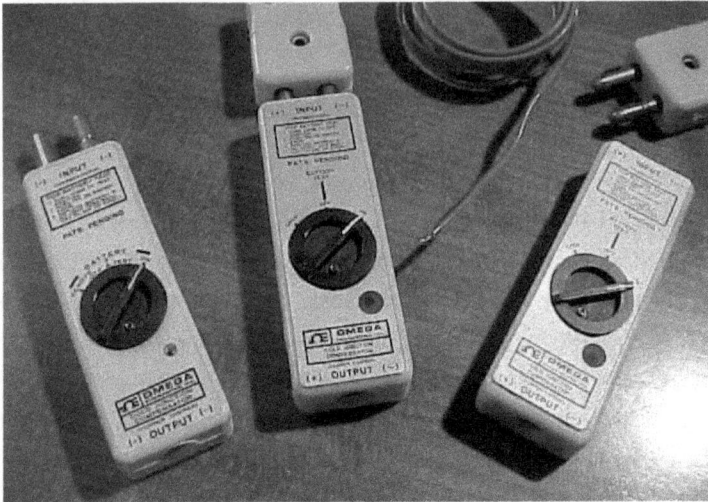

Figure 36 - Electronic ice bath references for Chromel-Alumel thermocouples.

The resulting output millivoltage value from the electronic ice bath can be taken to the temperature-millivolt data tables and the hot junction temperature can be read directly from the tables. In effect, the circuits perform the same function as immersing the thermocouple reference junctions into a mixture of 0° Celsius ice water as depicted in the following figure.

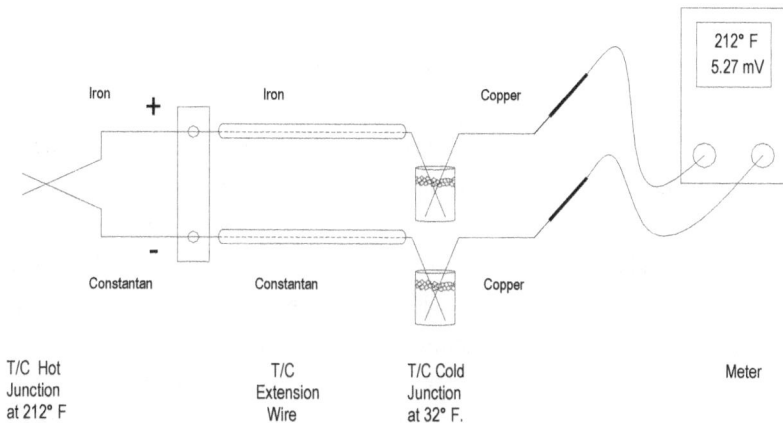

Figure 37 - The connection of thermocouple wires to electronic circuit wires always results in additional thermocouples being created at the reference junction.

As is demonstrated in the previous figure, the connection of the thermocouple extension wires to the electronic measuring instrument creates additional thermocouple junctions. The additional junctions form the reference junction and will always result in millivoltages that reduce the voltage generated by the hot junction. The diagram is provided to demonstrate the effect of an electronic ice bath in referencing the cold junction to 0° C so that hot junction temperature values can be read directly from the thermocouple millivolt-temperature tables.

Thermocouple Operation
Author's note - In thirty of studying measurement and control instruments, I've yet to find an explanation of thermocouple operation that applies fundamental electronic principles. Having worked extensively with thermocouples however, I can testify to their operation and the reproducible millivoltages associated with thermocouple temperature measurements. The following provides a bit of thermocouple history and the most commonly accepted explanation of thermocouple operation.

Although academically trained as a medical doctor, Thomas Johann Seebeck (1770-1831) preferred to study Physics. Recognized as the founder of the thermocouple, Seebeck researched the relationship between magnetism and electricity in metals when temperature is applied. During one of his experiments, Seebeck noted that when two different types of wires were joined at each end and a temperature increase was applied to one of the junctions, a magnetic field was observed around the wires and the other junction appeared to cool. The magnetic field indicated the presence of a current flowing between the junctions. (The following figure illustrates this effect.) Interestingly, Seebeck focused the subsequent research upon thermally induced magnetic effects instead of the temperature-millivoltage effects, deducing that the earth's magnetic field was the result of the temperature difference between the equator and the north and south poles.

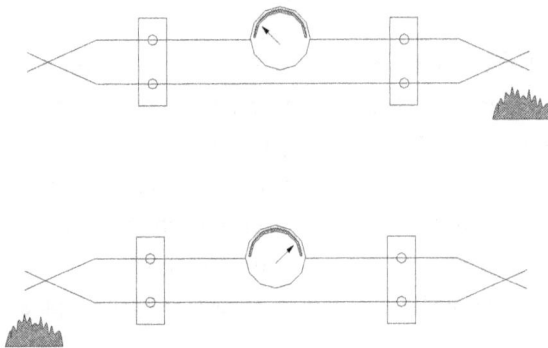

Figure 38 – An example of the Seebeck effect. Joining two dissimilar wires at each end and applying a temperature results in current flow. Reversing the applied temperature reverses the current direction.

The circuit provided in the previous diagram is representative of early thermocouple experiments. Industrial applications of thermocouples currently utilize half of the provided circuits in the diagram, preferring to utilize a generated millivoltage (EMF) instead of a circulating current between two junctions.

Although polynomial equations are available, thermocouple operation is understood to be empirical, or based upon scientific observation. Rather than predicting or proving with mathematics (other than millivoltage vs. Temperature reference tables), thermocouple operation requires coupling (connecting) two dissimilar metals together and applying a temperature to the joined end. The temperature difference between the 'hot' junction at the process and the 'cold' or reference junction (where the thermocouple output is located) generates a small, but very predictable millivoltage. The generated millivoltage is the result of the temperature difference between the two ends of the thermocouple at the process and at the measurement apparatus or transmitter. The greater the temperature difference, the greater the millivoltage generated - the smaller the temperature difference, the smaller the millivoltage. Should the two temperatures be equal with no temperature difference, the resulting millivoltage would be zero.

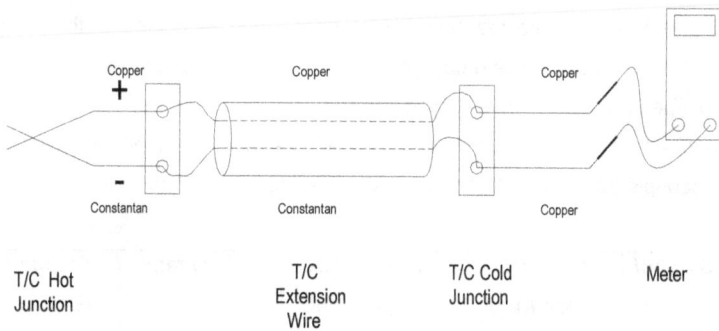

Figure 39 - The common components of a thermocouple-based temperature measurement system.

In a circuit operation sense, thermocouples act as a temperature controlled direct-current millivoltage source. Consequently, the meter in the previous figure is usually replaced with a high gain instrumentation amplifier or transmitter when applied in an industrial or experimental temperature measurement setting.

The most commonly accepted explanation of operation assigned to the thermocouple is centered on temperature effects upon electron activity. It is believed that the application of a temperature differential between the hot and cold junctions causes electrons activity to increase at the warmer junction. The electrons throughout the thermocouple assume varying degrees of thermally induced motion depending upon their location between the two junctions. The resulting variation in electron activity establishes a potential difference similar to the manner with which temperature would be conducted along the length of the thermocouple and corresponding extension leadwires. The potential differential or gradient along the length of the thermocouple circuit results in an electromotive force (EMF) or millivoltage appearing at the reference junction. The reference junction is also called the cold or measurement junction.

A testament to the thermocouple's industrial popularity, the ISA (Instrument, Systems and Automation society) and the ASTM (American Society of Testing Materials) has standardized upon color codes and symbols for the various thermocouple types. Thermocouple tables clearly indicate the generated millivoltage in relation to the applied temperature for the various types of sensor composition. Type 'J' (iron/constantan) and a type 'T' (copper/constantan) thermocouple tables have been included in the appendix of the chapter.

ISA Symbol	Materials (Pos-Neg)	Temperature Range – Ref. Jct. At 0 °C.	Millivoltage Range (Approx.)	Extension Wire Color Code (Pos – Neg)
B	Platinum (30% Rhodium) – Platinum (6% Rhodium)	0 to +1800° C.	0.0 to +14 mV.	Gray - Red
C	Tungsten - Rhenium	0 to +2300° C.	0.0 to +37 mV	White - Red

E	Chromel - Constantan	-250 to +1000° C.	-10 to +76 mV.	Purple - Red
J	Iron - Constantan	-200 to +1200° C.	-10 to +70 mV.	White - Red
K	Chromel - Alumel	-250 to +1350° C.	-6.5 to +55 mV.	Yellow - Red
N	Nicrosil - Nisil	-250 to +1300° C.	-4.5 to +47.5 mV.	Orange - Red
R	Platinum (13% Rhodium) - Platinum	-50 to +1750° C.	-.2 to +21.0 mV.	Black - Red
S	Platinum (10% Rhodium) - Platinum	-50 to +1750° C.	-.2 to +19.0 mV.	Black - Red
T	Copper - Constantan	-250 to +400° C.	-6.0 to +21.0 mV	Blue - Red

Figure 40 - Thermocouple symbology, output millivoltage, color-code and temperature measurement ranges.

Non-Contact Temperature Measurement

Pyrometry

Occasionally temperatures must be measured that are not practical for an insertion form of measurement instrument such as a sheathed RTD or thermocouple probe. In these cases the process temperatures are excessive, the object to be measured is in motion or cannot be safely or practically contacted. High voltage connections, large volume steam boilers, food processing and metal forging are examples of where temperature measurement deviates from traditional insertion practices. For these applications a non-contact class of temperature sensors called *pyrometers* has been developed. Loosely defined, Pyrometer means to fire-measure, or temperature measure. Specifically, pyrometer is a term assigned to a class of measurement instruments that observe the radiant infrared properties of a heated object to yield a temperature measurement through non-contact means.

Older forms of pyrometers utilized the radiation characteristics of objects to emit light when heated above the temperature of approximately 850° Fahrenheit. The early instruments commonly used the color hue, or tone of the heat source or heated object to determine the temperature. More recent devices take advantage of the infrared energy radiated from objects at temperatures above absolute zero temperature. Realistically, pyrometers can commonly measure temperatures down to near zero degrees Fahrenheit by detecting the infrared energy radiant from the object being measured. To understand the operation of pyrometers an investigation into the principles of infrared energy is in order.

Infrared light is the portion of the electromagnetic spectrum that is just below visible light. As can be seen in the following figure, The term "infrared" refers to a broad range of frequencies, beginning at the top end of those frequencies used for radio and television communication and extending up to the low frequency end of the visible spectrum.

The Electromagnetic Spectrum

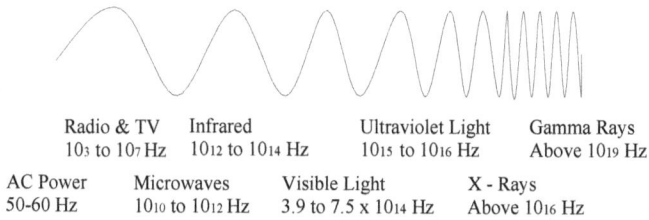

Radio & TV	Infrared		Ultraviolet Light	Gamma Rays
10_3 to 10_7 Hz	10_{12} to 10_{14} Hz		10_{15} to 10_{16} Hz	Above 10_{19} Hz

AC Power	Microwaves	Visible Light	X - Rays
50-60 Hz	10_{10} to 10_{12} Hz	3.9 to 7.5 x 10_{14} Hz	Above 10_{16} Hz

Figure 41 - The electromagnetic spectrum is normally expressed in wavelengths but is offered here in frequencies as a convenience.

As mentioned earlier in the chapter, all objects above absolute zero temperature exhibit a degree of electron activity as the electrons orbits the nucleus. The introduction of heat causes electrons to move in higher energy orbits. As the object is cooled and the electrons return to a normal orbit the excess energy is released as vibrating heat particles, or photons of light. The produced photon will exhibit a vibration frequency appearing as a specific color representative of the amount of energy released by the electron as it returns to its normal orbital path. Photons released in the spectrum just below visible light frequencies fall into the infrared category.

The infrared spectrum is composed of three frequency bands, near infrared, mid-infrared and thermal infrared. Near infrared (near-IR) is adjacent to visible light exhibiting wavelengths that range from 0.7 to 1.3 microns (micro-meters), or about $3x10^{14}$ Hertz. Mid-infrared (mid-IR) exhibits wavelengths ranging from 1.3 to 3 microns. Near-IR and mid-IR light are commonly used by electronic devices for communication such as handheld remote controls. Thermal-infrared (thermal-IR) occupies the largest part of the infrared spectrum with wavelengths of 3 to over 30 microns and is the only IR band resulting from photon emission, created as the electron returns to its normal orbit. Near and mid-IR is the result of light reflection.

Heat causes atoms to release photons in the thermal-infrared spectrum. The frequency of the thermal-IR energy varies in proportion to the heat source temperature. The hotter the object, the shorter the wavelength (the higher the frequency) of the infrared photon released. Objects between about 850° F and about 3000° F. will emit photons in the visible spectrum. A "warm" object appears red and as the object increases in temperature the color varies from red through orange, yellow, blue and white before becoming invisible in the ultraviolet portion of the spectrum.

Humans can't see thermal-IR energy but can feel IR whenever one stands near a heat source or fire. An IR temperature instrument receives and focuses the radiant thermal-IR energy upon a sensor or sensing array as with thermal imaging. Signal conditioning circuitry associates the specific wavelength, or frequency of the detected thermal-IR with temperature. IR instruments cover a range from approximately -4° F. to over 3500° F.

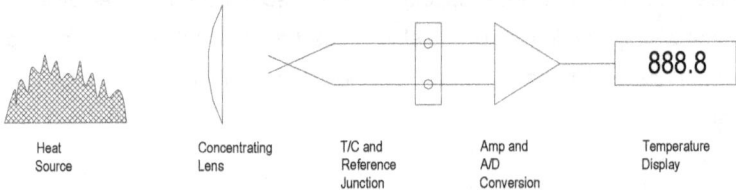

Heat	Concentrating	T/C and	Amp and	Temperature
Source	Lens	Reference	A/D	Display
		Junction	Conversion	

Figure 42 - Components of a typical portable pyrometer.

Two classes of pyrometers exist, portable and fixed. Handheld portable pyrometers require the instrument to be directed at the heat source. The IR energy is concentrated on a temperature-sensing element such as a thermocouple or a series configuration of multiple thermocouples called a thermopile. More elaborate sensors such as charge-coupled camera arrays are also used. The resulting signal from the sensor is conditioned and displayed as in the previous figure. Three types of portable infrared pyrometers are currently available, single wave, dual wave and multi-wave. Single wave instruments observe only a single portion of the thermal-IR frequency spectrum from the heat source. Dual and multi-wave pyrometers provide a more accurate indication by observing the radiant energy of the source at more than one frequency within the infrared spectrum.

In older "Optical" pyrometers the heat source was detected visually by looking through the instrument. The color of the heat source was manually matched to a colored lamp filament within the instrument. The filament color was manually adjusted to match the color of the heat source. When the tone of the filament matched the tone of the heat source, the temperature was read from an associated scale on the instrument. Optical pyrometers were limited to a narrow temperature range within the lower visible range of thermal energy between approximately 900° and 200° F. Optical pyrometers have been made obsolete by the appearance of portable IR pyrometers on the market.

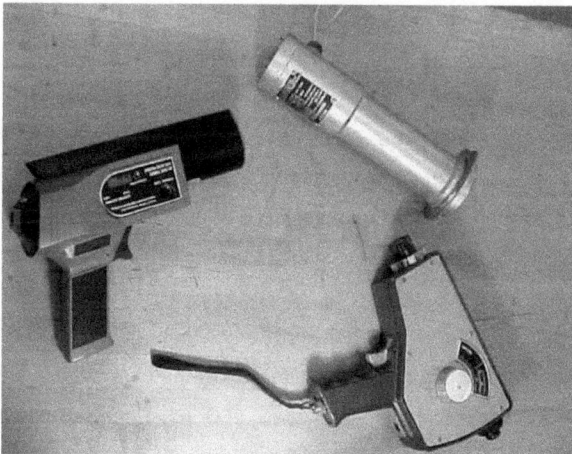

Figure 43 – IR, fixed and optical (lower right) pyrometers.

Fixed pyrometers, occasionally called radiation pyrometers are placed into a fixed location, directed at the heat source and generate an output signal that is used to control the temperature of the source. Fixed pyrometers commonly involve the use of thermopiles. The multiple thermocouple junctions increase the millivoltage resulting from the focused thermal-IR energy. The increased millivoltage provides greater sensitivity to temperature variation.

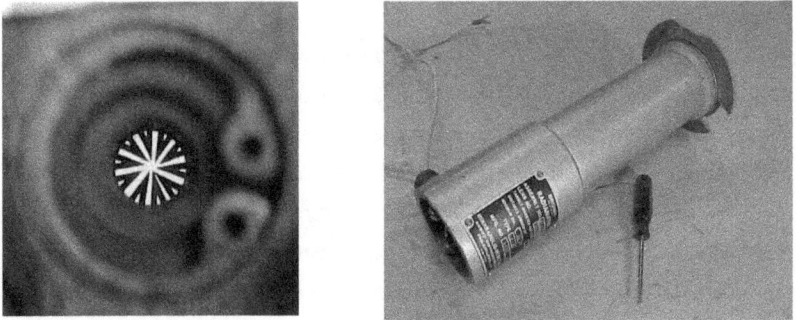

Figure 44 - A radiation pyrometer (right) and the sensing thermopile (left). For maintenance reasons, the thermopile is easily observed when the cover is removed from the pyrometer.

Often in fixed pyrometer applications the precise value of process temperature is not determined, as temperature value is not as important as maintaining a constant temperature. Precise values of process temperature are measured using other means or by the quality of the manufactured product. Of greater importance is having a sensor that once focused on a fixed location, is capable of providing a signal of significant resolution or sensitivity.

The current generation of pyrometers is extremely sensitive to thermal-IR energy and allows temperature measurements well below the temperature where heat radiates light in the visible spectrum (approx. 850° F). Handheld Infrared thermometers similar to the following figure are capable of measuring from 0° to 400 ° Celsius.

Figure 45 - A handheld infrared thermometer capable of measuring from 0 to 260 degrees Celsius.
Photo courtesy of Omega Instruments.

Infrared Thermocouples

Infrared thermocouples employ a lens to concentrate thermal-IR energy on a common single-element thermocouple. The generated thermocouple millivoltage is connected to a process controller and used as an analog to measure and control the process temperature. The objective behind the operation of the infrared thermocouple is to produce a signal that varies with temperature.

Figure 46 - An infrared thermocouple. Photo courtesy of Omega Instruments.

The generated millivoltage may NOT be referenced to a data table to determine the object temperature, as the purpose behind the device is to produce an analog of temperature for the purpose of temperature control and not precise temperature measurement. Due to the uncompromising nature of the locations where infrared thermocouple temperature measurements are performed, units can be purchased with air purges and cooling jackets to stabilize the output signal.

Emissivity coefficient

When applying pyrometers the nature of the temperature source, the measuring instrument and everything in between should be concerned. For an accurate temperature measurement the application should consider the emissivity coefficient of the heat source. The emissivity coefficient is a value between zero and one, and represents the thermal absorption characteristics of the heat source. Thermal sources can reflect, transmit or absorb heat. To be accurate in measuring the temperature of the source the measuring instrument needs to be sensitive to the absorption qualities of the source. Higher values of emissivity are associated with materials and objects that absorb heat where lower values are representative of shiny objects with reflective surfaces. Dark surfaced objects such as oxidized iron and copper absorb heat and are associated with emissivity coefficients of .7 to .9. Objects with shiny surfaces such as polished gold, copper, lead or cobalt reflect heat and commonly assume emissivity coefficients of .2 or under. Filtering is usually required for opaque objects that transmit thermal energy. On most of the more expensive instruments an emissivity adjustment is available.

Field of Vision

Since optics is always employed with pyrometers, the measuring instrument's field of view needs to be considered. The smaller the field of view the smaller angle of visual spread seen by the instrument. Small fields of view allow for an accurate measurement of a small area within the heat source. Larger fields of view yield more of an averaging effect.

Thermal Imaging

This material is currently under development.

Thermal imaging and thermal enhancement

creates pictures based on heat energy emitted by the viewed scene rather than light reflected off of it. Heat in the infrared region of the electromagnetic spectrum behaves much like visible light and can be optically focused and collected. Infrared thermal imagers employ various materials whose electrical properties change when heated to perform the transformation from infrared energy to electronic signal levels. These changing signal levels are translated into video signals in which a different shade of gray on the TV monitor is assigned for each detectable temperature (i.e. emitted energy) level. Because the energy being sensed is heat and not light, the TLV is totally unaffected by the amount of light in a scene. The ability to sense differences in temperature can often provide valuable information above and beyond the ability to see things in the dark.

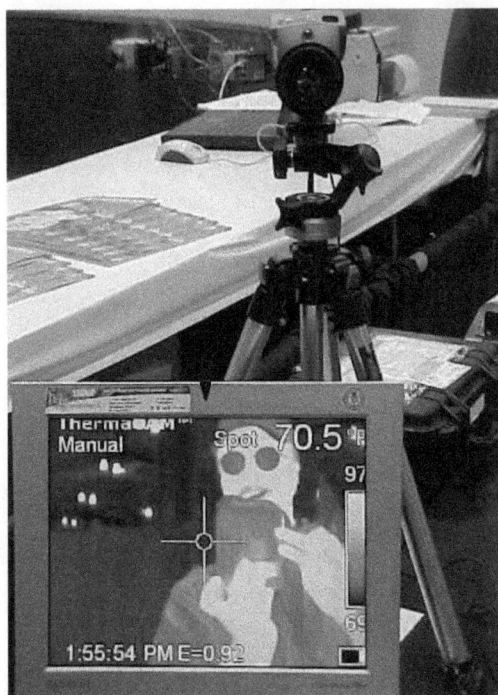

Application Notes

1 - Insertion forms of temperature sensors indicate the temperature of the probe and not necessarily the temperature of the process being controlled. For this reason the location of the sensing element within the process is critical. When measuring flowing fluids, assure that the sensor is placed into the path where the greatest amount of fluid is flowing.

2 – Speed of response should always be considered when applying a temperature sensor. The smaller the mass associated with the sensor, the greater the speed of response. Insertion forms of temperature measurements often require thermowells when placed into vessels to facilitate sensor removal and maintenance. The type of connection between the sensor and the thermowell determines the depth of the thermowell into the vessel.

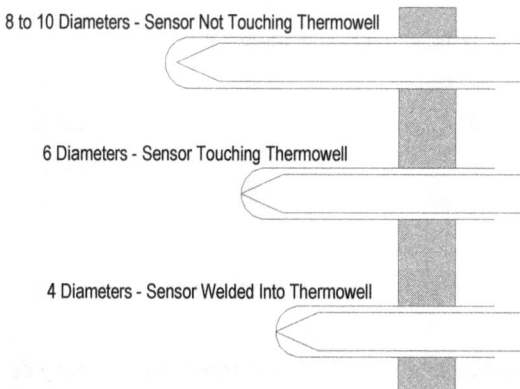

8 to 10 Diameters - Sensor Not Touching Thermowell

6 Diameters - Sensor Touching Thermowell

4 Diameters - Sensor Welded Into Thermowell

Figure 47 - The method of connection of the sensor to the thermowell determines the depth of the thermowell into the vessel.

- Future recc's -

Include T K and J tables in application notes?

Homework

Temperature Conversions

1 – Convert the following degrees Fahrenheit into degrees Celsius.

-100° F

98.6° F

200° F

1000° F

2 - Convert the following degrees Celsius into degrees Fahrenheit.

-200° C

-100° C

500° C

1000° C

3 – Convert the following into absolute temperature.

-100° F	-200° C
98.6° F	-100° C
200° F	500° C
1000° F	1000° C

4 – A *change* in temperature of 100° C corresponds to how many degrees F?

5 - A *change* in temperature of 100° R corresponds to how many degrees K?

Mechanical Thermometers

6 – A thermometer is calibrated to measure from 200° to 800° F. Determine the range and span of the measurement.

7 – A thermometer rated at 1% accurate is to measure a 1000° temperature span. Determine the probable error in degrees at any temperature within the span. If the current reading is 104°, determine the possible range of actual temperature values. Given this example, what happens to the percentage of error at lower temperatures? (Note – This material was not covered in the chapter but can be resolved by applying basic mathematical principles.)

8 – To complete an electrical circuit, a section of copper 6 inches in length must expand .01 inches beyond its normal length at 25° C. Determine the amount of temperature change required to complete the circuit.

9 – An iron rail is 33 feet in length. Assuming the rail is unrestrained, how much will its length change over the course of a year when subjected to temperatures of –20° to +150° F. Assuming a number of rails are connected to extend over a length of 100 miles, determine the resulting overall change in length for the same temperature variation.

10 – A petrochemical exhibits a specific gravity of .847 at 20° C. Determine the specific gravity if the product is heated to 70° C. Assume a cubical expansion coefficient of .000899 $in^3/in^3/C°$.

11 – A mechanical thermometer utilizes a gas-filled capillary system, bellows and spring to provide an indication of the temperature applied to the sensing bulb. Assuming the pressure within the capillary is 30 psig at room temperature (75° F), determine the pressure at 100° F and 300° F. A diagram of the thermometer is provided in the following figure.

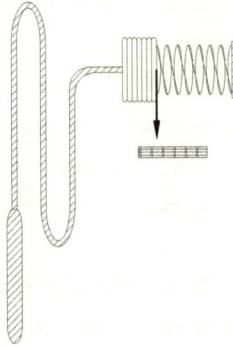

Figure 48 - The mechanical thermometer of problems #11 and #12 is composed of a sensing bulb, capillary, bellows and force-balancing spring.

12 – Using the mechanical thermometer of the previous figure and the pressures determined in the previous problem, determine the amount of indicator deflection as the temperature varies from 75° F to 100° F and from 75° F to 300° F. Assume the bellows surface area is 2 in^2 and the spring rate is 50 lbs/inch. (Answer = .75" at 300° F).

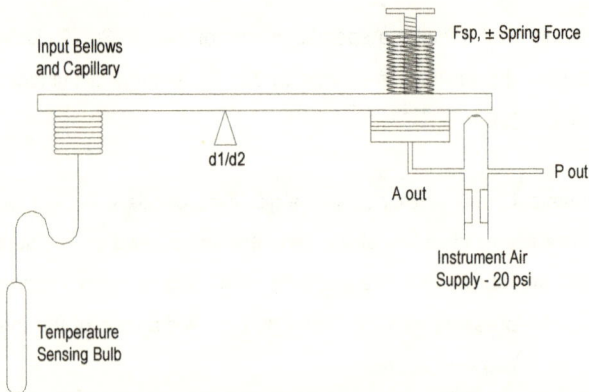

Figure 49 – The temperature transmitter of problem #13.

13 - Given mechanical temperature transmitter of the previous figure, determine the output pressure at -50° C and 150° C. Assume the capillary pressure is 10 psig at room temperature (25° C); the input bellows and output diaphragm (Aout) areas are 1 in², the d1/d2 gain is .725:1 and the spring force is .25# down. (The answer should approach 3-15 psi from -50° C to 150° C.)

Thermocouple Reference Tables
14 – Locate a set of type "T" thermocouple reference tables. Determine the millivoltage created when the hot junction is exposed to a range of temperature between 32°F and 212°F and the reference junction is at 75°F.

15 – Locate a set of type "J" thermocouple reference tables. Determine the millivoltage created when the hot junction is exposed to a range of temperature between 32°F and 212°F and the reference junction is at 75°F.

16 – Locate a set of type "K" thermocouple reference tables. Determine the millivoltage created when the hot junction is exposed to a range of temperature between 32°F and 212°F and the reference junction is at 75°F.

17 – Perform a web search to compare thermocouple millivoltage and temperature characteristics. Based upon your search, what thermocouple measures the widest range of temperature? What thermocouple generates the greatest millivoltage?

18 – Perform a web search for a set of thermistor resistance-temperature characteristic curves. Did the curves appear linear? If so, why?

This page retained intentionally blank.

Chapter 9 - Displacement, Velocity and Acceleration

Note – The material in this chapter is currently under development.

This chapter presents displacement-related processes and the commonly associated sensors. Sensors and physical variables common to the machining and metal working industries are investigated. Basic electronics is reviewed in the context of sensing technologies utilized in the measurement of position, velocity and acceleration. Included subjects follow.

Objectives:
Upon completion of this chapter, you should be able to:
- Explain the operational characteristics of common displacement sensors
- Explain the operational characteristics of velocity and acceleration sensors
- Interpret the common specifications of displacement sensors
- Interpret the common specifications of velocity sensors
- Interpret the common specifications of acceleration sensors
- Explain routine and unique sensor signal conditioning concerns
- Suggest calibration procedures for displacement, acceleration and velocity sensors

Introduction
Displacement Measurements
Position/Distance
Velocity
Acceleration
Displacement Application Notes:
Homework

Introduction

ET 203 is a study in sensors and the processes they detect. Among the processes discussed are position (or distance), velocity, acceleration, pressure, level, flow and temperature. As will be seen, many of the sensors associated with these processes utilize similar operating principles. Most basic operating principles require a solid understanding of fundamental electronics since variable capacitance, variable inductance and especially variable resistance compose the majority of process sensors.

Sensors are available in two forms, continuous (proportional) and discontinuous (switching). Continuous sensors provide an output that will change in a continuously variable manner with its process input. Discontinuous sensors, or process switches provide a contact closure, or similar discrete output variation at some pre-determined input value. The switch usually includes provisions to vary the set point and Hysteresis of the contact closure. These types of adjustments are common on pressure and temperature switches. More expensive and elaborate continuous sensors usually include provisions for gain and offset.

Continuous sensors can be further sub-divided into high level and low level, references to the output voltage levels. Low level sensors provide an output voltage value of millivolts since often composed of resistive strain gauge bridges internally. High level sensors output volts or milliamps and generally contain a millivolt sensor with a built-in amplifier. Span (gain) and zero (offset) adjustments are common with high level sensors. As might be expected, high level sensors are usually more expensive.

The processes and sensors discussed in this text will begin with position, velocity and acceleration since these are the most familiar yet often least understood.

Displacement Measurements

Position, velocity and acceleration have been included under displacement since each process represents an object being displaced. Position or distance is a single dimension measurement of length, (or distance) from a point to a reference location. Proximity is a discontinuous measurement of distance. A proximity detector determines if an object is within a given distance or zone and outputs a hi/lo indication accordingly. Proximity detection allows for determining the amount of liquid in a container as it is being filled, detecting people approaching an automatic opening door and counting the quantity of objects moving along a conveyor. In each case, determining the distance is not required. Merely determining whether an object is within a distance zone is of value.

Most non-contact methods of proximity detection are performed using acoustic pulses, magnetic fields or optical sensors and light beams. Simple proximity switches are assembled using a Microswitch® and a lever that drags along the device(s) to be detected.

Position/Distance

Continuous distance measurements are common in research, experimental testing and product development. Among the more simple devices used to measure distance is the 'Yo-yo.' Composed of a potentiometer and a retractable string, the yo-yo varies its resistance as the string is drawn (fig. 85). The gain, or transfer characteristic, of this and all other sensors are expressed as a function of the sensor's output in relation to its input. In the case of the yo-yo the gain would be in ohms per unit length, or if incorporated into an electronic circuit the gain might be volts per inch. Units of gain, when taken into consideration with the initial output offset of the sensor allow precise determination of the process being measured.

Figure 1 – The spring loaded Yo-Yo displacement transducer outputs a voltage in proportion to the length of line displaced. The metal assembly at the top is affixed to the device being measured.

LVDT

The Linear Variable Differential Transformer (LVDT) is a sophisticated and precise position sensor. The LVDT is used to measure position along a single linear axis within definite limits. The magnetic core of an LVDT is physically connected to the item position to be measured. The LVDT is constructed as a transformer with a movable core. As a transformer, the LVDT is composed of a single primary and two secondaries. An AC source voltage is applied to the primary, input frequencies of 400, 1000 and 10kHz are common. The primary flux couples both secondaries equally with the movable core centered or removed entirely. As the core is positioned above or below the center "Null" position, more magnetic field flux is coupled to the secondary winding, located above or below the center position, and a greater voltage is induced into that coil. Since the two secondaries are connected in series opposing, the secondary nearest the core and generating the greater induced voltage will cancel the lesser voltage associated with the remaining secondary.

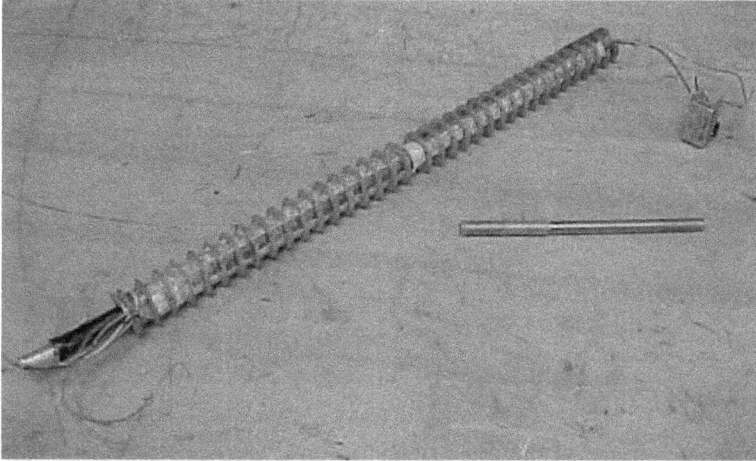

Figure 2 A "Baton" LVDT with the variable core removed and the signal conditioning integrated circuit obvious in the upper right corner. The length of this unit is approximately 12 inches, with a linear displacement range of about 4 inches. The two secondaries are obvious in the photo.

When observing the LVDT output from the series opposing secondaries, as the core is positioned from an extreme location through the center null into the other extreme, the output will be seen to change in amplitude and phase relation. The output is measured in relation to the input AC voltage. Phase sensitive discriminators, or demodulators are available in integrated circuit form to convert the amplitude and phase varying AC output into a linear bipolar (+ and -) DC signal if desired.

Review the LVDT specifications closely. Note such items as infinite resolution, nominal range, sensitivity and observe the output voltage linearity and phase as a function of core position plot for the Schaevitz spec. sheet graph (figure 2 on Schaevitz spec. sheet). Of greatest importance however is to develop a working knowledge of the sensor's internal operation as a variable transformer, so that the resulting specifications can be better interpreted.

Encoder

With the current growth of computer-based automation, many sensors are providing outputs in digital form. Among the more popular digital output sensors for measuring position is the encoder. Optical transmitters (light emitting diodes) and phototransistors are placed at each side of a slotted disk. As the disk is rotated, light is allowed to pass through the open slots on the disk generating light pulses incident on the phototransistor, and TTL logic pulses in the output of the phototransistor. Encoders are available in two forms, absolute and incremental. An *absolute* encoder provides a ten, twelve or sixteen bit Grey code output word as the input shaft is rotated. An absolute encoder has a specific and unique output word for each shaft position angle. Should power be lost, the output word will remain the same upon power-up. An *incremental* encoder output is merely a pulse train generated optically as the shaft is rotated. The incremental encoder's output appears the same at any arc of rotation. A counter is required to convert the pulses into an expression of rotational position, distance, location and so forth. Should the power be removed to an incremental encoder display, the unit will require calibration or "homing" upon power-up. Examples of both absolute and incremental units are included in the following specifications. Encoders of both types are commonly used with manufacturing, CNC and robotics.

Figure 3 – Although difficult to observe in the upper half of the photo, this absolute encoder resides on the end of a stepping motor and is used to feedback position information to a controller.

The following pages include specifications and descriptions of proximity sensors and displacement transducers. Review the specifications and attempt to determine the limitations of the individual sensors. Note the range of input and sensitivity, operation, output type and span, power supply requirements and expensive where possible. Especially review the LVDT and be prepared to address application concerns.

This text will expose the student to a quantity and variety of sensors, transmitters, actuators and related automation devices. In establishing familiarity with these and any other new component, attempt to establish the internal operation to the greatest extent possible. Approaching any new system or component from this perspective provides an intuitive understanding of the specifications and limitations of the device.

Velocity

Velocity is a rate measurement of distance in relation to time. Speed is another term for velocity. Units of feet per second, degrees per second, miles per hour, revolutions per minute and kilometers per hour are common velocity dimensions.

Two forms of velocity are commonly measured, linear and rotary. Conversion of one form into another for the purposes of measurement is common. The linear motion of a machine tool part might be converted through a rack and pinion gear into rotary motion for rotating an encoder or resolver to generate a velocity signal.

Velocity = Distance/Time

Following are specifications and explanations of linear and rotary velocity sensors. Note the gains, output values, power supply requirements and especially the internal operation of the sensors. Among the presented sensors is a variable reluctance transducer, or magnetic pickup. This widely used and easily adapted sensor generates pulses in proportion to its rotational velocity. The output frequency in Hertz is used to represent revolutions per minute.

LVT
The linear velocity transducer measures velocity within a short travel distance. A simple device, the LVT utilizes a fundamental concept in electronics of passing a magnet through a coil of wire to generate a voltage representative of velocity. As mentioned previously in this text, sensors provide a means of applying many of the most basic electronic concepts to the measurement of a physical process. Note the units of gain in volts per in/second.

Acceleration
Acceleration and the related physical quantity of vibration are common experimental measurements. Accelerometers are usually expensive and sensitive instruments. The output signal voltage is usually low-level, millivolts or commonly microvolts. Acceleration is the rate at which velocity is increased. Deceleration is the rate at which velocity is decreased. To best understand acceleration or deceleration, one may think of velocity changing in a period of time. For example, acceleration may be seen as an increase of velocity from 30

miles per hour (mph) to 40 mph in 30 seconds. Since by definition acceleration equals velocity divided by time, the computed acceleration becomes;

$$Acceleration = \Delta velocity / \Delta time$$

or,

$$Acceleration = \Delta (distance/time) / \Delta time$$

$$Acceleration = 10 \text{ mph} / 30 \text{ seconds} = .333 \text{ mph/second}$$

Acceleration and decelerations dimensions (units) are often defined as feet/second^2 , or feet per seconds squared. These units are best observed as feet per second per second (feet/second/second), or as described previously, a velocity change of feet per second every second.

Figure 4 – A 10g accelerometer. This unit is approximately the size of common cigarette lighter. Note the indication of the sensitive axis of motion.

Accelerometers are used to output a signal in relation to the amount of acceleration or deceleration to which the sensor is subjected. Accelerometers are capable of measuring accel/deceleration in one, two or three dimensions and are available in single, dual and triaxial versions.. Since accelerometers are often subjected to shock (defined as a Δacceleration/Δtime and often called a "jerk"), a specification of damping in CR units are also included. CR units refer to a time constant envelope surrounding the dampened periodic oscillation resulting from an applied shock.

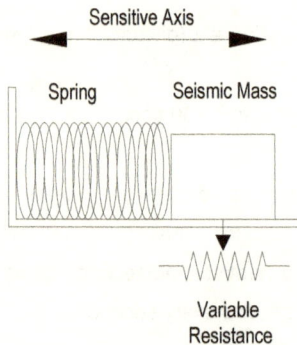

Figure 5 – Although the internal components of an accelerometer perform in a manner similar to the diagram, there exists little similarity in appearance. As the unit is accelerated left or right, the mass slides in the opposite direction pulling the spring and changing the resistance or output voltage if connected to a supply. As the mass is displaced the spring stretches causing a force balance condition between the accelerated mass and the opposing spring force.

Among the most common approaches to the internal construction of an accelerometer is a seismic spring-mass system attached to a frame as in the previous diagram. As the frame is accelerated the mass (allowed to move freely within the frame) is displaced and the attached spring stretches or compresses. Since the accelerated (or decelerated) mass exhibits a force (F = M*A), the mass force applied to the compressing spring results in a force balance measurement system. The displacement of the mass location generates a resistance change or piezoelectric effect in the accelerometer output. Resistive outputs require further conditioning with bridge completion resistors and/or amplification of the low-level voltage signal. A piezoelectric accelerometer signal is created when the test mass stresses a crystal to generate a charge. A "charge amplifier" accumulates (integrates) the charge and develops an output voltage in proportion to the applied acceleration.

Figure 6 – An internal view of an accelerometer, the seismic mass is hinged at the left end and observed in the center of the item in the lower right of the photo. It rests on fine wires, which can be seen in the photo. As the instrument is placed under acceleration, the wires lengthen or shorten depending upon the direction of acceleration. The change in resistance of the wire when strained and corresponding change in voltage becomes the output signal.

Piezoelectric and piezoresistive terms are often used to describe sensor internal operations. Piezo is a term used to describe an applied mechanical force. A (Piezo) force is created when pressure, vibration, acceleration or other physical variables are applied to a sensitive element (resistive or crystal) within a sensor. Piezoelectric applies a force to generate a voltage, usually a charge from a crystal. Piezoresistive applies a force to create a resistive variation, such as found with strain gauges.

When reviewing the enclosed accelerometer specifications, note the references to natural (resonant) frequency, sensitivity (gain), offset, internal circuitry and acceleration in "g's." If any of these quantities require clarification, ask..

Displacement Application Notes:

The following assists the student in describing the included sensor specifications, while providing an overview of typical applications. It is suggested that students closely review the specifications associated with each sensor and attempt to respond. The application notes are organized in a manner similar to the manuscript text with position sensing first, followed by velocity sensors, and accelerometers last.

Proximity Detection

A proximity sensor outputs an indication of whether or not an object is within a certain zone, area or distance from the sensor. Optical and inductive proximity sensors are most common. Inductive sensors require that the sensed object be ferrous, or capable of influencing magnetic fields. Optical proximity sensors utilize light and a reflection of the light to detect either an object, or a gap between objects.

When looking through the first seven sheets on proximity detectors, attempt to locate the following information:

-Basic operating principle; (inductive, optical, other)

-Types of available outputs; (TTL, 120VAC, Open collector, relay output. This is often the difference between a proximity sensor and a proximity control. In addition, the control generally has a set point adjustment of some form.

-Attempt to locate a specification representative of the distance between the sensor and the detected object; inches, fractions of an inch, feet, etc. What is the maximum detectable distance?

-Is the output span linear with the input? Discrete? Any contingencies?

-Power supply requirements; 24VDC, 120VAC, other

-Is the sensor compatible with all materials in all environments?

-Locate and determine typical applications.

Position/Distance Measurement

The previous proximity detectors would not completely qualify as sensors. A true sensor has an output that varies in some proportion to the physical variable at the input. The previous proximity detectors were merely switches. The output was a 2-state discrete representation of the input condition, in essence a single bit A/D converter. The proximity detector's output was indicating whether an object was within a specific distance, or not. Process switches exist for almost every sensor application with pressure, temperature, level and flow switches being commonly available.

The remaining specifications in this chapter and the text are taken from true proportional sensors, where the output varies with respect to the input. The output may not always have a linear relationship with the input, but will vary in some predictable manner with the input variable. For our purposes in this manuscript, the vast majority of sensors will exhibit a linear input/output characteristic.

The section on Position sensing begins with the Linear Displacement Transducer spec. sheet and continues to the Model MP-1 Magnetic Pickup spec. sheet. The Linear Displacement Transducer is another name for the Linear Variable Differential Transformer or LVDT. The LVDT is a transformer, where the sensor output is dependent upon the position of the transformer core. The sensor output is representative of core position and therefore representative of the core actuator's position. The actuator could be a hydraulic cylinder, linear motor, pneumatic positioners or any other linear motion device. Also included are pages on signal conditioning apparatus used to convert the phase and amplitude shifted AC output, into a bipolar varying DC output voltage.

Following the Linear Displacement Transducer, specifications are the Rotary Variable Differential Transformer or RVDT specifications. Note the similarities and differences between both. Both are precise and common among rotary and linear position measurements, especially in experimental or research and development applications. Also included are specifications and applications of Absolute and Incremental Encoders, digital position sensing devices containing outputs directly compatible with computer and digital circuits. When reviewing the specification and application sheets on all Position sensing, try to address the following considerations.

-Basic operating principle, how does it work?

-Types of available outputs? Low impedance, high impedance? Is additional conditioning necessary for the application?

-Physical considerations? Linear throughout the entire range of operation?

-Is the output range linear with the input? Any contingencies?

-Power supply requirements; 24VDC, 120VAC, Constant current?

-Compatible with all materials in all environments?

-Locate typical applications.

Velocity Measurement

The Variable Reluctance Transducer, or magnetic pickup, is a common and simple sensor for the measurement and control of rotary velocity, or speed. Mag. Pickups are available in two forms, active and passive. The passive form requires no power supply connection and is composed of a permanent magnet, a coil of wire and a gear or some other ferrous material which periodically cuts the magnet's lines of force. As the field is disturbed, the coil generates an EMF, which can commonly assume many volts of amplitude. In the case of a gear spinning within the magnetic field (usually within 50 thousandths of an inch from the mag. pickup) the output will be a periodic waveshape similar to a triangle wave. The waveshape frequency is directly related to the revolutions per minute of the gear. Often a 60-tooth gear is used since the frequency of generated output waveshape will match the revolutions per minute (RPM) of the gear.

Variable reluctance refers to the effort required to establish magnetic flux within space. As the gear tooth passes in front of the mag pickup, the reluctance decreases since flux can be established easier in iron than in air. As the gear teeth pass in front of the mag. pickup, the varying reluctance induces the voltage within the pickup coil. Passive mag. pickups are easily recognized by their two-wire lead connection.

Active magnetic pickups or "Zero Velocity Detectors" are capable of generating an output even at zero speed. Active pickups use the same components as the passive pickup but also include a Hall effect magnetic sensor and possibly an output drive circuit such as an open collector for TTL compatibility. The Hall efect sensor allows detection of magnetic field intensity without the normally required change in flux or reluctance. The Zero Velocity Pickup outputs a high level signal when a ferrous material is located in front of the pickup even at zero speed. A TTL level low state is output when no ferrous materials are positioned in front of the sensor. Consequently, Zero Speed Sensors are often used in metal detection and other low speed, rotary applications.

Both active and passive magnetic pickup specification sheets are included within the following information on velocity sensors. An additional specification on the Linear Velocity Transducer has also been included. When reviewing the specification and application sheets on Velocity sensing, try to address the following considerations.
-Basic operating principle, how does it work?
-Types of available outputs? Low impedance, high impedance? Is additional conditioning necessary for the application?
-Any physical or mounting considerations?
-Is the output range linear with the input? Any contingencies?
-Power supply requirements; 24VDC, 120VAC, Constant current?
-Compatible with all materials in all environments?
-Locate and revlew common applications.

Acceleration/Vibration Measurement

Uniaxial, biaxial and triaxial accelerometers have been included in the last pages of the following spec. sheets. Although minimal accelerometer operational material is provided, comprehensive specifications are included for all three types of accelerometers. A review of accelerometer operations and applications may be in order before investigating the specifications. It should also be noted that most accelerometers involve applying an acceleration-generated force against a silicon crystal and squeezing a "charge" out of the silicon. A "charge amp" is required to convert the very slight current from the high source impedance into a potential representative of acceleration. The included specification sheets do not use silicon crystals but semiconductor strain gauges attached to a flexible seismic mass. Likewise, the output circuit assumes the form of a half or full bridge circuit. Again, it should be understood the majority of accelerometers utilize a charge manipulation technique and NOT a resistive bridge.

When reviewing the specification and application sheets on accelerometers, it's beneficial to address the following considerations.
-Basic operating principle, how does it work?
-What do the specifications imply about internal operation?
-What types of outputs are available? Low impedance, high impedance? Voltage? Current? Standard signal spans?
-Is additional signal conditioning necessary for the application?
-What are the physical considerations?
-Is the accelerometer linear throughout the entire range of operation?
-Are any operational contingencies required?
-Power supply requirements; 5VDC, 24VDC, 120VAC, Constant current?
-Is the device compatible with all materials in all environments?
-How can the accelerometer measure vibration?

Homework

Unfortunately, homework for this chapter has yet to be developed.

Chapter 10 – Control Modes

The previous chapters in this course focused on mechanical methods of sensing and transmitting analogs of measured physical variables. This chapter introduces fundamental and advanced concepts employed in the automatic control of physical variables. Introductory concepts involve discrete switching concepts and progress into more sophisticated continuous control concepts. These chapters are under development.

Introduction

Automatic controllers utilize numerous control strategies to maintain a controlled variable at, or near the desired set point. All controller operational modes fall into one of two categories, discontinuous or continuous. Discontinuous control modes, or switching control, possess a finite number of output states. Most commonly the output will be only two extreme states, on or off. On/off and differential control are considered discontinuous since the output is only allowed to assume a 0% or 100% condition. This means the final control element will be either on or off, and incapable of any value in between the two extremes.

Discontinuous Control

With discontinuous control the process variable is always cycling within a tolerance around the set point. The controller and associated final control element (control valve) are continuously over-correcting and under-correcting for the always-present error. As the process deviates from set point, the controller output assumes an extreme condition of either 0% or 100%. The process error is corrected, but only briefly as the process moves through the set point value and into the opposite error polarity condition from the previous condition.

On/Off Control

The On/Off control mode cycles within a relatively narrow band around the set point but cycles at a high repetition frequency that usually causes excessive wear on the mechanical components within the system. The controller output is switched whenever the process passes through the set point value.

The specific output value is determined by the direction of the process deviation, and the reverse or direct action of the controller. With a direct acting controller, and a process value below set point, the controller output will be at 0%. As the process value increases (the result of the controller output being at 0%), the process error will move from negative to positive. As the process error changes polarity, the controller output will also change from 0% to 100%. The change in controller output will result in a corresponding change in the process error direction and polarity. In this manner the process continuously cycles around the set point.

Direct Action   Reverse Action
Controller Output   Controller Output
Max. Output   Max. Output

Min. Output   Min. Output
         Controller Input (Ep, Cm, C)   Controller Input (Ep, Cm, C)

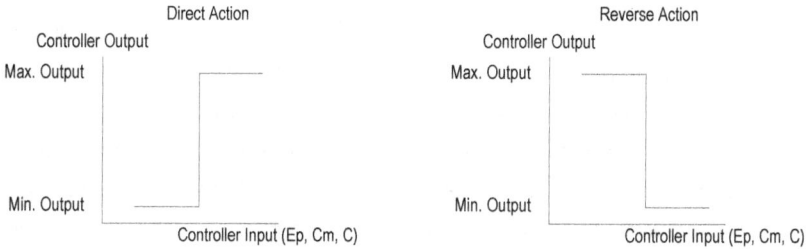

Figure 1 – On/Off control characteristics.

Differential/Hysteresis Control

The differential mode, also referred to as differential gap or Hysteresis control, broadens the tolerance around the set point causing the process to swing within a larger span of process values. Differential control is used to reduce the cycling frequency and improve the life expectancy of mechanical components. Differential control differs from On/Off control in that it contains two separate and distinct switching values each side of the set point. A differential zone or 'gap' is formed between the input/output switching values

equidistant around the set point. Assuming a direct acting controller and the process value below set point, the controller output might be either 0% or 100% depending upon if the process is increasing or decreasing in value. Assuming the output is low (0%) and the process is increasing towards the set point, the process will continue to increase up to and beyond the set point. Where the On/Off controller would cause an output change as the process passed through the set point, the Differential controller requires the process to move beyond the set point to the upper extreme of the input zone. Once the process continues beyond the upper switching point (or upper trigger point), the output will switch to a high (100%) state whereby the process will eventually reverse direction and begin decreasing in value. Again, the process value decreases passed the set point to the lower switching value where the output again changes reversing the process direction and continuing the process cycling.

Although discontinuous modes of control cause the process to continuously cycle around the set point, the resulting process deviation has a minimal effect upon the material being processed. In other words, Discontinuous control modes are employed where a process tolerance or deviation is acceptable. Such processes include air compressors, residential and commercial temperature applications, and some level control systems. Discontinuous control systems are also easier to assemble and set-up, occasionally requiring only a single component to perform as sensor and controller. Temperature switches, pressure switches and other process 'switches' often contain set point and differential gap adjustments internally.

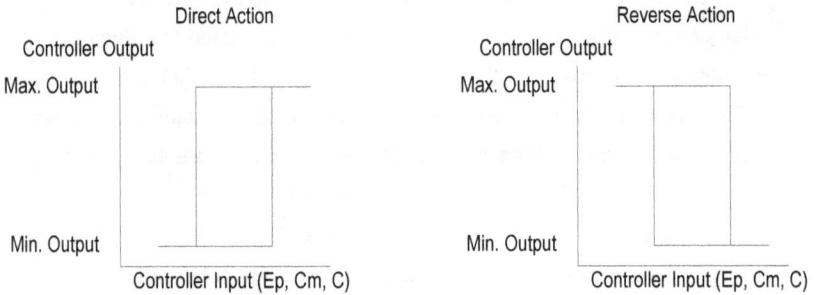

Figure 2 – Differential/Hysteresis control characteristics.

Continuous Control

For best results, continuous control loops should be established from a linear relationship between all components, from the process value through the controller output and on to the amount of corrective action introduced to the process. With each component operating in a linear manner, the performance of the control loop becomes far more predictable throughout the calibrated range of operating values. And since each of the components are connected together into a closed-loop configuration, should any single component exhibit a slight bit of non-linearity, meaning a curve somewhere on the input/output graph of any single component in the control loop, the control system will operate in either an overly responsive, oscillatory manner or a sluggish manner at or near the location of the curve on the component's I/O plot. To maintain consistent operation throughout the entire range of calibrated values, all non-linear components in the control loop are made linear with signal conditioning modules.

Proportional Control

The most basic form of continuous control is Proportional control. Proportional control establishes a predictable and proportional relation between the calibrated process input span, of say 100° to 500° F, and the controller output span of 0% to 100%. In simplest form, proportional control means that as the process temperature varies from minimum to maximum, a direct acting controller output will vary from minimum to maximum, or in the case of a reverse-acting controller, maximum to minimum.

Figure 3 - Four Yokogawa YS-170 process controllers - none are operating at the desired set point.

The proportional controller output (Pout) will equal the controller gain (Kp) multiplied by the error (Ep), plus (or minus) the controller's output bias (Po).

$$Or, \ Pout = Kp * Ep \pm Po$$
$$Where, \ Ep = C - SP$$

To predict the response of a proportional controller to a given process condition, one should visualize each of the involved stages between the process and the controller output. As an example, assume the process is the previously mentioned temperature span of 100 to 500 degrees Fahrenheit. From the process, the temperature sensor signal will be converted to a 4 mA to 20 mA transmitter output span. The transmitter output will be displayed at the controller as a 0% to 100% input span. At this time the process analog (4-20mA) is compared against the set point and an error is generated. The process error is amplified by the proportional controller gain (Kp) and offset by the controller offset-bias (Po). The output is then displayed as a 0% to 100% value at the controller. The actual controller output will most likely assume a value within a 4 mA to 20 mA span, and sent to a transducer or final control element such as a valve.

Temperature Fahrenheit Degrees 100 - 500	milliamps 4-20 mA	Process % 0 - 100 %	% Ep Sp = 300 C - SP	% Pout Po= 50% Kp= 1 KpEp+Po	Controller output current 4 - 20 mA	Valve 0 - 100 %
100	4	0	-50	0	4	0
120	4.8	5	-45	5	4.8	5
140	5.6	10	-40	10	5.6	10
160	6.4	15	-35	15	6.4	15
180	7.2	20	-30	20	7.2	20
200	8	20	-30	20	8	20
220	8.8	30	-20	30	8.8	30
240	9.6	35	-15	35	9.6	35
260	10.4	40	-10	40	10.4	40
280	11.2	45	-5	45	11.2	45
300	12	50	0	50	12	50
320	12.8	55	5	55	12.8	55
340	13.6	60	10	60	13.6	60
360	14.4	65	15	65	14.4	65
380	15.2	70	20	70	15.2	70
400	16	75	25	75	16	75
420	16.8	80	30	80	16.8	80
440	17.6	85	35	85	17.6	85
460	18.4	90	40	90	18.4	90
480	19.2	95	45	95	19.2	95
500	20	100	50	100	20	100

Figure 4 - Proportional control table of process to valve linear relation. The set point is set at 50% or 300° in this example and the output bias is set to 50% when the error is 0%.

In the end result, the controller output will assume a value in proportion to the process value. In proportional control systems the conversion of the process variable to a current span for transmission purposes, the comparison of the current signal against the set point for error generation, the amplification and biasing (offset) of the error, and the conversion of the computed controller output to a current signal are all proportional and predictable relations. In short, the controller output and final element will vary in a proportional manner (directly or inversely) with the process. The controller gain determines the dynamic relationship (controller output change per in relation to process change), and the controller offset determines the static relationship (a 50% value of the process range may result in a 75% controller output value). All of the previously mentioned proportioning concepts can be observed through the use of Y=MX+B characteristic tables and graphs, by plotting the % controller output against the process and controller inputs.

Figure 5 - The process variable (input), set (point) variable and manipulated variable (output) are displayed on the front panel of process controllers in numeric and graphical values.

Proportional Droop

Proportional control is the most logic attempt at eliminating the cycling associated with on/off and differential control. The result of using a proportional control mode is not what one might expect however. Most students of control assume that proportional control alone will be responsible for keeping the process at set point under all imposed conditions. Unfortunately, this is not the case.

Proportional control is said to "meet the load's demand." Meaning, should the load on the process increase, the controller output will increase or decrease in proportion depending upon whether the controller is set for reverse or direct action. With proportional control, the controller output will vary a proportional amount in responding to the load's demand. The process varies in conjunction with the load change and the controller output change. And although the proportional controller's output change will move the process toward the set point value, proportional control cannot completely correct a deviation of the process from set point. With proportional control, there exists a specific value of process variable for every value of the controller's output. This is obvious from the data table provided; note that each input from the process corresponds to a unique value of output. Varying the gain and offset bias of the controller can alter the relationship but will not change the basic concept; a linear and proportional relationship exists between the process value and the controller output. Likewise, *only a single value of controller output (the controller bias quantity, Po) will correspond to a process error of 0%, or set point.* This last statement is extremely important in understanding why proportional only control is incapable of maintaining set point under conditions of varying loads, varying disturbances or even varying set points.

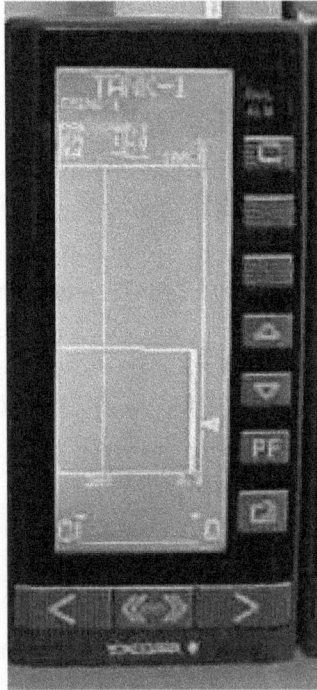

Figure 6 - A change in the process load can result in offset error, or droop in proportional controller loops.

Droop, occasionally called offset, is a reference to the inability of the proportional only controller to return a process to set point after a process deviation. The difference between the set point value entered into the controller and the actual process value is called offset or, since "offset" is used to describe an abundance of other items in controls the residual error resulting from proportional control is frequently called *droop*. After a load or disturbance change, the process will eventually settle into a value at some distance away from the set point. Whenever the process dampens into a final value away from set point, the difference between the set point and the process value, the process error, is the droop. Droop can be expressed as an absolute value in process units of psi, degrees, etc., or as the percentage, %Ep.

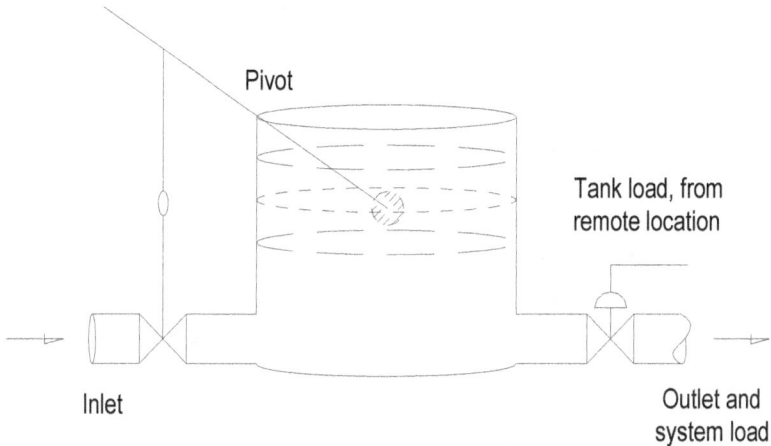

Figure 7 – Proportional control example of "Droop."

A simple, level control example can be used to demonstrate droop in proportional control. In the diagram, assume the two objects appearing as an 'X' shape are inlet and outlet valves to a holding tank. The outlet valve receives its position-input signal from a remote location while the inlet valve receives its positioning signal from the tank level directly. Initially assume both valves are positioned at 50% open. A signal then arrives at the outlet valve to assume a 75% open position. Since the inlet is at 50% and the outlet is now at 75%, the load on the system has increased and the tank level will drop. The resulting decrease in level causes the inlet valve to open further, which allows an increase in the input flow. If the inlet and outlet valves are matched, the tank level will decrease until the inlet valve allows an input flow equal to the outlet flow. When the inlet and outlet flows are matched, the tank level will stabilize and the entire system reaches equilibrium at a different tank level than that was observed at 50% valve positions. Assuming the tank level was at set point with 50% valve positions, the level will be operating with an offset error, or droop, at any other valve position. Again, there exists only a single value of controller output (inlet valve position in this example) that results in the process operating at the desired set point.

Systems like this are often referred to as surge tanks, and are used to eliminate large demands upon the main fluid supply, which could decrease the fluid pressure to all other systems supplied by the main or incoming line.

Proportional Gain and Band
The gain and offset bias in most industrial proportional controllers are usually made variable. These adjustments are available to fine-tune the operating characteristics of the control loop, to minimize the droop error and maximize the amount of response to a process upset. Proportional controller gain is occasionally referred to as *Proportional Band*, which is actually the inverse of proportional gain.

$$\% \text{ Proportional Band} = (1/ K_P) \times 100\%$$

With a proportional gain of unity (1), a 1% change in the input (process error) results in a 1% change in the output. Increasing the gain to a value of 10 will result in a 10% output change for a 1% input change. This condition would also be referred to as a 10% *proportional band*, meaning a process error variation of 10% would result in a controller output variation of 100%. Proportional Band is primarily a method of observing the amount of controller input change necessary to effect a 100% change in the output.

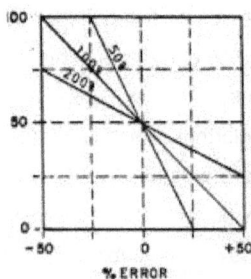

Figure 8 – Reverse acting proportional controller operating characteristics. Proportional band focuses on how much input, in percent, is required to change the output from 0% to 100%. A 50% prop band uses 50% of the input to generate 0% to 100% of the output range. Proportional bands greater then 100% result in a narrowed output range. Note the 200% prop band in the above graph results in an output range of 25% to 75%.

The best method of observing proportional gain and proportional band is to visualize a single linear plot with controller output on the y-axis and the controller input, *process error*, on the x-axis as in the previous figure. As the gain increases, the slope of the plot increases, requiring less of an input variation for a given output change. Decreasing the gain requires a larger input span variation to create a given output range. Focusing upon the input span is the essence of proportional band.

$$\% \text{ Proportional Band} = (1/ K_P) \times 100\%$$

In the previous figure a proportional band of 100% means the input error, E_P, must change from -50% to +50% (a 100% error variation) for the controller output to change from 0% to 100%. If the proportional band is decreased to 50% (or gain is increased to 2.0) the input must now change from -25% to +25% to cause an output change from 0% to 100%. Keeping in mind that the controller output range is limited to the standard 4mA to 20mA signal, errors outside of -25% to +25% results in the output holding at 0% or 100%. In a similar manner a proportional band of 200% requires an error variation of - 50% to +50% and results in a limited output range of 25% to 75%. Errors outside of -50% to +50% will not cause the output to move outside of 25% to 75%.

Regardless of whether proportional band or proportional gain is indicated on the controller, the gain is variable to optimize controller performance. It can be proven by rearranging the controller output equation to solve for Ep, that increasing proportional gain results in a decreased error. This is especially useful when applied to reducing droop; increased gain always reduces offset error. Increased gain cannot eliminate droop however, only minimize it. In addition, increasing proportional gain results in controller characteristic plot which approaches an on/off control mode. Too much gain will result in the control system becoming unstable and oscillatory. In fact, many control loops are "tuned" by increasing the proportional gain until the system oscillates and the n reducing the gain until the oscillation stops. It has often been said that the best-tuned control system is on the brink of becoming an oscillator. In any case, a continued increasing of the proportional gain (or decreasing the proportional band) will result in a control system becoming over-responsive, over-corrective, and very unstable.

Manual Reset by Biasing

The controller bias, or controller output offset bias, is often made variable to provide manual elimination of the droop error. As mentioned previously, proportional controller bias is primarily responsible for determining the controller output when the error is 0%, when the process is operating at set point. In a similar manner, if the gain has been optimized to just less than that which causes oscillation, the controller bias can be adjusted to cause slight changes in the controller output that will effect slight changes in the final element to move the process towards the set point. Biasing the linear controller input/output characteristic moves the plot vertically up or down by increasing or decreasing the bias respectively. The slope will not be affected; only the relative north/south location of the plot will be altered by a biasing variation. Aside from more sophisticated control modes, controller bias variation is the only method to eliminate droop.

Often the controller bias is made accessible and offered as an option called *Manual Reset*. The idea behind Manual Reset is to inch the process back to set point after a load or disturbance change by making slight variations to the controller bias adjustment manually by hand. This feature is only available on process controllers that provide access to the proportional offset. Most contemporary process controllers do not provide a manual reset feature preferring to provide automatic reset *proportional plus integral* control instead.

As was seen with purely proportional control, the process will assume an operating value equal to the set point at only a single value of load and subsequent controller output. Should the load, process disturbances or set point change when using proportional control the controllers' output bias will require a manual adjustment in order to return the process to the desired operating value, the set point. In order to automatically reset the process after a deviation, an *integrator* is required to accumulate the droop error and subsequently change the output until the error is eliminated and the process is returned to set point. The operation of the basic integrator will be described following the proportional problems.

Proportional Control – Chapter Problems

Equations:
%C = [(C- Cmin) / (Cmax - Cmin)] * 100%

%Ep = [(C-Sp) / (Cmax - Cmin)] * 100%

%Pout = (Kp*Ep) ± Po

Where,
        C = Process variable value
        Sp = Set point value
        Ep = Process error (C - Sp)
        Cmin = minimum calibrated process variable value
        Cmax = maximum calibrated process variable value
        Kp = Proportional gain
        PB or %PB = Proportional Band (1/Kp * 100%)
        Po or % Po = ± Proportional controller output bias

-Graph the following proportional controller characteristics; assume a set point of 50% where required:

#1 - Calibrated input span = 100 to 500 F
        Po = 50%
        Kp = +1 (DA)

#2 - Calibrated Input span = 0 psi to 15 psi
        Po = 75%
        Kp = +.5 (DA)

#3 - Same characteristics as in #2 except,
        Kp = +1 (DA)

#4 - Calibrated input span = 100 to 500 F
   Po = 50%
   Kp = -1 (RA)

#5 - Calibrated Input span = 0 psi to 15 psi
   Po = 75%
   Kp = -.5 (RA)

#6 - Same characteristics as in #5 except,
   Kp = -1 (RA)

-Determine the controller outputs given the following:
#7 - Calibrated Input span = 0 psi to 15 psi
   C = 9 psi
   SP = 7.5 psi
   Kp = 1 (DA)
   Po = 50%

#8 - Calibrated Input span = 0 psi to 15 psi
   C = 5 psi
   SP = 7.5 psi
   Kp = 1 (DA)
   Po = 50%

#9 - Calibrated Input span = 0 psi to 15 psi
   C = 5 psi
   SP = 7.5 psi
   Kp = 2 (DA)
   Po = 50%

#10 - Compute the offset error (droop)
   in problems 7, 8 and 9.

-From the previous problems,
   A) What effect does increasing proportional gain (Kp) have upon offset error (droop)?
   B) Why is proportional gain variable?
   C) Why is controller bias variable?

Author's note – the *gain*, or proportional band, in proportional controllers is frequently made variable to facilitate *tuning* the controller. Tuning allows the technician an opportunity to match the controller's response to the response characteristics of the process. Tuning employs only the proportional gain and determines how much response a proportional controller will provide to an error. If the gain is too high the controller will over-correct which will result in an error of an opposite polarity and ultimately causes an oscillatory or "hunting" type of response where the controller is constantly hunting for the set point, over-correcting and the process is cycling around the set point. Too little gain and the error will only be corrected a minimal and insignificant amount. The key is to find the specific gain value which minimizes the error without an oscillatory response.

When the *bias* is made variable it is usually labeled Manual Reset and allows the operator to manually adjust the controller output by varying the controller's output offset, or bias, which changes the valve position. While observing the process error the operator varies the Manual Reset in the appropriate direction to cause the process error to go to zero. The bias or Manual Reset needs to be adjusted whenever a difference between the set point and process, an error, appears. With proportional control an error will always be present after a load change, disturbance change or set point change.

Proportional Control – Chapter Questions

11 – In simplest form, explain proportional control.

12 – Explain why proportional control cannot return a process to set point after a load change.

13 – Name three examples where you have seen proportional control used. Hint, a governor is an application of proportional control.

14 – Explain "Droop" in proportional control systems. Where have you seen droop?

15 – Assume you're working in a power plant and currently looking at the proportional controller in the following figure where an error of 7.5 exists due to the set point being 9.8 with a process value of 17.3. Given access to the standard front panel controls of set point, automatic/manual, and manual output adjustment what could you do to make the process equal the set point value and make the process error equal zero?

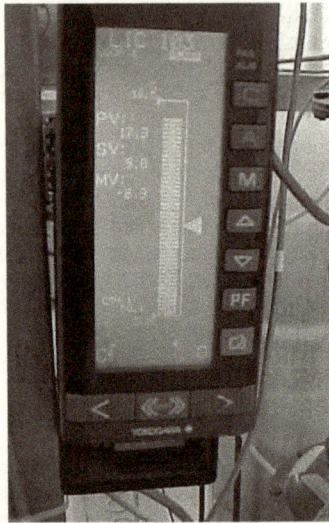

Figure 9 - A single loop controller where the set point (arrow at right) and the process (bar graph) are not equal. The difference is the process error.

Error, Integration and Integrators

The integrator is often referred to as a totalizer, accumulator or "area under the curve" computer. The idea behind these descriptions is that the integrator does not consider only the amplitude of an input voltage but also the time the input has been available at the integrator. Integrators are common in our daily lives but are often overlooked as mathematical functions. Water meters, electrical power and gas meters are all examples of totalizers or integrators. Among the most popular examples of an integrator is the common gasoline pump that everyone uses when filling the gas tank on their automobile.

As one fills the tank on their automobile, the nozzle is set into an open and locked, constant flow position. During this event two indicators on the pump display total flow and total cost. If an industrial transmitter were connected to the flowmeter measuring the gas flow into the vehicle, the transmitter would output a signal between 4 mA and 20 mA in proportion to the gasoline flow rate (gallons per minute, liters per second, etc.). As long as the flow into the automobile remains constant, the input to the integrator is constant and the output from the integrator is continuously increasing at a constant rate. Just as the total volume of the flow into your gas tank is accumulated over time, an integrator's output will be continuously changing, as long as a non-zero input is applied. Should the input go to zero, like when the tank is full and the pump stops, the output will hold the most recent value until reset, or until another input is applied. Should another non-zero input be applied, the integrator will either continue to build in the same direction on top of the last value, or in a reverse direction if the opposite polarity of input is applied – this would be effectively pulling gas from or tank. Likewise, an integrator takes a step function input and generates a ramping or continuously changing output.

The integrator is the rough equivalent of an RC time constant charge circuit, with a DC "step" voltage input. In a manner similar to the accumulation of charge on a capacitor, the integrator uses the input to continuously charge, or "build" the output. Unfortunately, the RC time constant circuit charges exponentially (non-linearly) since the charging capacitor voltage opposes the applied step voltage. The non-linear nature of the RC circuit and applied DC voltage would make a poor and somewhat unpredictable integrator. However, a constant current generator and capacitor form a simple and accurate integrator. As the current generator creates a constant value of current, the capacitor charges predictably with the voltage across the capacitor increasing linearly.

More than likely, the concept of integration is foreign to those students recently introduced to electronics. With most electronic circuits, a constant input value results in a constant output value. With an integrator, a constant DC-like input produces a changing output. See the included diagram representing one volt applied to an integrator for a twenty second time period. Note the input and output plots are both displayed in the graph.

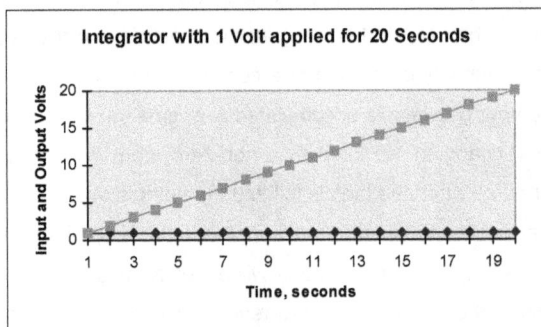

Integrator with 1 Volt applied for 20 Seconds

Figure 10 - Graph of a constant input voltage applied to an integrator. The ramping waveform is the resulting integrator output.

Integrators are understood to perform an "area under the curve" computation upon the input voltage waveform. At any point in time along the input wave shape one could compute the area under the curve by multiplying the input voltage by the time duration between time locations along the x-axis. As an example, the area under the input voltage at five seconds would be, 1 Volt * 5 seconds, which equals 5 with volt*seconds units. By moving from five seconds to the integrator plot, it can be seen that the integrator's output is also 5 Volts, representing an input of 1 Volt applied for five seconds. Had the input value and/or time been different, the integrator would still indicate the multiple of voltage multiplied by time at its output. The following example uses an input of -2 Volts applied to an integrator. Note the negative going waveshape from the integrator. Should the input move to zero while the integrator is ramping, the output will hold at the most recent value as in the included Integrator Ramping and Holding diagram.

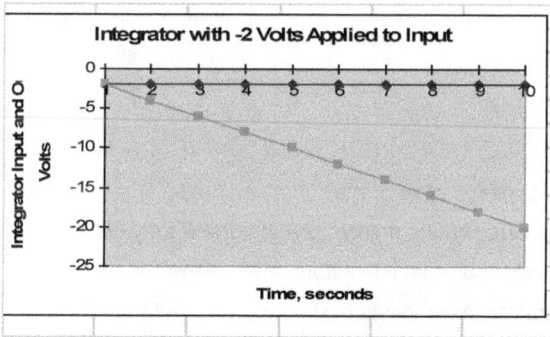

Figure 11 - A negative input to integrator results in a negative-going "ramping" integrator output voltage. Note the integrator output is the product of the input voltage (-2) multiplied by time.

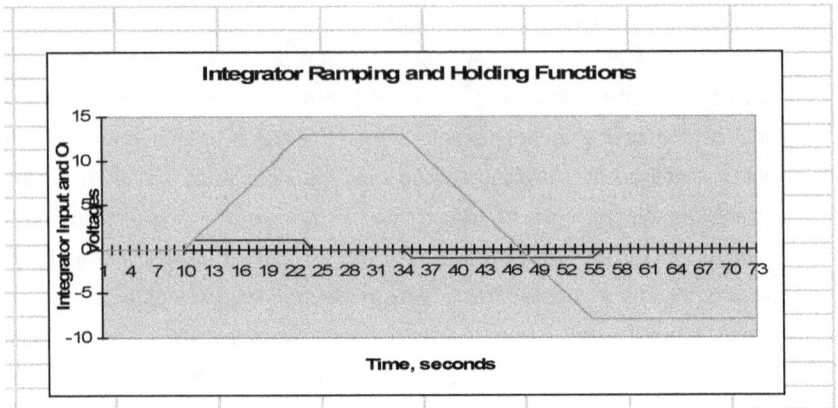

Figure 12 - The input and output from an integrator beginning at 10-seconds and accumulating to beyond 60-seconds, remember that the integrator responds to a DC appearing input with an accumulation of input versus time area or, a ramping output. Note the integrator output holding when the input goes to zero.

For each input voltage of zero, the integrator 'holds' at the last value prior to the input going to zero. In this manner the integrator may be started, stopped and even reset. Some industrial integrators, also referred to as timers, totalizers and 3-mode integrators, have inputs labeled as Run, Hold and Reset to perform the indicated characteristics.

Note also the effect of both positive and negative deviations on the integrator. As can be seen in the bipolar graphs, the integrator charges in a positive direction when responding to a positive input, then moves negatively in responding to the negative half-wave of the square wave input.

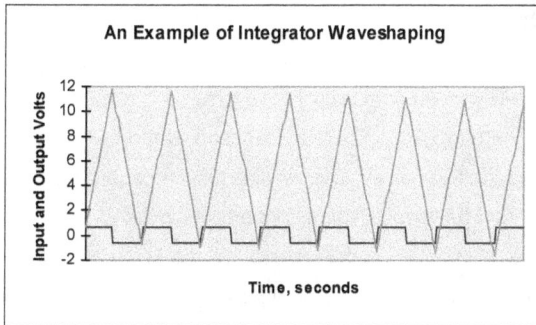

An Example of Integrator Waveshaping

Input and Output Volts

12
10
8
6
4
2
0
-2

Time, seconds

Figure 13 - An example of integrator wave-shaping, note the subtle negative-going motion of the triangle waveform peaks. What would cause this to appear at the integrator output? Where is the triangle waveform eventually headed?

The triangular waveshape at the output initially moved positive because of the appearance of the positive half-wave of the input first. During the brief period that the positive half-wave was at the integrator's input, the input and output appeared as in the first graph where a positive DC input was integrated into a ramping output. However, during the negative half-wave of the input, the negative half-wave of the input 'drives' the integrator in a negative direction from where it had last accumulated during the positive half-cycle. The process continues when the positive half-wave appears again during the next input cycle.

Astute observers will note the input is not perfectly symmetrical. Upon close inspection the output appears to be moving further negative with each cycle. This is due to the input containing slightly more area in the negative half-wave than in the positive. The input is not exactly referenced to or around zero volts but instead is slightly offset from zero in a slightly negative direction. If the input were precisely centered on zero, the integrator output waveshape would not appear to be incrementally moving negative.

Automatic "Reset"

An integrator, identical in operation to those previously described will be used in conjunction with a proportional amplifier to provide the ability of a controller to meet a load's demand (proportional contribution to the controller output) and to *return the process to set point after a deviation* (integrator contribution to the controller output). The bias function formerly associated with the proportional amplifier will be eliminated and replaced by the integrator. As a result, the integrator will be responsible for automatically resetting the process to set point after a process deviation and the controller's output will be the mathematical sum of the proportional response added with the integrator response. Hence, the term for the combined control modes is *Proportional plus Reset*, PI Control, 2-mode control or *Proportional plus Integral* Control. The proportional and integrating functions will be effectively paralleled and summed as can be seen in the equation following. The former proportional bias term, Po, is now performed by the integrator and has been slightly subdued in the following equation;

$$Pout = KpEp + Ki\!\int\!Ep\ dt + Pini$$

Where,

> Pout is the controller output percentage,
> Kp is the proportional gain,
> Ep is the process error, C – SP,
> Ki is the integrator gain,
> ∫Ep, the integral of process error, area of error vs. time
> dt represents the time period of the error
> Pini, initial controller output prior to the error occurrence

Note that the **Ki∫Ep dt + Pini** terms both occupy the location of the bias (Po) term in the purely proportional controller output equation. This indicates the integrator is now performing the controller biasing function. At first, the **Pini** term may seem somewhat abstract. **Pini** represents the ability of the integrator to retain, charge, or hold a controller output value in a manner similar to the examples described previously. In fact, if one were to investigate a proportional plus integral controller's output when the process is operating at the set point, one would note the output is coming solely from the integrator with nothing from the proportional portion since the error is zero. Like wise, **Pini** represents the controller output value just before a process deviation or the final controller output prior to the deviation. All of **Pini** is associated with the integrator and its ability to retain a value when the integrator input (process error) is zero.

The integrator gain, Ki, is an amplifying or multiplying function that provides a varying influence of the integrator on the controller output. Varying the integrator gain will vary the slope (rate of change) of the integrator's output that varies the speed of the integrator. For a given process error, a high integrator gain will cause the controller output to ramp rapidly, a low integrator gain will cause the controller output to change slowly. The idea behind "tuning" or setting the controller gains is to match the response of the controller output to the characteristics of the process and remaining control loop components. A controller responding too quickly will over-correct, causing the process to be very responsive and possibly even oscillate. Should the gain be too low, the process will return to set point slowly after a process upset and waiting for the process to creep back to set point is equally undesirable since the material being processed is manufactured out of specification during the time that an error exists.

Repeats/Minute - Minutes/Repeat

The units of integral gain are the elusive repeats/minute or minutes/repeat. Both units are a reference to the manner with which the *proportional* mode responds to an error between the process variable and the set point. Imagine a constant DC appearing error, the proportional control mode would respond by generating a constant DC appearing output quantity. The units of integral gain, repeats/minute, refer to the subsequent number of proportional output quantity changes created by the integrator.

As an example, if the proportional response to a given error is 10% and the integrator is set to 3 repeats/minute, in the next minute the integrator will change the controller by an additional 30% on top of the proportional response. Should the integrator be set to 3 *minutes/repeat*, to acquire an additional 10% change on top of the proportional response of 10%, or one repeat of the proportional response, would require 3 minutes. Upon inspection it can be assumed that the two units of R/M and M/R are inversely related. Numerically, as the R/M quantity increases, the integral gain and subsequent response increases. As the M/R increases the integrator response decreases. Integrator gains of repeats/minute are typically found on smaller processes with relatively quick reacting process changes whereas minutes per repeat are found on larger processes with relatively large transit times.

When working with PLC's and PC-based software control algorithms providing integral control action, look closely for the units of integral gain or tuning the control loop will be unnecessarily complicated. In the author's experience the same model and manufacturer of PLC, such as an Allen-Bradley PLC 5/15 will offer both repeats/minute and minutes/repeat to implement integral control action. In short, *don't assume that an increase in integral gain will result in increased speed of controller response.*

P+I Example Problem 1

Given the following Error (Ep) waveshape, determine the response of a PI controller with the following operational characteristics:
Kp =.5
Ki = .1
Po = 10%

- Error Data:

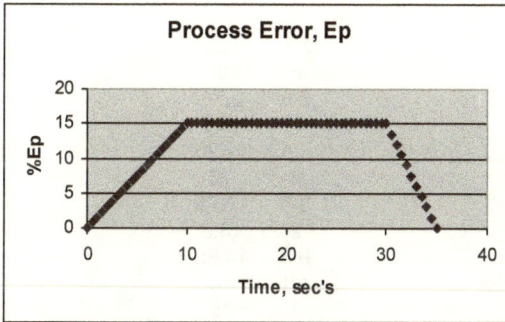

Figure 14 - An example of process error versus time during a load change or a disturbance change.

Data from the figure 12 error versus time plot follows. Time is in the left column versus Error, Ep, in the right column.

Time	Ep
0	0
0.5	0.75
1	1.5
1.5	2.25
2	3
2.5	3.75
3	4.5
3.5	5.25
4	6
4.5	6.75
5	7.5
5.5	8.25
6	9
6.5	9.75
7	10.5
7.5	11.25
8	12
8.5	12.75
9	13.5
9.5	14.25
10	15
10.5	15
11	15
11.5	15
12	15
12.5	15
13	15
13.5	15
14	15
14.5	15
15	15
15.5	15
16	15
16.5	15
17	15
17.5	15
18	15
18.5	15
19	15
19.5	15
20	15
20.5	15
21	15
21.5	15

22	15
22.5	15
23	15
23.5	15
24	15
24.5	15
25	15
25.5	15
26	15
26.5	15
27	15
27.5	15
28	15
28.5	15
29	15
29.5	15
30	15
30.5	13.5
31	12
31.5	10.5
32	9
32.5	7.5
33	6
33.5	4.5
34	3
34.5	1.5
35	0

- Error Equations:
- Determining Error (Ep) equations from previous graph:

- From 0s to 10s:
$Ep_A = 15\%/10s\ T + 0\%$
$\therefore Ep_A = 1.5T$, from 0s to 10s

- From 10s to 30s:
$Ep_B = 0\%/20s\ T + 15\%$
$\therefore Ep_B = 15$, from 10s to 30s

- From 30s to 35s:
$Ep_C = -15\%/5s\ T + 0\%$
$\therefore Ep_C = -3T + 105\%$, from 30s to 35s

- Output Determinations:
- Finding the Pout equations and values for the previous error waveshape, given:

Pout = KpEp + Ki∫Ep + Pini
Pout = .5Ep + .1∫Ep + 10%

- From 0s to 10s, Ep_A = 1.5T

Pout = Kp (1.5T) + Ki ∫ (1.5T) + 10
Pout = .5(1.5T) + .1∫(1.5T) + 10

- Subbing 0s, 5s and 10s for T, integrating the E_P (1.5T) equation and setting the limits of Integration at 0s and 5s, then 5s and 10s yields:

Pout @ 0s -
$Pout_{0s}$ = .5(1.5%/s*0s) + .1s∫(.75T²) + 10%
$Pout_{0s}$ = .5(0%) + .1s(0%s) + 10%
$Pout_{0s}$ = .5(0%) + .1s(0%s) + 10%
$Pout_{0s}$ = .5(0%) + (0%) + 10%
$Pout_{0s}$ = 10% at 0s, Pini = 10%

Pout @ 5s -
$Pout_{5s}$ = .5((1.5%/sT+ 0%)*5s) + .1s∫(.75T²) + 10%
$Pout_{5s}$ = .5(7.5%) + .1s(18.75%s − 0%s) + 10%
$Pout_{5s}$ = .5(7.5%) + .1s(18.75%s) + 10%
$Pout_{5s}$ = (3.75%) + (1.875%) + 10%
$Pout_{5s}$ = 15.625% at 5s, Pini = 11.875%

Pout @ 10s -
$Pout_{10s}$ = .5((1.5%/sT + 0%)*10s) + .1s ∫ (.75T²) + 11.875%
$Pout_{10s}$ = .5(15%) + .1s(75%s − 18.75%s) + 11.875%
$Pout_{10s}$ = .5(15%) + .1s(56.25%s) + 11.875%
$Pout_{10s}$ = (7.5%) + (5.625%) + 11.875%
$Pout_{10s}$ = 25% at 10s, Pini = 17.5%

Pout @ 10s, where Ep_B = 15, from 10s to 30s
$Pout_{10s}$ = (.5* (0T+15)) + .1s∫(15T) + 17.5%
$Pout_{10s}$ = (7.5%) + .1s(150%s − 150%s) + 17.5%
$Pout_{10s}$ = (7.5%) + .1s(0%s) + 17.5%
$Pout_{10s}$ = (7.5%) + (0%) + 17.5%
$Pout_{10s}$ = 25.0% at 10s, Pini = 17.5%

Pout @ 15s -
$Pout_{15s}$ = (.5* (0T+15)) + .1s∫(15T) + 17.5%
$Pout_{15s}$ = (7.5%) + .1s(225%s − 150%s) + 17.5%
$Pout_{15s}$ = (7.5%) + .1s(75%s) + 17.5%
$Pout_{15s}$ = (7.5%) + (7.5%) + 17.5%

$\text{Pout}_{15s} = 32.5\%$ at 15s, Pini = 25%

Pout @ 20s -
$\text{Pout}_{20s} = (.5* (0T+15)) + .1s\int(15T) + 25\%$
$\text{Pout}_{20s} = (7.5\%) + .1s(300\%s - 225\%s) + 25\%$
$\text{Pout}_{20s} = (7.5\%) + .1s(75\%s) + 25\%$
$\text{Pout}_{20s} = (7.5\%) + (7.5\%) + 25\%$
$\text{Pout}_{20s} = 40\%$ at 20s, Pini = 32.5%

Pout @ 25s -
$\text{Pout}_{25s} = (.5* (0T+15)) + .1s\int(15T) + 32.5\%$
$\text{Pout}_{25s} = (7.5\%) + .1s(375\%s - 300\%s) + 32.5\%$
$\text{Pout}_{25s} = (7.5\%) + .1s(75\%s) + 32.5\%$
$\text{Pout}_{25s} = (7.5\%) + (7.5\%) + 32.5\%$
$\text{Pout}_{25s} = 47.5\%$ at 25s, Pini = 40%

Pout @ 30s -
$\text{Pout}_{30s} = (.5* (0T+15)) + .1s\int(15T) + 40\%$
$\text{Pout}_{30s} = (7.5\%) + .1s(450\%s - 375\%s) + 40\%$
$\text{Pout}_{30s} = (7.5\%) + .1s(75\%s) + 40\%$
$\text{Pout}_{30s} = (7.5\%) + (7.5\%) + 40\%$
$\text{Pout}_{30s} = 55\%$ at 30s, Pini = 47.5%

Pout @ 30s, where $Ep_C = -3T + 105\%$, from 30s to 35s
$\text{Pout}_{30s} = (.5* (-3\%/sT + 105\%)) + .1s\int(-3T + 105) + 47.5\%$
$\text{Pout}_{30s} = (.5* (-3\%/s (30s) + 105\%)) + .1s\int(-1.5T^2 + 105T) + 47.5\%$
$\text{Pout}_{30s} = (.5*15\%) + .1s[(-1350 + 3150) - (-1350 + 3150)] + 47.5\%$
$\text{Pout}_{30s} = (7.5\%) + .1s(1800-1800) + 47.5\%$
$\text{Pout}_{30s} = (7.5\%) + .1s(0) + 47.5\%$
$\text{Pout}_{30s} = 55\%$ at 30s, Pini = 47.5%

Pout @ 32.5s -
$\text{Pout}_{32.5s} = (.5* (-3\%/sT + 105\%)) + .1s\int(-3T + 105) + 47.5\%$
$\text{Pout}_{32.5s} = (.5* (-3\%/s (32.5s) + 105\%)) + .1s\int(-1.5T^2 + 105T) + 47.5\%$
$\text{Pout}_{32.5s} = (.5*7.5\%) + .1s[(-1584.375 + 3412.5) - (-1350 + 3150)] + 47.5\%$
$\text{Pout}_{32.5s} = (3.75\%) + .1s(1828.125 - 1800) + 47.5\%$
$\text{Pout}_{32.5s} = (3.75\%) + .1s(28.125\%s) + 47.5\%$
$\text{Pout}_{32.5s} = (3.75\%) + (2.8125\%) + 47.5\%$
$\text{Pout}_{32.5s} = 54.0625\%$ at 32.5s, Pini = 50.3125%

Pout @ 35s -
$\text{Pout}_{35s} = (.5* (-3\%/sT + 105\%)) + .1s\int(-3T + 105) + 50.3125\%$
$\text{Pout}_{35s} = (.5* (-3\%/s (35s) + 105\%)) + .1s\int(-1.5T^2 + 105T) + 50.3125\%$
$\text{Pout}_{35s} = (.5*0\%) + .1s[(-1837.5 + 3675) - (-1584.375 + 3412.5)] + 50.3125\%$
$\text{Pout}_{35s} = (0\%) + .1s(1837.5 - 1828.125) + 50.3125\%$
$\text{Pout}_{35s} = (0\%) + .1s(9.375\%s) + 50.3125\%$

$Pout_{35s} = (0\%) + (.9375\%) + 50.3125\%$
$Pout_{35s} = 51.25\%$ at 35s, $Pini = 51.25\%$

- Excel Computations:
$Ep_A = 1.5T$, from 0s to 10s in .5s intervals

Time, sec's	Ep Values	Kp Only	Integ Only	.1 * Integ	Pini	P+I
0	0	0	0	0	10	10
0.5	0.75	0.375	0.1875	0.01875	10.01875	10.39375
1	1.5	0.75	0.75	0.075	10.075	10.825
1.5	2.25	1.125	1.6875	0.16875	10.16875	11.29375
2	3	1.5	3	0.3	10.3	11.8
2.5	3.75	1.875	4.6875	0.46875	10.46875	12.34375
3	4.5	2.25	6.75	0.675	10.675	12.925
3.5	5.25	2.625	9.1875	0.91875	10.91875	13.54375
4	6	3	12	1.2	11.2	14.2
4.5	6.75	3.375	15.1875	1.51875	11.51875	14.89375
5	7.5	3.75	18.75	1.875	11.875	15.625
5.5	8.25	4.125	22.6875	2.26875	12.26875	16.39375
6	9	4.5	27	2.7	12.7	17.2
6.5	9.75	4.875	31.6875	3.16875	13.16875	18.04375
7	10.5	5.25	36.75	3.675	13.675	18.925
7.5	11.25	5.625	42.1875	4.21875	14.21875	19.84375
8	12	6	48	4.8	14.8	20.8
8.5	12.75	6.375	54.1875	5.41875	15.41875	21.79375
9	13.5	6.75	60.75	6.075	16.075	22.825
9.5	14.25	7.125	67.6875	6.76875	16.76875	23.89375
10	15	7.5	75	7.5	17.5	25

$Ep_B = 15$, from 10s to 30s in .5s intervals

Time, sec's	Ep Values	Kp Only	Integ Only	.1 * Integ	Pini	P+I
10	15	7.5	0	0	17.5	25
10.5	15	7.5	7.5	0.75	18.25	25.75
11	15	7.5	7.5	0.75	19	26.5
11.5	15	7.5	7.5	0.75	19.75	27.25
12	15	7.5	7.5	0.75	20.5	28
12.5	15	7.5	7.5	0.75	21.25	28.75
13	15	7.5	7.5	0.75	22	29.5
13.5	15	7.5	7.5	0.75	22.75	30.25
14	15	7.5	7.5	0.75	23.5	31
14.5	15	7.5	7.5	0.75	24.25	31.75

Time, sec's	Ep Values	Kp Only	Integ Only	.1 * Integ	Pini	P+I
15	15	7.5	7.5	0.75	25	32.5
15.5	15	7.5	7.5	0.75	25.75	33.25
16	15	7.5	7.5	0.75	26.5	34
16.5	15	7.5	7.5	0.75	27.25	34.75
17	15	7.5	7.5	0.75	28	35.5
17.5	15	7.5	7.5	0.75	28.75	36.25
18	15	7.5	7.5	0.75	29.5	37
18.5	15	7.5	7.5	0.75	30.25	37.75
19	15	7.5	7.5	0.75	31	38.5
19.5	15	7.5	7.5	0.75	31.75	39.25
20	15	7.5	7.5	0.75	32.5	40
20.5	15	7.5	7.5	0.75	33.25	40.75
21	15	7.5	7.5	0.75	34	41.5
21.5	15	7.5	7.5	0.75	34.75	42.25
22	15	7.5	7.5	0.75	35.5	43
22.5	15	7.5	7.5	0.75	36.25	43.75
23	15	7.5	7.5	0.75	37	44.5
23.5	15	7.5	7.5	0.75	37.75	45.25
24	15	7.5	7.5	0.75	38.5	46
24.5	15	7.5	7.5	0.75	39.25	46.75
25	15	7.5	7.5	0.75	40	47.5
25.5	15	7.5	7.5	0.75	40.75	48.25
26	15	7.5	7.5	0.75	41.5	49
26.5	15	7.5	7.5	0.75	42.25	49.75
27	15	7.5	7.5	0.75	43	50.5
27.5	15	7.5	7.5	0.75	43.75	51.25
28	15	7.5	7.5	0.75	44.5	52
28.5	15	7.5	7.5	0.75	45.25	52.75
29	15	7.5	7.5	0.75	46	53.5
29.5	15	7.5	7.5	0.75	46.75	54.25
30	15	7.5	7.5	0.75	47.5	55

$Ep_C = -3T + 105\%$, from 30s to 35s in .5s intervals

Time, sec's	Ep Values	Kp Only	Integ Only	.1 * Integ	Pini	P+I
30	15	7.5	0	0	47.5	55
30.5	13.5	6.75	7.125	0.7125	48.2125	54.9625
31	12	6	6.375	0.6375	48.85	54.85
31.5	10.5	5.25	5.625	0.5625	49.4125	54.6625
32	9	4.5	4.875	0.4875	49.9	54.4
32.5	7.5	3.75	4.125	0.4125	50.3125	54.0625
33	6	3	3.375	0.3375	50.65	53.65
33.5	4.5	2.25	2.625	0.2625	50.9125	53.1625

34	3		1.5	1.875	0.1875	51.1		52.6
34.5	1.5		0.75	1.125	0.1125	51.2125		51.9625
35	0		0	0.375	0.0375	51.25		51.25

Response Plots:

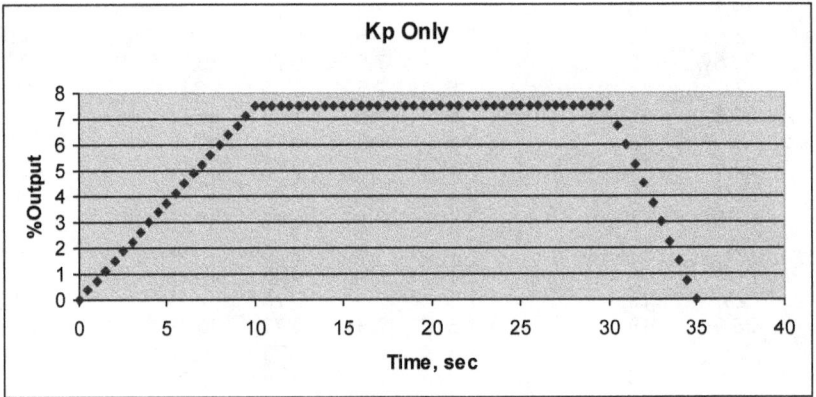

Figure 15 - Proportional only response to the figure 12 error, Ep.

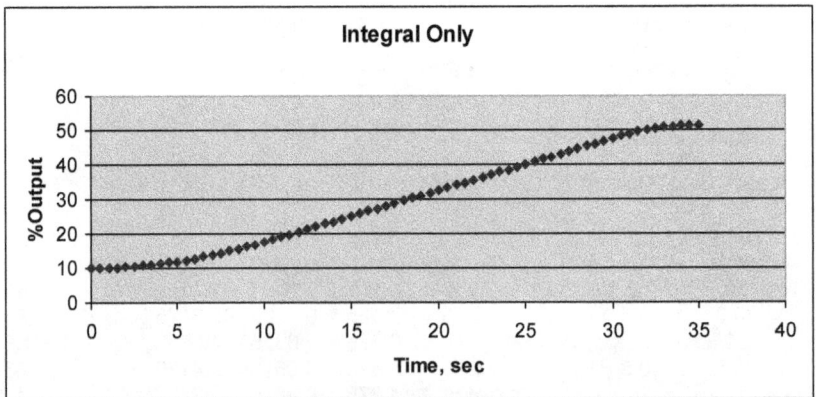

Figure 16 - Integral only response to the figure 12 error, Ep.

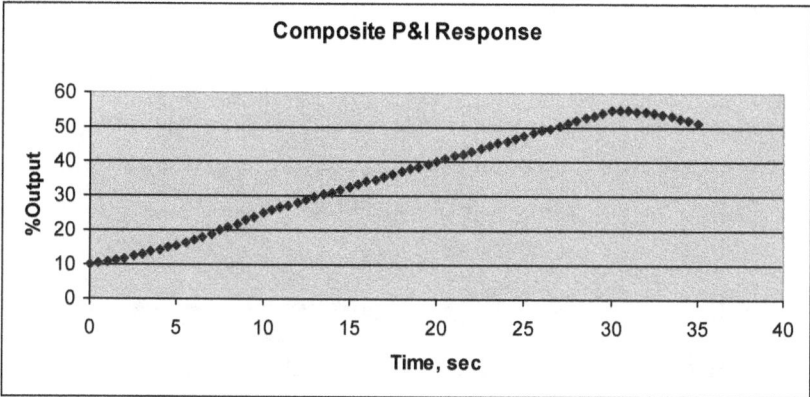

Composite P&I Response

Figure 17 - Combined proportional pus reset, PI, 2-mode or proportional plus integral response to a load change or disturbance change induced process error, Ep.

Integral Control Problems

1 - Given the following error (Ep) waveshape, determine the controller's response at 0s, 5s and 10s. Assume Kp = .5, Ki = 1 and Pini = 10%. Include an approximate waveshape of the controller output in the graph provided.

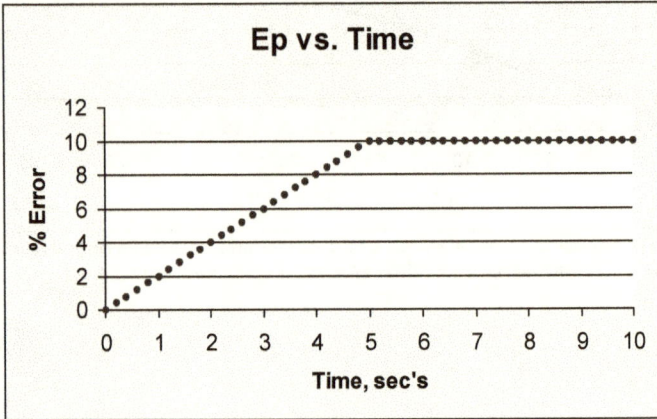

Ep vs. Time

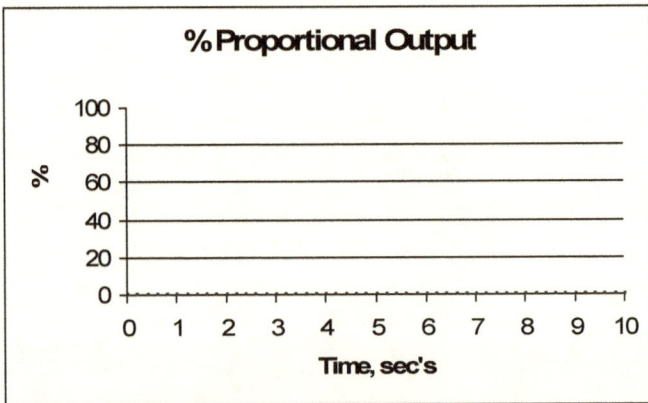

% Proportional Output

Approximate the proportional only contribution to the output here.

% Integral Output

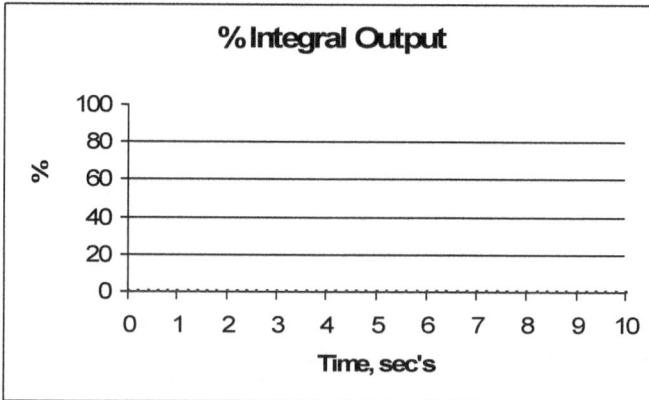

Approximate the Integral only contribution to the output here.

% Proportional Plus Integral Output

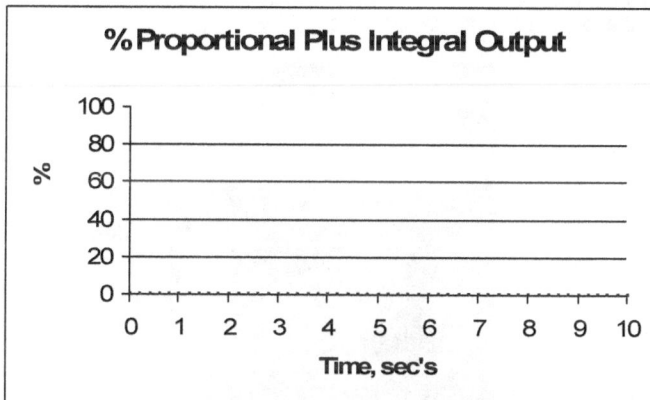

Approximate the combined P&I response here.

Integral Control Questions

2 – What does the proportional control mode provide or what feature does it add to the control system?

3 – Specifically, what does the proportional control mode respond to in the error?

4 – What is the perceived deficiency with proportional only control?

5 – How does the integral control mode correct the deficiency that's associated with proportional control?

6 – Specifically, what does the integral control mode respond to in the error?

7 – How does the "area under the curve" influence integral control action?

8 – What is a perceived deficiency with integral control?

9 – Assume you're working in a power plant and currently looking at the P+I level indicating controller (LIC 103) in the following figure where an error of 7.5 exists because the set point is 9.8 and the process is 17.3. Since the controller is using proportional plus integral, or "Proportional plus Reset" control, what would you expect to observe in the next few minutes? Note: The output valve controls flow from the tank. At the bottom of the controller is a horizontal bar graph indicator of controller output and specifically, valve position. In the lower left corner is the letter "C" representing a closed valve and in the lower right corner is the letter "O" representing an open valve. The controller output bar graph is currently barely visible at the far left near the "C" indication.

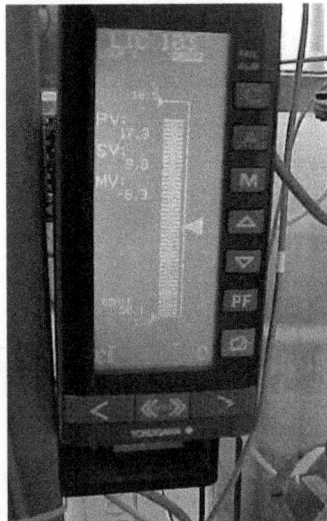

Figure 18 - A single loop controller where the set point (arrow at right) and the process (bar graph) are not equal. The difference is the process error.

Derivative Control Action

One might think the combined actions of proportional plus reset (integral) require no assistance in providing optimum control but in reality the combination has a characteristic flaw. Integrators by their nature are slow in responding to an error. As can be seen on any Integrator response plot, the value of the integrator's contribution to a controller output increases significantly with time. Yet in the processing industries materials are being manufactured out of specification during the time the process is deviating from set point. An additional control mode is required to "speed-up" the controller response to an error. Enter the derivative mode.

Mathematically, the derivative mode is a rate of change computation on the process error. As a process deviates from set point, the error seen by the derivative amplifier appears as a signal gradually moving away from zero, either positive or negative. The derivative is unlike the integral or the proportional in that the derivative is sensitive to only the rate of change. Therefore a constantly changing (ramping) error produces a constant value, DC appearing output from the derivative.

The following example demonstrates this concept. In the graph, the triangular appearing waveshape represents the process error. Note the error begins at zero, increases at a constant rate (slope), peaks, then begins decreasing at a constant slope and finally holds at a constant value. For each of the error segments a corresponding derivative, or rate of change value is presented. The shapes and amplitudes appear unrelated. But remember, unlike the proportional mode the amplitude of the error is inconsequential. Only the slope (in percent/second) is significant.

Derivative Action vs. Ramping Error

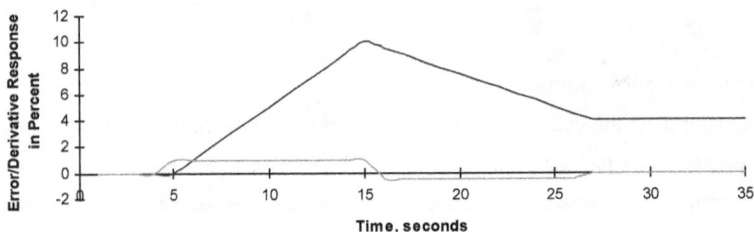

Referring to the diagram, the value of each wave is zero, prior to 5 seconds. As the error ramps positive just after 5 seconds, the derivative waveform assumes a value equal to the rate of change of the error. In the diagram, the derivative output is 1, indicating an error rate of change of 1% per second. The derivative output is computed as any other common slope, $\Delta Y / \Delta X$. The derivative output is usually amplified before being summed with the proportional, or proportional and integral modes. Therefore the derivative mode is represented by the equation;

$$Pout, Derivative\ Mode = Kd * \Delta Y / \Delta X$$

At approximately 15 seconds the error peaks and moves negative. Upon close inspection the derivative output moves negative to a value of -.5%, indicating a negative-going error slope of .5% per second. At about 17 seconds the error becomes constant at 4%. Note the output moves to zero when the error becomes flat. This is due to the loss of the slope. Even though the error is a non-zero quantity, the derivative is only sensitive to *change*. Since the error isn't changing, the derivative output becomes zero. The constant error amplitude will be returned to zero by the integral mode, if available.

Because of the derivative sensitivity to error variation, the derivative mode is often referred to as 'Rate' action. Proportional plus integral plus derivative controllers are also known as 'PID', 'three mode' and 'proportional plus reset and rate' controllers. In a controls context, the idea behind the derivative mode is to immediately respond to an error signal to minimize the process error, Ep, before it is allowed to develop. In essence, the derivative mode provides *anticipation* of the error signal since the derivative mode is sensitive to the direction and *time rate of change* of the error signal. As the error is just developing (before the proportional or integral can amplify or accumulate any significant quantity of error), the derivative mode detects the slope of the developing error and responds by rapidly positioning the final control element. The derivative is thereby attempting to minimize the deviation of the process from set point. In short, it should be understood that the greater the slope of the error (the faster the process change), the greater the derivative contribution to the output. In this manner the derivative mode provides a correcting action immediately upon process deviation while the error quantity is very small. For this reason the derivative mode of control is often referred to as *anticipation.* The derivative attempts to minimize an error in its' early period of development.

Due to its rapidly responding nature derivative control action is never used with noisy input signals to the controller or with flow systems. Noisy inputs will cause the controller output to experience large and violent output transitions which can destroy control valves and other final control elements which are often the most expensive components in the control loop. Derivative is never used with flow control loops for the same reason; suddenly closing a large valve in a water line can cause serious damage to hangers, buildings and related hardware – consider the amount of mass moving through large diameter piping systems and you'll understand why derivative control action isn't used in flow systems.

Derivative Gain
Derivative gain is measured in units of time, or seconds. The units represent the amount of time a proportional only controller would require to develop the contribution the derivative mode generates immediately. Sounding like more double-talk, the definition of derivative units relates to the ability of the derivative

amplifier to output a quantity equal to the error rate of change. Since the eventual contribution of the derivative mode will be summed with proportional or proportional plus integral modes, the derivative modes' influence on the overall output will require varying to increase or decrease the derivative effect on the controller output. The variation is provided by amplification, or derivative gain, symbolized Kd. When the derivative action is combined with the other available control modes, it is summed as a percentage. To convert the derivative action of error rate of change in percent per second, into percent for summing, the derivative action must be amplified by the units of seconds. Multiplying percent per second by seconds will equal percent, with the units of seconds canceling.

Mathematically,

$$Pout\ derivative = (Derivative\ gain) * (Error\ rate\ of\ change)$$
$$Pout\ derivative = (seconds) * (percent / second)$$
$$Pout\ derivative = percent$$

Derivative Control Problems

1 - Given the following error (Ep) waveshape, determine a Proportional plus derivative controller's response at 0s, 5s and 10s. Assume Kp = .5, Kd = 1 and Po = 50%. Include an approximate waveshape of the controller output in the graph provided.

Ep vs. Time

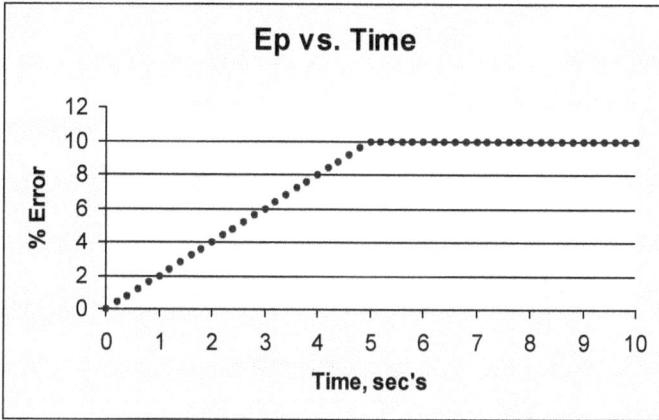

% Proportional Output

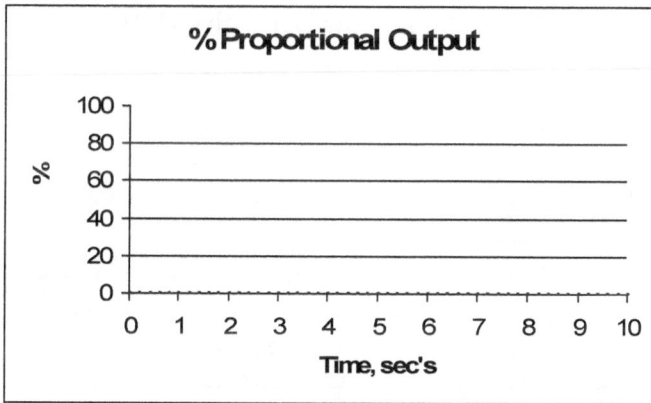

Approximate the proportional only contribution to the output here.

% Derivative Output

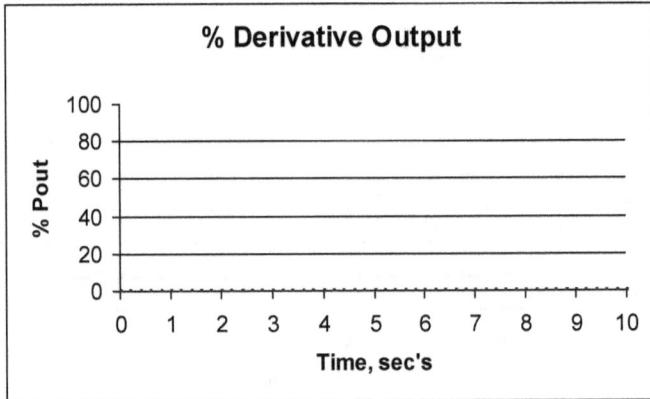

% Pout vs Time, sec's

Approximate the derivative only contribution to the output here.

% Proportional Plus Derivative Output

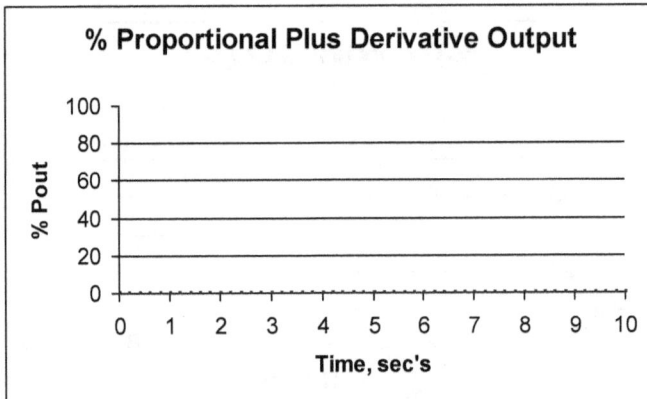

% Pout vs Time, sec's

Approximate the combined P&I response here.

Derivative Control Questions

2 – What does the derivative control mode provide or what feature does it add to the control system?

3 – Specifically, what does the derivative control mode respond to in the error?

4 – Would a controller using only a derivative mode of control be able to control a process?

5 – How does the derivative control mode correct or improve the deficiency that's associated with proportional or integral control?

6 – Specifically, what does the derivative control mode respond to in the error?

7 – How does the error amplitude influence derivative control action?

8 – What is a perceived deficiency the derivative mode of control?

9 – Assume you're working in a power plant and currently looking at the PID level indicating controller (LIC 103) in the following figure. In PID controllers the derivative action responds to the rate of change of the error which can be the result of a changing process or a change in the set point value since both quantities are included in the error calculation. In most industrial process control systems the set point can be changed much more rapidly than the process. Violent process upsets, often called "Bumps," can be induced by the derivative action responding to rapid set point changes. Explain how the effects of derivative action on set point adjustments could be eliminated? The following diagrams of figures 21 and 22 may prove beneficial in finding a solution to this question.

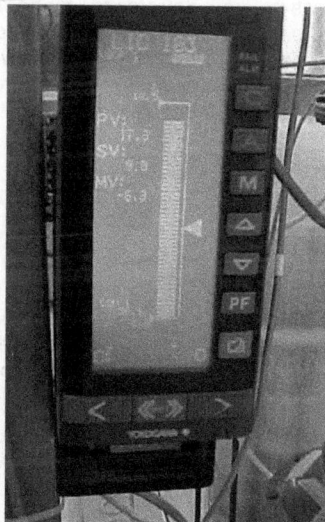

Figure 19 - A single loop PID controller where the set point (arrow at right) and the process (bar graph) are not equal. The difference is the process error.

Combined P+I+D Control

Of greatest importance in understanding the operation of three-mode PID control is the contribution provided by each of the individual control modes. Keep in mind the purpose of each control mode – proportional control meets a load's demand but cannot maintain set point if the load is continuously varying or if the process disturbances are varying. Integral action was added to return, or reset, the process to set point after a load or disturbance change.

Unfortunately, integrators tend to operate quite slowly. Increasing the integrator gain will increase the integrator's speed but it also results in over-correction of the error and over-correction results in process oscillation where the process continuously cycles around the set point. In some processes a subtle and slow oscillation can be tolerated but in others oscillations are absolutely intolerable. Try to imagine an automotive cruise control system that oscillates around the set point speed or a temperature control loop baking bread or Twinkies where the oven temperature is in oscillation. The finished product would be leaving the oven where half are over cooked and half are under cooked with very few being cooked adequately. To enhance the relatively slow PI controller output response to an error derivative or rate action was added to the controller.

Rate action provides a pulse or burst of manipulated variable to the process in an attempt to catch the process before it deviates too far from set point during a load change or temporary disturbance. The amplitude and direction of the pulse is proportional to the rate of change of the process. Derivative action observes the speed and direction of the error and adjusts the final control element rapidly in an attempt to maintain set point during a process interruption.

Figure 20 – Using PID control and pneumatic disk brakes this reel-tension-paster unit senses and controls web tension as paper is pulled from these rolls individually into the press above. As the press speed increases derivative control action detects the rate of speed increase and removes a constant amount of brake preventing the paper from tearing out. As the press speed decreases the opposite applies, the rate of change of press speed is determined and the brake is increased by a constant amount equal to the rate of change to keep the web from going slack and stopping the press.

In combining the individual PID control modes two basic configurations are employed. The following figure illustrates a single-gain approach where tuning the controller involves increasing the gain, Kc, until the process oscillates and then reducing Kc until the oscillation stops. This form of tuning is common in many programmable controllers (PLC's) and inexpensive single-loop three-mode controllers. By varying Kc the error to all three modes is increased or decreased equally which results in all three controller modes increasing or decreasing their output contributions proportionately. The control modes cannot be tuned individually which usually results in some degree of system instability or a less-then-desirable response to an upset.

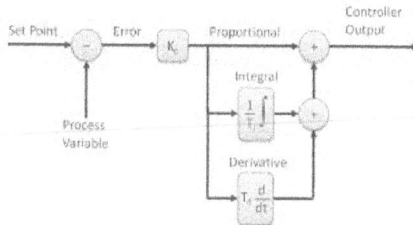

Figure 21 – This diagram of a single gain determining PID controller is a non-interactive form of PID controller algorithm. Controllers employing this combination of PID control modes are tuned by varying a single gain coefficient, Kc.

The second form of PID control has individual gain controls for the proportional, integral and derivative modes. Tuning a PID control loop is performed to optimize the response of the loop. A well tuned loop won't oscillate more than one to three cycles after a load variation or set point change returning to set point in a minimum of time. More importantly, a skilled student of PID loop tuning can customize the response of the controller to an error for a given application.

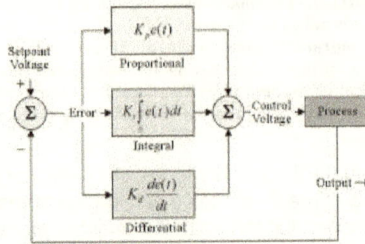

Figure 22 – Tuning independent PID gains as seen in the above diagram requires knowledge of the interaction between the control modes and an ability to set the individual gains by observing the contributions of the proportional, integral and derivative modes to the controller output.

Numerous techniques are available to tune control loops, several will be investigated in the lab however, all tuning techniques involve some degree of fine tuning after the P, I and D gain values have been determined. This will also be demonstrated in the lab experiments.

For additional information on PID control and loop tuning consider visiting the following sites or doing a Google search. If you find some decent PID or tuning demonstrations, please send me the website.

PID demonstrations and explanations:
http://blog.opticontrols.com/archives/344

http://137.132.165.75/matlab/Smit/theory1.html

http://www.fcet.staffs.ac.uk/sow1/control_resources.htm

http://blog.analogmachine.org/2012/02/04/pid-control-demonstration/

Recorded PID and related control lectures (most of this is typical university engineering stuff):
https://controls.engin.umich.edu/wiki/index.php/Recorded_Lectures

Loop Tuning Information:
http://www.controleng.com/new-products/hmis-and-operator-interfaces/single-article/loop-tuning-fundamentals/43a6422d48.html

http://ww2.nscc.edu/dean_t/EETH2370/tuning_rules.htm

http://blog.opticontrols.com/archives/131

Contemporary Control Automation

Current technology in process automation is a combination of systems developed over the past two decades, and systems being currently developed. Most current technologies are developed around the common personal computer. Unfortunately, the common PC is net perceived by the manufacturing and process industries as being "industrial strength." To a great extent the perception is accurate. Industrial components such as semiconductors, printed circuit boards, connectors, switches and the like are designed, manufactured and selected to operate accurately and repeatedly in extreme environments, often under less than desirable conditions. Similarly, industrial components such as I/O modules and programmable controllers (PLC's) usually are employed under rigid limitations with little variety or flexibility. Especially when compared to the extensile, multi-tasking abilities of the contemporary PC.

However, PC's are becoming more accepted in supplemental industrial applications. In the office domain the PC has proven to be a valuable asset requiring little support or assistance. As such, a blend of current industrial strength components and commonly available PC's is the current state-of-the-art in process automation. Personal computers are most popularly employed as data display and control facilitation devices. To perform in this capacity several items are integral to the overall cohesion and operation. Essential items include data communications, software to support communications, and hardware at the process and the operator interface.

Evolution

To best understand the current application of PC's in industry, a brief description of industrial control history is in order. In the 1950's, analog systems using vacuum tubes and a combination of electronic and mechanical components were used to sense, transmit, compare, correct and control a process variable. Many of these components are still in use today as a result of industry's' meticulous research and development processes resulting in reliable, long-life components. Transistors were developed in the fifties and deployed extensively during the sixties in analog control components. Integrated circuits began appearing in the early seventies in analog and eventually in discrete digital forms. Controllers, transmitters and transducers began to move hard in the direction of integrated electronics during this decade. Microprocessors were developed in the seventies and appeared in hybrid form, along with analog electronic components, by the end of the seventies. During the mid to late sixties computers appeared in manufacturing capacities. The most common application of industrial computers was in the manufacturing industries where NC machines were used. Numerical Control (NC) computers used punched paper tape to record and playback the program to direct cutting tool operations. NC machines were large, slow and required constant maintenance, although in time became ubiquitous.

Around this time computers, some analog, appeared for the first time in process automation capacities. The first applications of computers in process automation were in direct digital capacities where the computer replaced the computer. Computers were not reliable in the late sixties to early seventies and were discarded almost as rapidly as they appeared. As direct digital systems, a computer failure would result in process downtime, an industrial Cardinal sin. When computers returned in the eighties, industry was skeptical.

General Motors was retooling for new production runs every year, which probably took all year. Every year dozens of new models would roll off the assembly lines at GM, Ford and Chrysler. Supporting the assembly of new models required new controls, or complete modification of existing controls. New wiring, control cabinets, new or relocated transducers and actuators, welders, assembly lines, new assembly processes and procedures, each year. In an attempt to minimize the retooling efforts, GM realized many of the discrete (switching) processes could be consolidated into a re-programmable computer-like device. This device was the predecessor of the programmable controller, the PLC. GM eventually entered into an agreement with Allen-Bradley to develop the PLC. Allen-Bradley fitted the PLC with analog I/O, counters, timers, basic mathematical functions and more sophisticated control features such as PID functions. Industry accepted the PLC since Allen-Bradley, Square D, Siemens, Modicon and other controls manufacturers had been making controls components such as relays, motors and motor controls for decades. Early programming devices for PLC's were hand-held programmers and proprietary dumb terminals. Eventually RS-232 or RS-485 communications, proprietary software and computer to PLC communication boards, modules and cables allowed programming to be performed from a PC.

Human Machine Interface

As mentioned, computers have become accepted but as ancillary, operational enhancement devices. Currently PC's are to display process characteristics, work-in-process status, material inventories, finished goods and shipping information. Depending upon the software and hardware sophistication, some PC's can also direct switching, set point and control information to the process components. In short, the PC is a process interface, or Man-Machine Interface (MMI), often called Human-Machine Interface (HMI). The HMI exchanges information with the process through data communications with PLC's or other controls components configured as a supervisory control system. Communication between the PLC and the PC can be performed in a number of ways. Communication buses such as Fieldbus, Prophybus, Canbus, and numerous others are often proprietary, requiring all components be acquired from a single controls supplier. Modbus for example is a data communications system for Modicon PLC components. Others, more universal in their application are called open-architecture buses. Common computer networks such as Novell, IPX, Ethernet and NetBios are also used to communicate between the HMI, PC and the control system devices. Many recent control system installations are connecting all control components on a two-wire, half-duplex, high-speed RS-485 network. Having all components on a network facilitates troubleshooting, calibration and information access by the HMI software.

Finally, after the controls communication has been established between the process and a PC, the appropriate MMI software package is acquired. Wonderware, Intellution, Cimplistic and various others offer MMI applications and usually support additional features such as pre-developed graphics and animation, control functions, statistical analysis, Internet access and recently maintenance assistance from a growing number of vendors via the World Wide Web.

The Human Machine Interface is a valuable addition to Supervisory Control and Data Acquisition (SCADA) applications. Currently available HMI packages provide far more than data display features, and are capable of acquiring information from a number of sources concurrently. It is suggested the student search the Web for HMI vendors such as GE/Fanuc's Cimplisitic, Wonderware and Intellution to become more familiar with HMI applications and abilities.

HMI Software

The software used to implement a Human-Machine Interface application is relatively easy to assemble, but usually suggests a working knowledge of C+ or Visual Basic. A working knowledge of Visual software is to better apply and modify the operation of the HMI package, and to assist in the development, repair or modification of any required hardware drivers within the HMI. If nothing else, previous visual programming provides insight into how the HMI software is assembled and operates.

Programming of the HMI software is performed by manipulation of objects and icons. Each object is assigned a unique tag name, or tag ID. The tag ID object might be representing a pump, valve, tank, or instrument in the process system. Tag ID objects are interconnected to perform informational and control functions such as displaying pressure, level, flow or temperature information, or starting/stopping motors, opening/closing solenoids and valves. In many applications complete manufacturing operations are represented by tag-identified objects. Security levels are provided to allow access and operation control.

An advantage to networked process information is to allow multiple users in remote locations to access process information and control functions.

Selecting the Right Fieldbus

The following information was located on the Fieldbus website. It has been included here to introduce some of the more common data communication buses and standards available in the control industry today. It is suggested that the student perform a search on industrial data networks for additional related material. - JL

Click to jump to the descriptions of FILBUS, BITBUS, WorldFIP, Profibus, or CAN
Click here to jump to the FILBOX and NANOBOX remote I/O systems

Introduction to Fieldbus

Fieldbus or Fieldbuses are a special form of local area network dedicated to applications in the field of data acquisition and the control of sensors and actuators in machines or on the factory floor. Fieldbuses typically operate on low cost twisted pair cables and unlike traditional networks, such as Ethernet, where performance is measured in throughput when transferring large data blocks, fieldbuses are optimized for the exchange of short point to point status and command messages. GESPAC began working on fieldbuses in 1989 and has developed considerable expertise in this field with products now available for use with the Filbus®, Bitbus®, FIP, CAN and Profibus standard networks. GESPAC is capable of offering a complete solution, including network controllers, I/O modules, distributed microcomputers, blind nodes and software for most fieldbus standards. Host controllers are available for today's most widely used system buses, including G-64/G-96, VME, PC, iSBX, and PC-104. GESPAC has a long-term commitment to provide field bus connectivity to future standards and is active in a number of Fieldbus standardization bodies.

Many fieldbus standards exist in the market today. Each of them has been invented at a different time, by a different company and for different purposes. The right fieldbus standard for you is a matter of requirements and preferences. The tables below are designed to help you understand the differences between the various fieldbuses supported by GESPAC and guide you in your selection.

FILBUS

FILBUS was developed by GESPAC as a comprehensive remote I/O system, based on distributed intelligence and peer-to-peer communication. Firmware functions are built into each FILBUS I/O module and allow basic capabilities such as pulse count, delay before action, and sending/receiving messages to/from other modules on the network. Host adapters for the G-64/G-96, VME and PC bus are available.

Fieldbus	**FILBUS**
Speed	375 Kbits/s
Max # nodes with repeaters without repeaters	250 32
Max distance with repeaters without repeaters	13.2 km max 1.2 km max
Arbitration	Master/slave
Cable type	Twisted pair

Header/Data size	1 to 256 bytes
Major benefits	Event driven I/O software in modules
Primary applications	Remote I/O Data acquisition

FILBUS compatible products available from GESPAC:

- Couplers for PC/ISA bus, G-64/G-96 bus, and VME bus
- FILBOX family of remote I/O modules
- Starter kits

BITBUS

BITBUS was originally introduced by Intel as a way to add remote I/O capability to Multibus systems. This original fieldbus is one of the most mature and most broadly used networks today. BITBUS allows programs to be downloaded and executed in a remote node for truly distributed system configurations.

Fieldbus	BITBUS
Speed	375 Kbits/s
Max # nodes with repeaters without repeaters	250 32
Max distance with repeaters without repeaters	13.2 km max 1.2 km max
Arbitration	Master/slave
Cable type	Twisted pair
Header/Data size	1 to 13 or 52 bytes
Major benefits	Large user base Nodes programmable
	Intelligent I/O modules

Primary applications	Process control

BITBUS compatible products available from GESPAC:

- Couplers for PC/ISA bus, iSBX bus, G-64/G-96 bus, VME bus
- Intel compatible remote I/O modules

FIP

FIP provides a deterministic and reliable scheme for communicating process variables (generated by sensors and executed by actuators) and messages (events, configuration commands,) at up to 1Mbit per second on inexpensive twisted pairs cables. FIP uses an original mechanism where the bus arbitrator broadcasts a variable identifier to all nodes on the network, triggering the node producing that variable to place its value on the network. Once on the network, all modules that need that information "consume" it simultaneously. This concept results in a decentralized database of variables in the nodes and remarkable real-time characteristics. This feature eliminates the notion of node address and makes it possible to design truly distributed process control systems. Click here to access the WorldFIP organization home page

Fieldbus	**WorldFIP**
Speed	1 Mbit/s
Max # nodes with repeaters without repeaters	256 64
Max distance with repeaters without repeaters	>10 km 2 km
Arbitration	Bus Arbiter
Cable type	Twisted pair
	1 to 128 bytes

Header/Data size	
Major benefits	Distributed data base Very deterministic
Primary applications	Real-time control Process/machine

WorldFIP compatible products available from GESPAC:

- Couplers for PC/ISA bus, PC/104, and G-64/G-96 bus
- FILBOX family of remote I/O modules

PROFIBUS

PROFIBUS is a fieldbus network designed for deterministic communication between computers and PLCs. Based on a real-time capable asynchronous token bus principle, PROFIBUS defines multi-master and master-slave communication relations, with cyclic or acyclic access, allowing transfer rates of up to 500 kbit/s. The physical layer 1 (2-wire RS-485), the data link layer 2, and the application layer are all standardized. PROFIBUS distinguishes between confirmed and unconfirmed services, allowing process communication, broadcast and multitasking. GESPAC provides PROFIBUS hardware and software for connecting G-64/G-96 PCs or 68K real-time systems to third party PLCs and supervisors on PROFIBUS networks.

Fieldbus	PROFIBUS
Speed	500 Kbit/s
Max # nodes with repeaters without repeaters	127 32
Max distance with repeaters without repeaters	800 m 200 m
Arbitration	Token passing
Cable type	Twisted pair
Header/Data size	250 bytes

Major benefits	Powerful messaging
Primary applications	Inter-PLC communication Factory automation

Profibus compatible products available from GESPAC:

- Coupler for G-64/G-96 bus

CAN

Controller Area Network (CAN) is a fast serial bus that is designed to provide an efficient, reliable and very economical link between sensors and actuators. CAN uses a twisted pair cable to communicate at speeds up to 1Mbit/s with up to 40 devices. Originally developed to simplify the wiring in automobiles, its use has spread in machine and factory automation products thanks to its outstanding features:

- Any node can access the bus when the bus is quiet
- Non-destructive bit-wise arbitration to allow 100% use of the bandwidth without loss of data
- Variable message priority based on 11-bit packet identifier
- Multimaster, peer-to-peer, and multi-cast reception
- Automatic error detection, signaling and retries
- Data packets 8 bytes long

CAN is the basis of several sensor buses such as DeviceNET of Allen Bradley, CAN Application Layer (CAL) from CAN in Automation, or Honeywell's SDS. GESPAC offers a CAN host adapter on the PC-104 for its family of PC Compatible G-64/G-96 single board computers.

Fieldbus	CAN

Speed	up to 1 Mbit/s
Max # nodes with repeaters without repeaters	N/A 30
Max distance with repeaters without repeaters	N/A 40m-1Mb/s, 1km-20 kb/s
Arbitration	CSMA
Cable type	Twisted pair
Header/Data size	8 bytes fixed
Major benefits	Low cost Efficient for short messages
Primary applications	In sensors/actuators Automotive

CAN compatible products available from GESPAC:

- Couplers for PC/ISA bus, and PC/104 bus
- NANOBOX family of remote I/O modules

About the author –

Jonathan Lambert is a professor emeritus of engineering technology at Black Hawk College in Moline, Illinois. As a senior member of the International Society of Automation (ISA) and a thirty-year member of the American Society for Engineering Education (ASEE), Jon holds an AAS degree in Instrumentation Engineering Technology from Black Hawk College, a BS degree in Electrical Engineering Technology (EET) from Bradley University and an MS degree in Industrial Engineering with a Design of Experiments (DOE) emphasis from the University of Iowa. He began a career in electronics, sensors and controls repairing traffic signals for the City of Quincy, Illinois in 1970 and has held positions as a test engineer for Motorola Corporation, a chief broadcast engineer for Black Hawk College Educational Television/WQPT Quad-Cities Public Television and served for 35 years as a professor, lead instructor and department chairman at Black Hawk College. Jon has provided technical assistance, engineering consulting and new product testing services for Iowa-Illinois regional employers including the Harvester and Seeding divisions of John Deere Worldwide Product Development, Roth Pump, Small Newspapers, the Moline Dispatch, Martin Engineering, Chemplex Corporation, US Army/Edgewood Chem/Bio Center, River Stone Group, Moline Consumers Company, Ipsco Steel, Kewanee Boiler Manufacturing, Chrome Locomotive, National Railway and numerous other regional firms.

Jon is currently providing contract engineering services and adjunct engineering technology instruction at Black Hawk College in Moline, Illinois. He can be reached at LambertJ@bhc.edu.

Many thanks to my wife, Antoinette, for allowing me to work on this manuscript during "our" time and especially to my mentor Richard Henry, without whose patience, guidance, understanding and support this text and my career would never have happened.

This page retained intentionally blank.

www.ingramcontent.com/pod-product-compliance
Lightning Source LLC
Chambersburg PA
CBHW021022210326
41598CB00016B/890